Countering Insurgencies and Violent Extremism in South and South East Asia

This volume of case studies examines the rise in violent extremism, terrorism and insurgency in South and South East Asia, and subsequent state responses.

The South and South East of Asia has experienced various forms of extremism and violence for years, with a growing demand for academic or policy-relevant work that will enhance understanding of the reasons behind this. The violent challenges in this area have taken a variety of forms and are often exacerbated by lack of governance, tie-ins to existing regional criminal networks, colonial legacies and a presence of international terrorist movements. Written by experts with field experience, this volume analyzes the key element of successful response as the appropriate application of doctrine following nuanced assessment of threat. In practice, this often means counterinsurgency doctrine. The essays also analyze the need for irregular war practitioners to systematically examine the changing character of intrastate violent irregular challenges. The volume fills a gap in the understanding of patterns, drivers, organizations and ideologies of various insurgent and terrorist groups, and state responses. It also provides a set of recommendations for addressing the unfolding situation.

This book will be of much interest to students of terrorism and political violence, counterinsurgency and counterterrorism, Asian politics and security studies in general.

Shanthie Mariet D'Souza is Founder and President of Mantraya and Member of the Research and Advisory Committee, Naval War College, Goa, India. She has a PhD in International Relations from Jawaharlal Nehru University, New Delhi, India.

Cass Military Studies

For more information about this series, please visit: www.routledge.com/Cass-Military-Studies/book-series/CMS

Countering Insurgencies and Violent Extremism in South and South East Asia

Edited by Shanthie Mariet D'Souza

Routledge
Taylor & Francis Group

LONDON AND NEW YORK

First published 2019
by Routledge

2 Park Square, Milton Park, Abingdon, Oxfordshire OX14 4RN
52 Vanderbilt Avenue, New York, NY 10017

Routledge is an imprint of the Taylor & Francis Group, an informa business

First issued in paperback 2020

British Library Cataloguing-in-Publication Data
A catalogue record for this book is available from the British Library

Library of Congress Cataloging-in-Publication Data
A catalog record has been requested for this book

ISBN: 978-1-138-61555-7 (hbk)
ISBN: 978-0-367-66249-3 (pbk)

Typeset in Times New Roman
by Wearset Ltd, Boldon, Tyne and Wear

Contents

PART II

Cautious optimism – or false dawn?   107

PART III

Quagmires   223

PART IV

Victory?   285

Illustrations

Figures

Tables

Contributors

Mohamed Bin Ali is Assistant Professor at Studies in Inter-Religious Relations in Pural Societies (SRP) Programme, S. Rajaratnam School of International Studies (RSIS), Nanyang Technological University (NTU), Singapore. His latest book is entitled *The Roots of Religious Extremism: Understanding the Salafi Doctrine of Al-Wala' wal Bara*, Imperial College Press (Insurgency and Terrorism Series Volume 9), World Scientific Publishing Company, November 2015.

Dawood Azami is an award winning journalist and academic. He works as a Multi-Media Editor at the BBC (London), leading a team of journalists to cover international news. Azami was a Visiting Scholar at the University of Westminster, London, and the Ohio State University, Columbus, USA. Countering the Islamic State Group in Asia.

Kirklin J. Bateman is the Chair, War and Conflict Studies Department at the College of International Security Affairs of the National Defense University in Washington, DC. A retired U.S. Army officer, he completed his graduate studies in history at George Mason University.

Shanthie Mariet D'Souza is Founder and President of Mantraya; Member of Research and Advisory Committee, Naval war College, Goa; Board of Directors, Regional Centre for Strategic Studies, Colombo; Associate editor for the Journal of Asian Security and International Affairs; Editorial board member of Small Wars and Insurgencies; Expert and Contributor to the Middle East-Asia Project (MAP) at the Middle East Institute, Washington DC; Senior analyst for the South Asia desk with the Wikistrat Analytic Community, and Adviser for Independent Conflict Research and Analysis in London. Dr. D'Souza has previously been a Visiting Research Associate (2017) at Murdoch University, Perth, Australia; Research Fellow (2010–2014) at the Institute of South Asian Studies (ISAS), National University of Singapore; Associate Fellow (2006–2010), Institute for Defence Studies and Analyses, New Delhi; Fulbright Fellow (2005–2006) and Visiting Research Associate at South Asia Studies, The Paul H Nitze School of Advanced International Studies, Johns Hopkins University, Washington DC. As Adviser at the Independent Directorate of Local Governance (2015–2016), Government of the Islamic Republic of Afghanistan, Senior Transition Consultant (2013), United Nations Mine Action Service and External Reviewer and Consultant (2011) for Action Aid International, Afghanistan, she has carried out strategy reviews, organizational planning and field research in various provinces of Afghanistan. She has been an International Election Observer for the audit and recount of Afghanistan's Presidential Runoff elections in 2014. She has also conducted field studies in Pakistan, China, Africa, Canada, United States, Jammu and Kashmir and India's North East. Among her most recent published work is an edited book titled *Afghanistan in Transition: Beyond 2014?*, co-edited books, *Perspectives on South Asian Security* and *Saving Afghanistan*.

Antonio Giustozzi is currently Visiting Professor at King's College London. He is the author of several articles and papers on Afghanistan, as well as of five books, *War, politics and society in Afghanistan, 1978–1992* (Georgetown University Press), *Koran, Kalashnikov and laptop: the Neo-Taliban insurgency, 2002–7* (Columbia University Press), *Empires of mud: war and warlords in Afghanistan* (Columbia University Press), *Policing Afghanistan* (with M. Ishaqzada, Columbia University Press, 2013) and *The army of Afghanistan* (Hurst, 2016). He is currently researching a project on the Islamic State in Central and South Asia.

Richard Javad Heydarian is a Manila-based academic, columnist and author. He has taught political science at De La Salle University and Ateneo De Manila University in Manila, the Philippines, and is the author of, among others, *Asia's New Battlefield: US, China and the Struggle for Western Pacific* and *Duterte's Rise: A Populist Revolt against Elite Democracy*. He has authored academic journal articles on Philippine politics, South China Sea disputes, and Islamism and Islamic movements. He is a columnist for *Manila Bulletin* and resident political analyst for GMA Network.

Sameer P. Lalwani is a Senior Associate and Co-Director of the South Asia Program at the Stimson Center where he researches nuclear deterrence, interstate rivalry, and civil conflict. Lalwani is also an Adjunct Professor at George Washington University's Department of Political Science and was previously a Stanton Nuclear Security Postdoctoral Fellow at the RAND Corporation. Lalwani holds a PhD from MIT's Department of Political Science.

Anatol Lieven is a professor in Georgetown University's School of Foreign Service based in Doha, Qatar and a senior fellow of the New America Foundation in Washington DC. His book *Pakistan: A Hard Country* was published in 2011. He was previously correspondent for *The Times* (London) in Pakistan and Afghanistan.

Geoffrey Macdonald is a Professorial Lecturer in George Washington University's Department of Political Science and the Principal Researcher for Democracy and Governance at the International Republican Institute. His research focuses on political violence, extremism, democratization, and conflict management. He has written widely on violent extremism and South and South East Asian politics.

Thomas A. Marks is Distinguished Professor and MG Edward Lansdale Chair of Irregular Warfighting Strategy at the College of International Security Affairs (CISA) of the National Defense University (NDU) in Washington, DC. He was promoted to the position after 12 years as Department Chair, War and Conflict Studies (WACS), CISA.

Rhonda Mays is Regional Deputy Director for Asia at the International Republican Institute. Her professional and academic experience focuses on democratization, human rights, and political history and contemporary politics in Southeast Asia.

Samir Puri lecturers at the Department of War Studies, King's College London. His books include *Fighting and Negotiating with Armed Groups*, published by the International Institute for Strategic Studies in its Adelphi series (2016), and *Pakistan's War on Terrorism* (Routledge, 2011).

Andrin J. N. Raj is the Southeast Asia Regional Director for the IACSP-Centre for Security Studies based in Kuala Lumpur Malaysia. He is also a Visiting Lecturer at the MARSEC COE, Akzas Naval Base in Turkey and at the Institute for Diplomacy Foreign Relations, Malaysia. He was a former Visiting Fellow and an ASEAN Scholar at the Japan Institute for International Affairs and his publication *Japan's Initiatives in Security Cooperation in the Straits of Malacca on Maritime Security in Southeast Asia: Terrorism and Maritime Security* was published by JIIA and Militant Islam in Malaysia: Synergy between Regional and Global Jihadi Groups.

Bibhu Prasad Routray is Director, Mantraya. He has been a Visiting Professor, Murdoch University, Perth; a Deputy Director in the National Security Council Secretariat, New Delhi; Director, Institute for Conflict Management, Assam; and a Visiting Research Fellow, S. Rajaratnam School of International Studies, Singapore. Routray specializes in decision-making, counter-terrorism, and dissent articulation in South and Southeast Asia; and is the author of *National Security Decision Making in India* (RSIS, Singapore).

Luke Waggoner is a Senior Fragility and Resilience Specialist at the International Republican Institute, where he oversees programming on countering violent extremism. He has managed CVE research programs in Indonesia, Tunisia, Kosovo, and elsewhere.

Marvin G. Weinbaum is professor emeritus of political science at the University of Illinois at Urbana-Champaign, and served as analyst for Pakistan and Afghanistan in the U.S. Department of State's Bureau of Intelligence and Research from 1999 to 2003. He is currently a scholar-in-residence and Director of Afghanistan and Pakistan Studies at the Middle East Institute in Washington, DC.

Introduction

Understanding insurgencies and violent extremism in South and South East Asia

Shanthie Mariet D'Souza

South and South East Asia have long been recognized for the complexity of their violent internal challenges, with a growing need for academic or policy-relevant work to enhance comprehension. This is all the more important since these challenges have taken a variety of forms and have regularly leaped beyond national boundaries, exacerbated by contextual factors that include lack of governance, support not only from within but from without, tie-ins to existing regional criminal networks, steady-state conflict economies, colonial legacies, multi-ethnic structures, and presence of international terrorist movements. Appropriate response to such challenges necessarily extends beyond tactical considerations to address the drivers that propel the resultant insurgencies, terrorism, and violent extremism.

Still, progress has remained elusive. Instances abound of cases that have gone on for nearly three-quarters of a century. In the interim, not surprisingly, new factors have impacted legacy roots of conflict, creating still further demands upon policymakers. To understand these dynamics, this volume brings together a wide array of case studies by practitioners and academics with rich field experience in South and South East Asia. The objective is to produce a nuanced understanding of state responses, past and present, thus to provide a better sense of what appears to work and what does not.

Understanding insurgencies, terrorism, violent extremism and irregular warfare

Necessarily, the contributions of the authors bring into focus ongoing debate within irregular warfare strategy and scholarship. The long history of what some have incorrectly termed "low-intensity conflict" reveals not only how ubiquitous irregular challenges have been but also how often their importance has been ignored, setting the stage for future strategic setbacks – certainly not victory. In particular, more often than not, what are violent political challenges are met with a response that is overwhelmingly military. This can be successful – witness the Chinese elimination of local society in order to incorporate Tibet within the larger Han polity, as well as the ongoing effort to do the same in Xinjiang – yet it is certainly not the doctrinal approach of democratic polities. Regardless, something short of victory or even resolution has been a hallmark of sorts as illustrated by the cases in this study.

It is noteworthy that even the powerful United States has often been faulted for its dismaying record of failing to adapt to "small wars," despite its considerable experience fighting "Native Americans, Philippine insurrectos, the Vietcong, al Qaeda, the Taliban, and numerous other irregulars."[1] Such shortfall invariably seems to stem in large measure from a lack of understanding of just who these irregulars are, their objectives and methods, and hence what an irregular counter should consist of. The central challenge, then, is to find a framework of analysis that defines the dynamics of these armed struggles. Thomas A. Marks and Paul B. Rich, drawing upon people's war, offer a useful approach.[2] They highlight that focus must first be upon the challenge to the existing order, assessing the kinetic/nonkinetic synthesis that operationalizes the way forward to the objective. In the imperative to mobilize manpower and resources required, agency interacts with structural contradiction to build a new world that challenges the existing world. Through formation of a counter-state the old-order ultimately is destroyed and replaced.[3]

Far from being an approach unique to a particular time or place, much less culture, this is a generic methodology. The existing order is challenged politically, with violence utilized as conceptualized by the strategic vision of the initiator. One challenge may see bombs accompanied by propaganda tracts as the way forward, while another may emphasize mass mobilization through demonstration of resistance. Still another may carefully mobilize through political action in order to field violence through a swelling force of armed manpower representing the counter-state. Variants are as numerous as the imagination. The error for the state, though, is in conceptualizing this as a form of warfare as opposed to a form of violent politics. The semantic difference may seem slight, but in reality is overwhelming. The first deploys specialists in violence to counter violent challenge; the second recognizes that violence is but the enabler for political solution to political challenge.

In this project, terminology becomes salient. If traditional rebellion informed by ideology privileges violence, the attack upon the innocent for political reasons gives us terrorism rather than insurgency. The latter avoids foregrounding violence in favor of mass mobilization through appeal to greed and grievance. Whether such appeal is real or contrived is a matter for analysis. The key, as noted by any number of scholars, is that what was once called "pure terrorism" eschews the work of alternative state-building that is the essence of insurgency. Both efforts are inspired by ideology (most simply, political theory wedded to dictates for action), which invariably, where rebels are concerned, is labeled "extremist," but the manner in which they proceed is very different. This, as discussed in this volume's contributions, has profound implications for counter.

Put most simply, those who challenge the existing order not surprisingly engage in different approaches that are themselves reflections of different assessments of both situation and necessary course of action. The term CVE, countering violent extremism, has been much in vogue over the last decade yet goes wide of the reality that leaders and followers – trailblazers and joiners might be as useful a formulation – are not the same thing. The mere act of mobilization stemming from alienation and frequently marginalization cannot be explained simply by looking to something called "radicalization," which is another way of saying "embracing extremism." This may be the case for a small band of committed individuals who privilege violence to the near exclusion of all else in their assault upon the society they have marked for remaking. Yet for those who do not conflate violence and political action into a unified whole, the label is wide of the mark. In reality, those who initiate the revolutionary project are acting upon their subjective assessments of societal imperfection, while those who join the project largely, certainly until indoctrinated, seek improvement of their own slice of that imperfection. They seek mediation while their leaders seek to alter societal terms of reference, to alter the very structure of society. The leaders' mindset may certainly be labeled extremism, accompanied as it is by violence, but the followers' orientation is best understood through the prism of recruitment as opposed to conversion.

As major episodes of violence invariably are labeled warfare, it stands to reason that political violence of the sort under discussion will be irregular in its conflict form. Most famous, of course, is guerrilla warfare. Yet all irregular warfare is not guerrilla warfare, albeit the frequent conflation of the two. Modern terrorism, which attempts to bring terror to the confines of one's privacy, is waged differently. While guerrilla tactics have not changed, goals have turned political. The Naxalites in India amply demonstrate this case. The distinction between the present and the past is the emergence of ideological agendas. Boot observes:

> The precise nature of the ideological agendas being fought for has changed over the years, from liberalism and nationalism (the *cri de coeur* of guerrilla fighters from the late eighteenth century to the late nineteenth century), to socialism and nationalism (which inspired guerrillas between the late nineteenth century and the late twentieth century), to jihadist extremism today. All the while, guerrilla and terrorist warfare have remained as ubiquitous and deadly as ever.[4]

The challenge, therefore, is about adopting an appropriate state strategy of response that is built upon careful differentiation between those components of rebellion that define it and hence provide it with its label. Responding to a primary group for which attacks upon the innocent have become the whole of the political effort ("propaganda by the deed," in the original formulation) is very different from responding to a mass-based upheaval driven by ideology but for which violence is only an enabling mechanism. Necessarily, the security forces, which exist to manage violence (either for or against), normally are placed in a position of primacy, at least in implementing the solution decided upon. Yet this highlights the danger involved, since the more political the challenge, the less appropriate is both the security force mindset and its tools.

It hence is imperative for political leaders and practitioners to understand the kind of war upon which they are embarking. Marks notes:

> If insurgency is an armed political campaign – mass mobilization of a counter-state to challenge the state for political power – then it is intuitively obvious that counterinsurgency is intended to prevent this. As a strategic category, the goal of counterinsurgency is always legitimacy.[5]

Implied, though not stated directly, is the corollary. To the extent violence has become the armed political campaign – that is, terrorism – legitimacy remains central, but as an enabling mechanism rather than a goal. A viable state effort cannot proceed in the absence of sufficient support.

Tragically, in some states (witness China or Russia), there remains a notion that "kill them all" is both a viable and preferred strategy, though one may add that China – if we compare Xinjiang to Tibet – has decided that neutralization is preferable to genocide, if only for the sake of keeping up appearances. Regardless, for most states, most certainly democracies, there is a recognition that neither killing nor emphasizing control works particularly well in the long-term and must be tactically a part of something larger, strategically. This can be readily understood through the cases just noted. For Beijing, resistance within Tibet or Xinjiang to Han domination means the indigenous populations are the problem – terrorists – and must be dealt with. In contrast, were Beijing a democracy, the first question would be, "What problem is causing resistance?" This indeed is the approach that generally informs responses laid out in the chapters of this book.

They make clear that just as all politics is local, the security of the populace remains the touchstone for democratic process. Although territorial control can be temporarily achieved by use of force, it can be maintained only if the government and security forces have some degree of popular legitimacy. This involves the proverbial "winning of hearts and minds" – a phrase certainly used most memorably in the early modern era by John Adams – "The Revolution was effected before the War commenced. The Revolution was in the Minds and Hearts of the People"[6] – then popularized by Sir Gerald Templer during the Malayan Emergency, in the late 1940s and 1950s.[7] Though simple enough in theory, operationalizing the sentiment has proved rather more difficult in fact. Foreign support for a local regime necessarily remains an expeditionary force, cut off by culture and language. Local regimes, however, do not get a free ride in this respect. The case of Sri Lanka's Sinhalese-dominated forces in this volume come immediately to mind. Similarly, Taliban has proved quite adept at tarring Kabul as "a puppet regime," lacking legitimacy and backed by the infidels (foreigners), to whip up anti-government sentiment in the tribal and rural areas of south and east Afghanistan.

Afghanistan further highlights the point, regardless of the "rural" nature of a society today – Nepal also comes immediately to mind – most states now experience a fusion of rural to urban created by both physical and virtual linkage. This creates a hybrid battle space demographically and spatially even as the same phenomenon occurs in conflict reality. Isolated rural communities have largely become relics of the past, while modern methods of counterinsurgency and counterterrorism, to include satellite surveillance and drones, have led some analysts to conclude that people's war-type insurgencies (or "Maoist") and the creation of secure base areas are problematic.[8] As David Kilcullen, among others, has argued, insurgencies in the future are increasingly likely to take an urban form.[9] Though there is something to be said for the idea of an increasingly transnational form, with coastal cities (such as Mumbai in India) vulnerable to coastal attack, of greater moment strategically for our consideration is the sort of post-Maoist movements illustrated by the Nepali and Bolivian cases. These erase the boundaries between violent and non-violent political action intent upon seizure of power, and are dangerous precisely because they are grounded in objective considerations of the past but explicit consideration for leveraging the changed context of the twenty-first century. Their decision to emphasize what certainly were "supporting forms" at one point in the Maoist trajectory has generally left states countering them either clueless or simply befuddled. Nepal survived in its democratic form due to a serendipitous combination of circumstances; Bolivia did not. Therein lies continuity with the past and a warning for the future.[10]

Current debates have largely missed the emergence of new radical forms of violent politics that privilege the nonkinetic strategically while deploying the kinetic tactically as enabling mechanisms. The armed strike is a premier example and plays to the weaknesses of democracy much in the manner Putin and Russia have played Trumpian America like a fiddle through perception management. Once termed simply "political warfare" by the Chinese, later "unrestricted warfare," leveraging such nonkinetic weapons at the operational level has for at least a century been doctrine in communist forms of irregular warfare. The approach, though, has moved to guide strategy itself.

Rather than focusing upon such pressing concerns, the current debate on counterterrorism and counterinsurgency in the United States has gone off into the weeds. On the one hand, though terrorism remains a pressing concern, dealing with it has been bifurcated into a domestic component, driven by the increasingly politicized dictates of homeland security, and a foreign component, met with a combination of intelligence-driven targeted killing (largely through drones) and building partner capacity (largely through special operations forces). To the extent coherent mesh is achieved, it occurs only in particular theaters. Afghanistan, once the shining symbol of American determination in a global war on terrorism, has been shunted to the side as an economy of force mission dedicated to staving off an embarrassing defeat rather than achieving a strategically meaningful victory. The result has been to increasingly divide consideration of counterterrorism and counterinsurgency, which in reality have much in common, as noted in previous discussion.

Given the extent to which the U.S. has dominated recent theory and implementation of counterterrorism and counterinsurgency, it is worthwhile to devote a moment to exploring just how such a point was reached. Much of the present controversy and aversion to anything to do with irregular warfare derives from the American refighting of the Vietnam War in Afghanistan and then Iraq. Despite protestations to the contrary, as well as a conscious effort to do correctly what was widely perceived as not just done incorrectly, but botched by the Vietnam era practitioners, the final product was not only remarkably the same in result but in execution, with only the "contents of the boxes" altered in any strategic diagram of conflict. "Night raids," to use a salient illustration, differed from "search and destroy" only in their vaunted intelligence-driven effectiveness. Their measure was still body count and had as little impact upon the strategic dynamic of the war(s) as did their kinetic predecessor. Drones are simply the logical next step in the process. The flaw, as it was in Vietnam, remains the lack of a political solution to what remains a political challenge.

The American approach – which its partners and allies used to conceptualize their own strategies and tactics – as is well known, derived from the work of a group of public relations-minded counterinsurgency enthusiasts who emerged around then-LTG David Petraeus at Fort Leavenworth and helped formulate the widely read doctrinal publication, FM 3–24 *Counterinsurgency Field Manual*. This group of followers of what some have called "neoclassical COIN" came to be termed *COINdinistas*, a coterie of highly educated and qualified military intellectuals who set out on the ambitious task of transforming the wider mindset of the American (and Western) military establishment so that it grasped the centrality of counterinsurgency in future U.S. expeditionary warfare.[11] This was done, somewhat paradoxically, as ostensibly a part of the larger war against terrorism, even as the widespread use of terrorism, as one method by insurgents, frequently led to a fusion of statistics, thus negating their meaning and utility.

The ideas and ability of the *COINdinistas* to re-define U.S. military doctrine ultimately unleashed extensive criticism,[12] with the debate noteworthy for its use and abuse of history, since a series of (often colonial) examples were essentially the basis for the doctrinal principles.[13] Porch and Gentile were particularly trenchant critics,[14] though both in turn were laid low by Ucko.[15] Still, the exchange had value. Porch, for instance, provided a corrective to the historical distortions that undergird much recent counterinsurgency thinking, arguing that "modern counterinsurgency proponents ransack[ed] history to unearth primordial versions of key COIN concepts with which to decorate their doctrine's pedigree."[16] In particular, Porch faulted practitioners for honing in on the tactical learning of military forces, as if tactics might hold the "key to victory."[17] This point is important and speaks to the over-militarization of counterinsurgency (and, by implication, counterterrorism) scholarship. As most of those writing on counterinsurgency appear to be militarily oriented – even as much counter-terrorist research displays the opposite trend – there has been a focus on the battlefield experiences of security forces rather than on the political heart of the matter. Whereas counterinsurgency texts – to include Galula and FM 3–24 – are keen to highlight the political essence of the endeavor, they devote far less attention to actually understanding and studying what politics is, what it means, and how it works.[18] The disappointing results in the field should therefore not surprise.

This deficiency is unfortunate but also avoidable. In the study of counterinsurgency (past, present, and future) and counterterrorism, it indeed is critical that "military history becomes an aspect of total history; not to 'demilitarize' it, but because the operational aspect of war is best studied in terms of the multiple political, social and cultural context that gave, and give, it meaning."[19] Similarly, despite counterinsurgency's emphasis on language and cultural training, its prescriptions are also assumed to be applicable across time and place,[20] as well as in counterterrorism. If in its present form, counterterrorism is often considered to be an overwhelmingly domestic concern, there is an implicit assumption that there thus adheres a certain home-field advantage that is not the case in expeditionary mode abroad. That such is often not the case (consider, for instance, the Chinese example above) tends to be overlooked.

What makes counterinsurgency and counterterrorism all the more difficult for democracies, it has long been recognized, is that there are few *quick* victories to be had. Since 1775, the average insurgency has lasted seven years (and since 1945, it has lasted almost ten years). Attempts by either insurgents or counterinsurgents to short-circuit the process usually backfire.[21] The experiences of the United States in Iraq in 2007–2008, Israel in the West Bank during the second intifada, the British in Northern Ireland, and Colombia in its fight against FARC (Revolutionary Armed Forces of Colombia), all show that it is possible for democratic governments to fight insurgents effectively if they pay attention to what the U.S. military calls "information operations" (also known as "propaganda" and "public relations") and implement some version of a population-centric strategy. But these cases are warnings not to take lightly entry into irregular conflicts. Future foes are unlikely, in other words, to repeat the mistake of nineteenth-century colonial subjects who often fought European invaders in the preferred Western style. Irregular tactics, on the other hand, have proven effective, even against superpowers.[22]

A quick survey of historical response to what increasingly has been termed simply "irregular challenge" demonstrates that even now, after much thought, emphasis remains upon military tactics or peacebuilding, rarely upon the trajectory of insurgency and the concomitant state responses necessary to address the political roots of the problem. This is particularly the case for South Asia, which may in some sense be considered one of the most analytically rich regions available for study. Regardless, there remains a need for analysis of the efforts to counter violent extremism (CVE) as well as on counterinsurgency (COIN) and counterterrorism (CT) in such manner as to move beyond the descriptive. This is a challenge since matters pertaining to national security are highly politicized and to a large extent bureaucratized, thus remaining officially confidential and closely guarded (as witnessed in inter-departmental and inter-agency rivalry where information is not shared and at times even distorted).

Of a different sort is another challenge, one of greatest importance: the need for field-based research at a time when "large-N studies" seem again to be all the rage. Even the term "civil war" has lost its plain-sense meaning and become but a cipher for a particular level of body count. This is far from useful. Indeed, little is said now, irrespective of the veneer of sophisticated terminology and bean-counting, which was not said in more straight-forward manner during the Cold War and counted better in the Vietnam War era and immediately thereafter. This reality inhibits thought-provoking analyses that could aid policymakers devise effective strategies to addresses and prevent future threats. Moreover, there is very little that is innovative in terms of methodology as concerns irregular war.

This volume

To understand the changing character of irregular warfare, of violent politics – insurgencies, the rise in violent extremism, and terrorism – as well as state response, this volume brings together a number of case studies by experts. Each of these essays provides a nuanced understanding of the strategies of the groups and responses of the states concerned. Each addresses a fundamental reality: what accounts for success or failure – as seen from the ground. Can successes and failures be understood in terms, for the first, of luck or design, of the second, in terms of bumbling or false doctrine?

This raises a host of subsidiary questions at both the theoretical and empirical levels. Is there a unique national approach to dealing with various forms of extremism? How is the relevant doctrine/strategy arrived at? Have there been any lessons learnt from previous operations that inform the present strategy/doctrine? Does the reluctance to use or the actual use of excessive force itself limit the effectiveness of the strategy? Have dialogue, peace talks, or negotiations been used as tools in strategy? Are there unique principles in implementing the COIN/CVE strategy that sets the country apart from the rest of the world? What new methodologies could a specialist researcher adopt in the coming years that will enrich the field contributions to policy? The answers unfold in the volume's four parts.

Emerging challenges

In the first part, "Emerging challenges," five authors examine countering the Islamic State in Asia and violent extremism in Indonesia, Philippines, Myanmar, and the Maldives.

Dawood Azami's chapter, "Countering the Islamic State in Asia," examines the emergence and expansion of the Islamic State (IS) in the region with a special focus on South and Southeast Asia. He analyzes how IS impacted the dynamics of militancy as it tried to co-opt or challenge the existing militant groups on both the military and ideological fronts. The contribution looks at state response as well as the impact of rivalries between certain state actors and the lack of a multilateral mechanism to tackle regional violent extremism. Drawing on field research, Azami points to the challenges IS faces in the future and its evolving tactics and strategy in Asia.

Richard Javad Heydarian, in "The Philippines' counter-terror conundrum: Marawi and Duterte's battle against the Islamic State," provides an analysis of the implications of the Battle of Marawi, where Islamic State-affiliated groups managed to endure a five months-long siege of the Philippines' largest Muslim-majority city. He examines the Rodrigo Duterte government's specific counter-terrorism doctrine, how it evolved over the decades, and its degrees of success (and lack thereof) under various administrations. Ultimately, there emerge policy options to deal with CT and CVE challenges in the country.

In tracing "The evolution of violent extremism and state response in Indonesia," Geoffrey Macdonald, Rhonda Mays and Luke Waggoner treat both sides of the title's dyad as deeply rooted in historical contestations over definitions of the Indonesian nation and the government's central objective of ensuring the continued existence of a unified Indonesian state. Two key periods in the history of violent extremism in Indonesia are examined: the rise of Jemaah Islamiyah and other terrorist organizations (1999–2009), and the ascendency of decentralized terrorism inspired by the Islamic State (IS) (2016 to the present). Indonesia's approach to countering violent extremism has evolved and adopted new, more effective tactics. The decentralized nature of IS-inspired terrorism will require that counter move beyond traditional hard-power, security-based operations toward a more holistic counterterrorism strategy that includes governance improvements at both the local and national level.

In the chapter, "The Rohingya and Myanmar's counter terrorism approach," Bibhu Prasad Routray details that the rise of resistance in the form of the Arakan Rohingya Salvation Army (ARSA) is a consequence of the years of discriminatory policy by the military and its harsh, dysfunctional policies as it responded to tactical self-defense by at least some element among the victims. For the Rohingya and the other various ethnic groups who inhabit Myanmar's periphery have indeed been at the receiving end of the military's scorched-earth policy and disproportionate response to conflicts. Far from providing a solution to the crisis, the "kill them all" approach under the false rubric of "counter-terrorism" is in all likelihood preparing the ground work for a sustained and more problematic phase of violence.

Kirklin J. Bateman, in "Emerging violent radical Islamism in the Maldives," observes that despite a homogeneous ethnic and religious population, as well as geographic isolation from sources of radicalism and extremism in South Asia and the Middle East, the Maldives is experiencing challenges that threaten to unsettle the archipelago and disrupt its thriving tourism industry. From the explosion of a terrorist bomb in the capital city, Malé, a decade ago, the once tranquil, tropical paradise has increasingly been forced to grapple with the reality that it is not immune to the same extremism that has produced other regional trouble-spots. Nevertheless, response is hampered by ongoing debate as to the causes of the violent radical Islamism and appropriate state response.

Cautious optimism – or false dawn?

The second part, "Cautious optimism – or false dawn?," examines whether "peace" in some of the erstwhile conflict theatres is indeed real.

In "Back to the future: Nepali people's war as 'new war'," Thomas A. Marks reveals that though the last two decades of strife in Nepal (1996–2017), neatly divided as they are by the November 2006 peace agreement, have seen two variants of people's war played out. In each of the two decades examined, an insurgent strategic approach comprised of violent and non-violent components astutely emphasized the particular lines of effort appropriate to the moment in order to continue the revolutionary struggle even as international and national contexts shifted. In the first decade, successful focus upon violence resulted in domination of 75–80 percent of the population. Thereafter, in the second decade, a revised approach blended political warfare with terrorism to continue the quest for power. It is the specifics of how this was done that make the case illustrative of what has been called "new war" – waging war despite a formal context of peace.

In responding to what has all the hallmarks of a less astute Maoist movement, Samir Puri assesses that "India's two-track response" to dealing with its "Naxalites" has neglected political process. Ostensibly a mix of the kinetic and nonkinetic, of security and development, in reality state strategy, which has included dialogue, has been minimally effective in bringing conflict to a resolution. Progress has been made through containing rather than solving the conflict. Hence its lethality in annual deaths has declined in the years since 2004, when the insurgency consolidated under the CPI-Maoist banner, but this has had the pernicious consequence of engendering a disincentive to engage in meaningful negotiation. The resulting underdevelopment of the political track necessarily means there are few prospects of the conflict ending.

Bibhu Prasad Routray takes such assessment still further in "Delimiting an Indian strategic approach to counterinsurgency," arguing that New Delhi's persistent failure in resolving some of its longstanding conflicts is linked to inability to implement necessary counterinsurgency practices. Extrapolating four key rules of engagement from four cases of success, he observes that sub-optimal results have stemmed from implementation of diverse policies in different theatres. The variation has become more acute with different political parties in power attempting to implement individual, parochial models of counterinsurgency.

A challenge of a different sort absorbs Mohamed Bin Ali in the chapter, "Countering violent extremism: the Singapore experience," wherein he examines the Singapore experience in dealing with the threat of violent extremism posed by Jemaah Islamiyah (JI) organization and the Islamic State (IS). Of particular note have been the efforts undertaken by local Muslim organizations, such as the Religious Rehabilitation Group (RRG) and the Islamic Religious Council of Singapore (MUIS), to counter extremism and prevent radicalization within the Muslim community. These exemplify Singapore's unique community-based approach to counter extremist threat. Important lessons emerge for the regional and international communities based on Singapore's experience in dealing with the threat of extremism and violence.

Neighboring Malaysia draws on a more traditional emphasis on law enforcement and intelligence-gathering in fashioning a twenty-first century approach to its present threat of religious violent extremism. Andrin J. N. Raj, in "Challenges in counter terrorism and countering violent extremism in Malaysia," examines the common drivers of violent extremism as well as challenges that must be met by capacity-building and best practices. He addresses countering terrorism from an alternative perspective using ethnographic and anthropological studies rather than relying solely on an ideological approach.

Quagmires

Part III of the volume treats the region's protracted conflicts. As indicated by the title, it examines the festering nature of struggles given sustenance by the very nature of state response (or lack thereof).

Antonio Giustozzi's "The counterinsurgency quandary in post-2001 Afghanistan" highlights the lack of understanding by key players of the changing character of the insurgency and hence the methods necessary for dealing with it. Afghan security forces have been unable to contain an internally divided Taliban, which in turn was unable to coordinate its own activities effectively in 2014–2015. The security forces steadily lost ground even as there was little effort to develop an indigenous Afghan counterinsurgency strategy with necessary doctrine. A decade-and-a-half of international military intervention with diverse foreign advisers contributed to disparate perceptions of appropriate response. A string of Taliban victories has reportedly spurred a sense of urgency, but it remains unclear if Kabul will break out of its state of strategic dependency in order to deal with a Taliban transition to maneuver warfare executed by regularized guerrilla units.

Examining the agents of insurgency that have originated in the tribal belt, Marvin G. Weinbaum's "Insurgency and violent extremism in Pakistan" traces the various campaigns and counterinsurgency strategies that have been employed to contain and defeat the threat. He examines the commonality between the violent extremism of Pakistan's jihadi and sectarian organizations and the insurgents in the Federally Administered Tribal Area (FATA) in terms of root causes, as well as the drivers and pathways for recruitment. Social and peer pressure and material incentives help explain individual mobilization. Both align in their rejection of the basic tenets of the country's democratic constitutional order and hence a willingness to use force to impose their vision of a political system compliant with *Sharia*. The state's counterinsurgency efforts, hampered by internal politics exacerbated by social division, have shied away from fully confronting insurgency on the country's Afghan frontiers and have failed to meet the security challenges posed by violent radical Islamist organizations. Nevertheless, regional strategic calculations and steadily growing domestic violence have forced many of Pakistan's military and civilian leaders to reassess their policies for dealing with terrorism and domestic extremism.

In tracing the recent history of Pakistani counterinsurgency, Anatol Lieven argues that the security forces have won their fight against the insurgency of Pakistani Taliban, though terrorism will remain a serious problem for the foreseeable future. As set forth in "Counterinsurgency in Pakistan: the role of legitimacy," victory resulted not from new tactics but the recovery of legitimacy. In Balochistan, the nationalist insurgency has been of a different form and weaker than that of the Taliban, but it may prove longer-lasting. A number of interconnected questions emerge from this internal effort. Has the Pakistani experience been *sui generis*, or does it form part of wider patterns of success in counterinsurgency? As part of their campaign, did the Pakistani security forces develop a new doctrine of counterinsurgency or did they make things up as they went along? And in general, what are the lessons of the Pakistani experience for other countries faced with insurgency?

A similar ending has not attended the ongoing struggle in the Southern Border Provinces (SBP) of Thailand. In "Thailand's south: roots of conflict," Thomas A. Marks highlights the local and separatist foundation which has served to give the violent radical Islamism of present upheaval its unique character. Global and regional patterns notwithstanding, the Southern insurgency, though privileging terrorism in its approach, has remained an essentially locally produced and sustained movement, committed to local solutions for perceived repression and marginalization of Malay-speaking Muslims within the larger Thai-speaking, Buddhist polity. Necessary in-depth examination, though, has been driven by external actors due to the physical and analytical restrictions attendant to political restrictions upon local assessment of the situation.

Victory?

Part IV includes in-depth examination by two scholars of one of the most discussed recent cases of "victory" in irregular warfare, that of Sri Lanka in some three decades of costly battle with the Liberation Tigers of Tamil Eelam (LTTE). In the end, the state indeed eliminated its foe. Yet the authors speak to the problematic nature of both the process and the end-state, thus leading to the necessary question mark following the term "victory."

The costly counterinsurgency effort is examined by Sameer P. Lalwani in his chapter, "Size still matters: explaining Sri Lanka's counterinsurgency victory over the Tamil Tigers." He argues that even in – perhaps, especially in – cases of counterinsurgency that rise to civil war (actual as opposed to the statistical placeholder), resource mobilization remains essential to explaining outcome. Moving beyond military effectiveness literature, which fails to take into account material preponderance and relies more on theories of strategic interaction, Lalwani demonstrates the explanatory role ultimately played in the Sri Lankan case by national mobilization and its resulting material preponderance. Using a new quantitative dataset assembled on annual loss-exchange ratios, he demonstrates the superiority of materialist explanations above those of skill, human capital, and regime type.

In conclusion, Thomas A. Marks also examines "Sri Lanka: state response to Liberation Tigers of Tamil Eelam." Though covering mobilization of both sides in-depth, he is particularly focused upon the strategic implications of terminological misuse. The state initially misdiagnosed insurgency as terrorism, then maneuver warfare as guerrilla action. Only when civil war emerged was Colombo able to land a knockout punch; but even then, it had failed to assess the manner in which post-heroic values had shaped the global context to such an extent that Colombo found itself, far from being cheered for victory, rather assailed as war criminals. This proved particularly unfortunate, for in terms of strategic, operational, and tactical adaptation, the Sri Lankans ultimately provided a superb illustration. More tragic, the contested end of the struggle caused shrill voices to hold up LTTE as something it never was, a desirable (laudable even) alternative to democratic societal imperfection.

Future pathways

Assessed in their totality, the contributions in this volume go far toward offering robust use of terminology and analysis appropriate for the present state of irregular challenge. For analysts, a plethora of issues remain. Salient is the reality that the lifeline of all forms of extremism remains the nexus of ungoverned spaces, weak state and governance mechanisms, external support stemming from globalization (whether from states, global markets, or alienated diasporas), and transnational linkages with international terrorist movements that manifest themselves in a variety of forms. Appropriate response to extremism therefore must address a highly complex challenge that extends well beyond national boundaries.

Whether South and South East Asian states should replicate the models of the West and other parts of the world in dealing with insurgencies and terrorism or should evolve their own strategies has remained a matter of debate but one that has been in a sense quite overcome in practice. The British effort in Malaya, for instance, remains one of the popular sources of counterinsurgency approaches among many of the states in the region, even as the case itself is ill-understood and even less relevant. Better, in this, the twenty-first century, to examine the articles herein. In their contributions, the authors begin to fill existing gaps in theory in such manner as to be of value both to analyst and practitioner.

Notes

1 Boot, "The Evolution of Irregular War."
2 Marks and Rich, "Back to the Future – People's War in the 21st Century," 409–425.
3 Ibid.
4 Boot, "The Evolution of Irregular War."
5 Marks, "Counterinsurgency in the Age of Globalism," 22–29.
6 From John Adams to Hezekiah Niles, February 13, 1818.
7 Boot, "The Evolution of Irregular War."
8 Joes, *Urban Guerrilla Warfare*, 5.
9 Kilcullen, *Out of the Mountains*.
10 For the former, Marks, "Terrorism as Method in Nepali Maoist Insurgency, 1996–2016"; for the latter, Spencer and Melgar, "Bolivia, a New Model Insurgency for the 21st Century: from Mao Back to Lenin."
11 Rich, "A Historical Overview of US Counter-insurgency," 5–40.
12 Ibid.

13 Jones and Smith, "Grammar But No Logic," 439–440.
14 Porch, *Counterinsurgency*, 434; Gentile, *Wrong Turn*, 208.
15 Ucko, "Critics Gone Wild," 161–179.
16 Porch, *Counterinsurgency*, 9.
17 Porch, *Counterinsurgency*, 322.
18 Ucko, "Critics Gone Wild."
19 Black, *Rethinking Military History*, 19.
20 Gventer, "Counterinsurgency and its Critics," 637–663.
21 Boot, "The Evolution of Irregular War."
22 Ibid.

Bibliography

Adams, John to Hezekiah Niles, February 13, 1818; available at: https://founders. archives.gov/documents/Adams/99-02-02-6854 (accessed April 2, 2018).

Black, Jeremy. *Rethinking Military History*. London: Routledge, 2004.

Boot, Max. "The Evolution of Irregular War: Insurgents and Guerrillas from Akkadia to Afghanistan," *Foreign Affairs* 92 (2) (March/April 2013) available at: www.foreign affairs.com/articles/2013-02-05/evolution-irregular-war (accessed March 12, 2018).

Gentile, Colonel Gian. *Wrong Turn: America's Deadly Embrace of Counterinsurgency*. New York: The New Press, 2013.

Gventera, Celeste Ward. "Counterinsurgency and its Critics," *Journal of Strategic Studies* 37 (4) (2014), 637–663.

Joes, Anthony James. *Urban Guerrilla Warfare*. Lexington, KY: University Press of Kentucky, 2007.

Jones, David Martin and M. L. R. Smith. "Grammar But No Logic: Technique is not Enough – A Response to Nagl and Burton," *Journal of Strategic Studies* 33 (3) (June 2010), 439–440.

Kilcullen, David. *Out of the Mountains: The Coming Age of the Urban Guerrilla*. London: Hurst, 2013.

Marks, Thomas A. "Counterinsurgency in the Age of Globalism," *The Journal of Conflict Studies* 27 (1) (Summer 2007), 22–29; available at: https://journals.lib.unb.ca/index. php/JCS/article/view/5936 (accessed March 4, 2016).

Marks, Thomas A. "Terrorism as Method in Nepali Maoist Insurgency, 1996–2016," *Small Wars and Insurgencies* 28 (1) (February 2017), 81–118; available at: www.tandf online.com/doi/full/10.1080/09592318.2016.1265820 (accessed April 2, 2018).

Marks, Thomas A. and Paul B. Rich. "Back to the Future – People's War in the 21st Century," *Small Wars and Insurgencies* 28 (3) (June 2017), 409–425.

Porch, Douglas. *Counterinsurgency: Exposing the Myths of the New Way of War*. New York: Cambridge University Press, 2013.

Roper, Col. Daniel S. "Global Counterinsurgency: Strategic Clarity for the Long War," *Parameters* 38 (3) (Autumn 2008), 100.

Rich, Paul B. "A Historical Overview of US Counter-insurgency," *Small Wars and Insurgencies* 25, no. 1 (May 2014), 5–40.

Spencer, David E. and Hugo Acha Melgar. "Bolivia, a New Model Insurgency for the 21st Century: from Mao Back to Lenin," *Small Wars and Insurgencies* 28 (3) (June 2017), 629–660 available at: www.tandfonline.com/doi/full/10.1080/09592318.2017.1 307617 (accessed April 2, 2018).

Ucko, David H. "Critics Gone Wild: Counterinsurgency as the Root of all Evil," *Small Wars and Insurgencies* 25 (1) (May 2014), 161–179.

Part I
Emerging challenges

1 Countering the Islamic State in Asia

Dawood Azami

Introduction

Along with the establishment of its caliphate on June 29, 2014, the Islamic State (IS) – also known as ISIS or ISIL – presented a grand ambition of removing secular governments and spreading its version of Islamic rule across Muslim lands. Within a week of declaring the caliphate, Abu Bakr Al-Baghdadi made his first public appearance at the famous Al-Nouri Mosque in the center of Iraq's second city, Mosul, in which he asserted his position as the "Caliph (*Khalifah*)," or political-spiritual leader of all Muslims and named the countries and regions where Muslims faced hardships along with the violation of Islamic sanctities. While speaking in a confident and defiant mode, Al-Baghdadi also referred to a number of Asian countries and regions including Afghanistan, Pakistan, and Iran, as well as India and Kashmir, Indonesia, Burma (Myanmar), the Philippines, China and East Turkestan (China's Xinjiang Uighur Autonomous Region).[1]

As the group continued to consolidate power in its headquarters in Iraq and Syria, it called upon Islamist groups elsewhere to join its caliphate under the leadership of Al-Baghdadi. As part of its expansionist vision, the Islamic State formulated an aggressive strategy to evict its opponents and absorb existing militant groups. From its inception, the Islamic State attempted to differentiate itself from established Islamist groups and attacked its opponents on both military and ideological fronts. The Islamist State's robust propaganda, its powerful narrative, and the simplicity of its message attracted the attention of many militants in various parts of the world to include Asia. The group cleverly exploited real and perceived injustices and grievances among different Muslim communities in different countries and offered individuals redress. It convinced many Muslims across the world that joining the Islamic State would enable them to participate in building a new home and a bright and just future for "true Muslims."

The Islamic State labeled existing militant groups as incompetent accusing them of lacking the right vision and a winning strategy. In contrast, its brand offered militants a new template and a compelling vision for waging violent jihad. The group not only offered many battle-wary militants inspiration; it promised to provide training and materiel support as well. Indeed, the Islamic State's promise of establishing the long-awaited pan-Islamic caliphate was a dream come true for many militants.

From the very beginning, the Islamic State made clear its aim of expanding its rule through a wider network of *wilayats* (provinces or regions) as part of a single caliphate. Within weeks of Al-Baghdadi's grand announcement, the group's supporters distributed a map online that detailed areas of the world the Islamic State planned to have under its control within the next five years. The map, with slight variations, identified areas the group viewed as its long-term targets. In addition to expanding its caliphate throughout the Middle East and North Africa, the map revealed the Islamic State's ambitions to extend into large parts of South, Central, and East Asia by 2020. The group's "ten-state solution" map divided the area of its future global caliphate into ten administrative units. According to this map, Afghanistan, the Indian subcontinent (including India, Pakistan, and Bangladesh), Central Asian States, parts of Iran, and potentially parts of China (especially Xinjiang) and Indonesia would all become parts of its Khorasan/Khurasan province.[2] Thus, Khorasan represents the Islamic State's largest province both in terms of area and population.

The establishment of this province on January 26, 2015 marked the first time the Islamic State had officially spread outside the Arab world. Khorasan, which means the Land of the Rising Sun, is an ancient name for Afghanistan and the surrounding parts of Pakistan, Iran, and Central Asia.[3] Khorasan became the second most important region for the Islamic State following its headquarters in Iraq and Syria for mainly four reasons. First, the presence of numerous jihadist groups and ungoverned spaces made it an ideal place for the group to recruit and establish safe havens. Second, high ranking Islamic State members who had previously lived in the region, mainly as part of Al-Qaeda, already had links with a number of local militants.[4] Third, the presence of various Salafi/Wahhabi and anti-Shia groups also enhanced the region's importance and the Islamic State's chances of success there. Finally, following the establishment of its first Asian base in the Afghanistan–Pakistan region in January 2015, the Islamic State planned to export its militancy to Central Asia, South Asia, Southeast Asia, and China.

Caliphate calling

The idea of a caliphate has significant appeal for all Islamist groups. Over the past century, various Islamist groups have longed for the establishment of a Muslim Caliphate (*Khilafat*) consisting of all Muslim regions. Current debates about the caliphate are in many ways linked to the policies of European imperial powers from a century ago. The revolts against the Ottoman Empire's Sultan, who also had the title of the Caliph, in the first quarter of the twentieth century, were a cause of major concern for many Muslims, especially in South and Central Asia. The Muslims of colonial India (today's India, Pakistan, and Bangladesh) extended great respect to the Ottoman Sultan and usually recited his name in Friday prayer sermons. Thus, the question of the caliphate emerged as a significant political issue when European powers began dismantling the Ottoman Empire and its Caliphate (headquartered in modern day Turkey) after the First World War (1914–1918).

As a reaction to the partitioning of the Ottoman Empire, Muslims in British India established the *Khilafat* (Caliphate) Movement, which attracted groups and individuals from various political and social backgrounds and thus became the most prominent global political campaign on behalf of the Ottoman Caliphate. The "Khilafat Manifesto," published in 1920, called upon the British to protect the Caliphate and encouraged Muslims of the Indian sub-continent to unite and hold the British accountable for this purpose.[5] In the same year, the Khilafat leaders and the Indian National Congress – the largest political party in India led by Gandhi – created an alliance to work and fight together for the causes of the Khilafat and *Swaraj* (Indian independence from foreign denomination). Although the aims, ideology, and tactics of the Khilafat Movement of the twentieth century were different from what the Islamic State group propagates, several Islamist groups and individuals in South, Central, and Southeast Asia share the Islamic State's main objective of establishing a "Global Khilafat."

In addition to exploiting the existing grievances among certain groups and communities in the region, the Islamic State evokes the nostalgia of and desire for the global Islamic Caliphate. The group has simultaneously targeted people it hoped to recruit in three ways. First, it contacted important commanders and influential individuals within existing militant groups mainly through emissaries and using the services of Islamic State members from Middle Eastern countries who had acquaintances with local militants. Second, it targeted members of the general public and lower ranking militants using different means and tools of communication including the Internet, radio, publications, and face-to-face preaching. Third, the Islamic State focused its efforts on the Muslim diaspora, both in Muslim and non-Muslim countries, in an effort to have these individuals influence relatives and acquaintances in their home countries. The later strategy also proved effective in the Gulf Cooperation Council (GCC) region, which is home to millions of expatriates from South and Southeast Asia, with Indians forming the largest community.

Within weeks of the Islamic State's announcement of its caliphate, a number of individuals and militant groups in different parts of Asia began pledging allegiance (*bay'ah*) to Al-Baghdadi, mainly through the internet and social media. A Pakistani jihadi group, Tehreek-e-Khilafat Wa Jihad (Caliphate and Jihad Movement) and Ansar al-Tawhid al-Hind (Supporters of Monotheism in India) were the first South Asian militant groups that pledged allegiance to Al-Baghdadi in July 2014.[6] Further, some Pakistani militants had already gone to Iraq and Syria with the help of those Arab fighters who had previously fought in the Afghanistan–Pakistan region.[7] Attracted by the new jihadi template the Islamic State offered, some former commanders of the Pakistani Taliban Movement (TTP) and several members of other Pakistani Salafi and sectarian groups also pledged their allegiances. Meanwhile, the Islamic State's establishment of its caliphate and its surprising successes on the battlefield in Iraq and Syria inspired many other groups and individuals across Asia.

In early 2016, reports about the presence of local Islamic State chapters and sleeper cells emerged in different parts of Pakistan.[8] In January 2016, Rana Sanaullah, the law minister in the provincial government of Pakistan's most populated Punjab province, disclosed that around 100 people, including a few women, had left Pakistan to join the Islamic State in Iraq and Syria.[9] A month later, the director general of Pakistan's Intelligence Bureau (IB), Aftab Sultan, stated that the Islamic State had emerged as a threat in Pakistan as several militant groups, specifically the Sunni sectarian groups such as Lashkar-e-Jhangvi (LeJ) and Sipah-e-Sahaba Pakistan (SSP), "had soft corner for it."[10] Meanwhile, Pakistan's Foreign Secretary also acknowledged the Islamic State posed a serious threat to Pakistan.[11] In September 2016, the Pakistani military spokesman revealed that the country's security forces had arrested more than 300 Islamic State members including a few Iraqi and Syrian nationals in Pakistan.[12] More importantly, despite frequent high ranking Pakistani official denials of Islamic State presence in Pakistan,[13] the military announced in July 2017 that it launched a major operation against the Islamic State in the north-western region's Khyber Agency along the Afghan border.[14] Subsequently, reports emerged that Islamic State operatives were recruiting inmates at the Central Jail in Karachi, the country's largest city.[15]

In Afghanistan, where the Islamic State established the de facto capital of its Khorasan province, a few former lesser cadres from the disgruntled Afghan Taliban group and some members of Hizb-e Islami (Islamic Party) – a former Afghan insurgent group led by Gulbuddin Hekmatyar – also announced their allegiance. There were also individuals who joined the Islamic State because of its monetary resources and seeking both money and power. Nevertheless, the bulk of Islamic State fighters who settled in some of Afghanistan's ungoverned spaces, especially eastern Nangarhar Province, came from neighboring Pakistan. According to US, NATO, and Afghan officials, about 70 percent of Islamic State fighters based in Afghanistan were previously part of the Pakistani Taliban group (TTP).[16] Other Islamic State members based in Afghanistan include Afghans, Arabs, Tajiks, Uzbeks, Kazakhs, Kyrgyz, Turkmen, Chinese (both Uighurs and Huis), Chechens, Iranians, and Indians.[17] Several Iranian recruits – mostly from its ethnic and Sunni minorities such as Kurds, Baloch, and Arab/Ahwaz – also went to Iraq and Syria.[18]

The Islamic State further strengthened its position in Asia when Usman Ghazi, the leader of the Islamic Movement of Uzbekistan (IMU) – Central Asia's largest jihadist group that acts as a bridge for a wider variety of Central Asian fighters including Uighurs from western China[19] – pledged his allegiance to the Islamic State in August 2015.[20] Outside the Middle East and North Africa, Central Asia contributed the most fighters to the Islamic State in Iraq and Syria. Thousands of volunteers from five Central Asian states – Kazakhstan, Kyrgyzstan, Tajikistan, Turkmenistan and Uzbekistan – went to the Middle East to fight for the Islamic State.[21] Several members of the Chinese Turkestan Islamic

Party (TIP) and East Turkestan Independence Movement (ETIM) also pledged loyalty to the Islamic State. According to a Chinese source, hundreds of Uighurs from China's Xinjiang Region as well as Huis (Chinese Muslims of Han ethnicity) either went to Iraq and Syria or began promoting the Islamic State cause in the region.[22]

While this was occurring, the Islamic State was also gaining recruits from other South Asian as well as Southeast Asian countries. In September 2014, the commander of US Pacific Command, Admiral Samuel Locklear, stated that approximately 1,000 recruits from India to the Pacific may have joined the Islamic State to fight in Syria or Iraq. Without specifying the countries or a time-frame, Admiral Locklear added the "number could get larger as we go forward."[23] In the same month, M. K. Narayan, India's former National Security Adviser, acknowledged that up to 150 Indians were fighting with the Islamic State and added that "the figure could be much higher."[24]

In November 2014, a senior Islamic commander in Iraq and Syria, Abu Hudhaifah, claimed that the group was in contact with Indian, Bangladeshi, and Pakistani organizations and individuals that supported its objective of establishing a global *khilafah* (caliphate).[25] Over the next two years, several Indians went to the Middle East to join the Islamic State while some of them surfaced in Afghanistan. Meanwhile, Indian security forces arrested a number of suspected Islamic State members in various parts of India.[26] Islamic State sources as well as Indian officials also confirmed the presence of and attacks by Islamic State militants in the Indian administered Kashmir.[27] According to an Indian government official, "we simply don't know how many Indian citizens have joined ISIS?"[28] Given India's Muslim population of approximately 200 million, the number of Islamic State Indian recruits remains low on a per capita basis. Nevertheless, the Islamic State has posed new security challenges to India's social stability and internal security. Even the world famous religious school and the great seat of Muslim learning in South Asia, Darul-Uloom Deoband (India) apparently panicked after the emergence of the Islamic State and several Muslim organizations across India held meetings to counter the group's propaganda.[29]

Neighboring Bangladesh was another major target for the Islamic State where it hoped to gain substantial support.[30] The group introduced Abu Ibrahim al-Hanif as its head in Bangladesh and published his interview in the April 2016 issue of the group's magazine, *Dabiq*. Al-Hanif claimed that he was "able to connect and cooperate" with fighters in various Islamic State provinces, including Khorasan.[31] Abu Ibrahim al-Hanif was the nom de guerre for Tamim Ahmed Chowdhury, a 30-year-old Canadian of Bangladeshi origin who allegedly orchestrated some of Bangladesh's worst militant attacks including the overnight storming and siege of Holey Artisan café in the capital Dhaka on July 1, 2016. The Islamic State attack killed 22 people, most of them foreigners. Bangladeshi security forces eventually killed Chowdhury during a raid in August 2016.[32] Although, Dhaka has repeatedly denied the existence of an organized Islamic State network in the country and frequently called such groups as "a few home-grown outfits,"[33] some Bangladeshi security officials acknowledge that hundreds of their compatriots appear to have joined the Islamic State. They have also expressed concerns that the arrival of hundreds of thousands of Rohingya refugees following tension and violence in neighboring Myanmar might give the group another opportunity to exploit the drivers of alienation among the refugee population and gain more recruits.[34] The Islamic State also made inroads into the Indian Ocean archipelago of the Maldives, known to millions of tourists as an idyllic holiday destination. On a per-capita basis, according to some measures, the strategically located island nation has produced more jihadis than any other country with a number of them ending up in Syria and Iraq.[35]

The Islamic State's declaration of its caliphate also energized a number of militants in Southeast Asia, a region that has historically featured Al-Qaeda activity. The region – comprised of three Muslim-majority states (Indonesia, the world's most populous Muslim country, Malaysia, and Brunei), and several other countries with substantial Muslim minorities (the Philippines, Thailand, Singapore, and Myanmar/Burma) – emerged as a major recruitment center for the Islamic State. Positive responses to the group's call posed new and heightened challenges for the governments in Southeast Asian region which was already home to a wide variety of violent Islamist groups ranging from those focused on domestic issues and separatist movements to international jihadists. While the links between Southeast Asian militant groups and those centered in the Middle East had been relatively weak, the emergence of the Islamic State led to a new phase of Islamist militancy in the region.

Several factions of the Abu Sayyaf Group (ASG),[36] which is the most terrifying extremist group in the Philippines, and the Mindanao based Bangsamoro Islamic Freedom Fighters (BIFF), a splinter group of the Moro Islamic Liberation Front (MILF), were among the first Southeast Asian groups that pledged support for the Islamic State in August 2014. A few other smaller groups in the Philippines including Ansar Khilafah in the Philippines, the Islamic State in Lanao, and Jamaat al Tawhid wal Jihad (a group formerly loyal to Al-Qaeda) also released videos pledging their allegiance to the Islamic State.

By 2017, the Philippines, which has a history of Islamist and separatist violence, emerged as the second largest Islamic State stronghold in Asia after the Afghanistan–Pakistan region. The Islamic State attack against government forces in the largely Muslim southern Philippine city of Marawi in May 2017 caught the Southeast Asian region by surprise and quickly became the country's most significant internal security crisis in years. Led by Isnilon Tontoni Hapilon – a former prominent commander of Abu Sayyaf who in 2016 emerged as the leader of all Islamic State aligned groups in the Philippines including the Maute group – the Marawi attack involved hundreds of local and foreign militants. Although Philippine security forces killed Isnilon Hapilon in October 2017,[37] his ability to capture parts of Marawi city and endure five months of ground and air strikes fueled wider concerns about the Islamic State's potential in Southeast Asia. In addition, the group's ability to control an urban center in the Philippines for months, especially at a time when the group was losing ground in Iraq and Syria, furthered its propaganda and publicity and had far-reaching ramifications across Southeast Asia. Despite the killing and capturing of a number of suspected militants, the threat from IS-linked groups in the Philippines remains. In early 2018, Philippine authorities announced the arrest of two foreign Islamic State suspects in Marawi from the Middle Eastern origin indicating the international component of the group and its links to the conflict in the Philippines and the surrounding region.[38]

While the Marawi siege was underway, the Islamic State was also making gains in other parts of the region. The Indonesian based Jemaah Islamiyah (JI) – Al-Qaeda's historical ally in Southeast Asia – suffered defections to the Islamic State. Shortly after its announcement of the caliphate in June 2014, Abu Bakar Bashir, the spiritual leader and co-founder of Jemaah Islamiyah, as well as the emir of its offshoot Jemaah Ansharut Tauhid, pledged allegiance to Al-Baghdadi. The cleric, while serving a 15-year jail sentence, called on his followers to support and fight with the Islamic State. Mujahideen Indonesian Timor (MIT), a US-designated terrorist group, also joined the Islamic State after its leader Abu Warda Santoso, swore allegiance to Al-Baghdadi. Thus, within months of the declaration of the Islamic State caliphate, hundreds of Southeast Asians, mainly from Indonesia, Malaysia, Singapore, and the Philippines joined the Islamic State, with a number of them including some women, traveling to the Middle East.[39]

Although Indonesia banned Jemaah Islamiyah (JI) in 2008, the Islamic State continued its activities in the Asia Pacific nation and influenced other groups. In July 2018, an Indonesian court disbanded a relatively new group, Jamaah Ansharut Daulah (JAD), widely considered as the country's largest Islamic State-linked group. Indonesian authorities accused JAD – formed in 2015 and designated by the US State Department as a terrorist group in 2017 – of being involved in a deadly 2016 Jakarta attack and a wave of suicide bombings in May 2018 in the country's second-biggest city Surabaya in which two families – including girls aged nine and 12 – separately blew themselves up at three churches and a police station, killing more than a dozen people.[40]

Islamic State's failure in Asia

Within a year of announcing its caliphate, the Islamic State managed to eclipse Al-Qaeda in size, strength as well as reach by finding affiliates and establishing ever-larger numbers of franchises. The group had the potential of achieving even greater successes in Asia due to a number of favorable factors including physical terrain (forests and mountains), the presence of a wide range of militant groups as well as the existence of ungoverned spaces and the relative ease with which militants in the region could cross borders and establish links with the outside world. Furthermore, the lack of fruitful diplomatic engagement among various countries in Asia has made it easier for the Islamic State to operate and take advantage of both inter-state and intra-state tensions. Nevertheless, the group has struggled to become a major force in the wider Asian region.

The Islamic State generally attracted groups with waning relevance and relatively little capability; thus, bringing little in the way of true support for the group. Most groups that pledged allegiance to the Islamic State hoped this would bring them materiel support and that the Islamic State's aura of success would infuse them with relevance. On the other hand, almost all major militant groups present in different parts of Asia opposed the Islamic State's emergence and expansion. In Asia, the group was unable to make significant headway due to religious, ideological, logistical and cultural impediments.

The rivalry between the Islamic State and some of the established militant groups in the region proved a major obstacle for Islamic State expansion in Asia. A number of these Islamist groups have ideological and political differences with the Islamic State, undermining its religious credentials. These groups consider the Islamic State as an organization, rather than a caliphate and argue that the process for appointing Abu Bakr al-Baghdadi as caliph was not legitimate as it did not take place before a religious council representing all Muslims.

The Islamic State also became the first major militant group that directly challenged the leadership of the Afghan Taliban group which is arguably the biggest militant group in Asia and calls itself the Islamic Emirate of Afghanistan.[41] The Islamic State rejected the authority of the Taliban's founding leader Mullah Muhammad Omar, whom his followers regarded as the *Amir-ul-Momineen* (Leader of the Faithful) of the Islamic Emirate of Afghanistan. Indeed, the Islamic State dedicated the cover article of the December 2014 edition of it magazine, *Dabiq*, to highlight contradictions in policies of both Al-Qaeda and the Afghan Taliban and accused the two groups of compromising the integrity of Islam for personal gain. The Islamic State criticized Mullah Omar for practicing a "distorted version of Islam" and "using a patriotic and nationalistic tone." It also condemned the Taliban leader for his "respect of international conventions and borders"[42] and for rejecting "any initiative to conduct any operation outside Afghanistan in an effort to appease the international community."[43]

In the July 2015 issue of *Dabiq*, the Islamic State published a detailed *Fatwa* (religious edict) in which the group derided Mullah Omar as an Afghan nationalist and someone who lacked global vision. Highly critical of him, the group stated Mullah Omar "was at most one day a former leader of one of the Islamic lands" and did not fulfill the conditions and qualifications of becoming an *Imam* (leader of global Muslim community).[44] The Taliban in Afghanistan fought against the Islamic State on both military and ideological fronts[45] calling it a deviation and sedition *(Fitna)*, a "Western conspiracy" to "defame Islam" and "weaken Islamist groups."[46] Indeed, CIA Director John Brennan told the US Senate in June 2016 that the Islamic State "has struggled to maintain its cohesion in part because of competition with the Taliban" in Afghanistan.[47] The Afghan Taliban and the Islamic State have frequently fought against each other, mainly in eastern Afghanistan. In July 2018, the Taliban managed to drive out the Islamic State group from the north of Afghanistan. Dozens of the group's fighters were killed in a month long onslaught by the Taliban in Jawzjan province while those who remained alive surrendered to the Afghan government.[48]

The Afghan Taliban has also assured regional countries including Russia, China, and Iran, that unlike the Islamic State, its armed struggle is limited to Afghanistan and that it will not allow the Islamic State to establish a base in Afghanistan and threaten their security.[49]

The Afghan Taliban also extended its influence when it renewed its alliance with the Islamic Jihad Union, a splinter faction of the Islamic Movement of Uzbekistan after the main faction joined the Islamic State in August 2015. Like the Islamic Movement of Uzbekistan, the Islamic Jihad Union was also based in North Waziristan Tribal Agency prior to the Pakistani military's offensive in mid-2014.[50] The main Pakistani Taliban group (Tehreek-e Taliban Pakistan), headed by Maulana Fazlullah, also rejected the Islamic State's self-professed caliphate and criticized its brutal tactics in a pamphlet released in May 2015.[51]

Al-Qaeda, which had already started its war against the Islamic State in Iraq and Syria, exported its animosity with the group to Asia. Most major Asian militant groups allied to Al-Qaeda, which has deep roots in the region, were unwilling to end their old ties established during the leadership of Osama bin Laden. Conversely, most of the Pakistani and Indian based Islamist militant groups have traditionally focused on India and Kashmir rather than the Middle East. These groups usually invoke the prophesies about *Ghazwa-e Hind* (Battle of India), a great battle in India between believers and unbelievers before the Koranic end times. Meanwhile, Al-Qaeda renewed its focus on India following the emergence of the Islamic State to retain the support of these groups through the formation of Al-Qaeda in the Indian Subcontinent (AQIS).[52]

The Islamic State also proved divisive in Southeast Asia, even among some of the most violent Islamist groups in the region. Jemaah Islamiyah (JI) rejected the Islamic State and dismissed its members as extremists. Indeed, when the group's spiritual leader, Abu Bakar Bashir, pledged allegiance to the Islamic State, many of his followers abandoned him and chose to remain with Al-Qaeda. Other groups, such as the conservative Indonesian Mujahideen Council (Majelis Mujahidin Indonesia), also questioned that legitimacy of the process under which the Islamic State declared its caliphate.[53]

Rifts and internal divisions within local Islamic State factions and affiliates have proved major obstacles for its wider expansion in Asia. Defections have further weakened the group as it has failed to meet the high expectations its propaganda and recruitment campaign created. Many militants in various parts of Asia have turned their backs on the Islamic State after initially joining and then withdrawing support. The first major rift in the Islamic State branch in the Afghanistan–Pakistan region occurred when Abdul Rahim Muslim Dost – a leading Islamic State figure and major recruiter – rebelled against Hafiz Saeed Khan, the founding governor (*wali*) of the Islamic State's Khorasan province, in October 2015. Muslim Dost, a native of Afghanistan's eastern Nangarhar province, accused Hafiz Saeed Khan of using cruel tactics, violating Islamic teachings, disrespecting Afghans and putting the outfit in a wrong direction.[54] In Southeast Asia, the Islamic State suffered a serious blow when in early 2016, Abu Bakar Bashir, the founder of the Al-Qaeda affiliated Southeast Asian militant network, Jemaah Islamiyah, reversed a previous pledge of allegiance to the Islamic State. Southeast Asia's most influential militant cleric said that he was pulling his support because his knowledge of the group was limited and that he had changed his mind about the Islamic State.[55]

Moreover, the Islamic State's lack of respect for culture and history and local traditions annoyed local populations. For example, the group banned wedding ceremonies (which are usually accompanied with music) and prevented people in areas under its influence and control in Afghanistan from holding traditional funerals calling them incorrect innovations in Islam. Local people also opposed the group's brutal tactics and extreme punishments such as beheadings, burning homes and crops, and abduction of people it viewed as opponents or non-supporters. In one notorious example, an Islamic State video released in August 2015 showed ten blindfolded community elders being forced to sit on holes in the ground which were filled with explosives. The men in Afghanistan's Nangarhar province were then blown to pieces. Like its counterparts in the Middle East, the Islamic State in Asia took violence to new levels. Attacks on mosques, educational centers, shrines, and other public gatherings killed numerous civilians including women and children and resulted in a wider resentment toward the group. While the Islamic State has been successful in creating fear, such tactics proved counterproductive in winning over the people and damaged the group's popular appeal.

The religious ideology of the Islamic State is alien to the majority of Muslims in the region. The group's practice of *Takfir* – declaring other Muslims apostates and non-believers due to their supposed apostasy, heresy, or deviation from the Islamic State's interpretation of Islamic teachings – has proved unpopular among the majority of Asian Muslims. The vast majority of Muslims in South, Central, and Southeast Asia are Sunnis who are unfamiliar with the Islamic State's austere Wahhabi/Salafi ideology, an ultra-conservative Sunni form of Islam that dates from the mid-eighteenth century. Although the Islamic State opposes Sufism (Islamic mysticism) and calls it un-Islamic and polytheism, this form of Islam is popular among Muslims across Asia. Overall, attacks on shrines of Sufi Muslim saints and the imposition of its ideology has alienated the general public from the Islamic State.

Finally, the long distance between local Islamic State groups and its headquarters in Iraq and Syria has proven a major hurdle for the group's further expansion in Asia. With the passage of time, regular communication with the Islamic State Central and the receipt of money, weapons, and other types of physical support and advice has become increasingly risky for local Islamic State leaders in Asia. More importantly, losing territory and power in Syria and Iraq was a significant blow to the morale and finances of local Islamic State affiliates in other parts of the world, including Asia.

Eliminating the Islamic State in Asia

The emergence of the Islamic State brought new security challenges to many parts of Asia. As a relatively new group, the governments in the region are often challenged with understanding the full scale and scope of the Islamic State's capabilities and activities in their respective countries. Since the group's emergence in Asia, government officials across the region have offered confusing and contradicting statements about the presence, strength, and elimination of the Islamic State's local chapters. Depending on political and security calculations and interests, officials usually understate the presence and strength of Islamic State affiliates as well as local support for the group in order to demonstrate control of the situation.

Although the exact number of people who joined the Islamic State group in Central, South, and Southeast Asia is not documented, it is at least several thousand. Government officials in almost all Asian countries have repeatedly emphasized that they would deny the group any space on their soil. Nevertheless, the Islamic State is not only active; it is evolving, emerging in a variety of ways in different parts of Asia and changing its tactics and strategy to ensure its survival and expansion.

State responses to counter the Islamic State in Asia have mostly involved kinetic application of instruments of national power through military operations and law enforcement activities. From 2015 to 2018, government officials in many Asian countries regularly claimed the defeat or near defeat of Islamic State affiliates. For example, in early March 2016, Afghanistan's President, Ashraf Ghani, announced that Afghan and International Security Assistance Forces had defeated the Islamic State in the eastern parts of the country and stated, "Afghanistan will be their graveyard."[56] In early 2017, General John Nicholson, the senior US commander in Afghanistan, announced "a series of operations designed to defeat ISIL-K in 2017 and preclude the migration of terrorists from Iraq and Syria into Afghanistan."[57] The next month, US and NATO headquarters announced a reduction in the number of Islamic State militants in Afghanistan to about 700 or "perhaps even less."[58] In early 2018, however, Afghanistan's National Security Adviser, Mohammed Hanif Atmar, said there were approximately 3,000 Islamic State members in his country, more than four times of the US/NATO's estimate of 700 fighters the previous year.[59]

The Islamic State suffered several serious setbacks in Afghanistan and other parts of Asia from 2016 onwards. In addition to a large number of fighters, four governors (*walis*) of the group's Khorasan province have been killed in Afghanistan in US air strikes since July 2016. Nevertheless, even as the Islamic State lost both men and territory, the group continued to attract new followers. In Afghanistan, its members consist mainly of foreign militants and former members of the Afghan Taliban and Hizb-e Islami (Islamic Party) as well as newly radicalized and mostly unemployed youth. Faced with military pressure, the group expanded to other parts of the country and attracted new members from within Afghanistan and the rest of Asia, especially Pakistan and neighboring Central Asian countries. From 2017 onward, the Islamic State group in Afghanistan also received fighters from the Middle East, mostly those Central and South Asian fighters who had gone to fight in Iraq and Syria. After losing some territory in Afghanistan's eastern Nangarhar province, it established bases in neighboring Kunar, Nooristan, and Laghman provinces. Islamic State also established pockets in a few northern provinces including Jawzjan and Faryab.

Other Asian countries have experienced similar outcomes despite official promises and counter-Islamic State strategies. While the lack of clarity surrounding Islamic State's strength and strategy is a serious obstacle in formulating an adequate response at the national level, the lack of mutual trust among several countries, especially in South and Central Asia, is a major hurdle in countering the Islamic State at the regional level. Instead of working on joint mechanisms to stabilize the broader regions, several nations accuse each other of having incompatible agendas.

Afghan authorities often blame Pakistan's military establishment for the rise of violent religious extremism including the Islamic State in the region, claiming that those fighting within the group are promoting the strategic cause of Pakistan.[60] Many officials in the Indian government also believe that the IS-Khorasan group is linked to Pakistan and is part of a broader Pakistani strategy in the wider region and its effort to keep Afghanistan weak and under pressure.[61] Conversely, many in Pakistan see the group operating under the name of the Islamic State in Afghanistan as part of a nexus between Afghanistan and India or even an Indian-sponsored entity.[62] These varying narratives with scant scholarly works that corroborate them – all portray a region facing numerous challenges with respect to various threats associated with the Islamic State.

Meanwhile, Russian officials and politicians have accused the United States and NATO of orchestrating the deterioration of security in Afghanistan and the expansion of the Islamic State to destabilize the region.[63] On the other hand, Afghan and US officials have accused Moscow of deliberately exaggerating the strength and threat of the Islamic State Khorasan group to justify its interference in Afghanistan and other Central Asian countries. Meanwhile, Iran has accused the United States of orchestrating the emergence of the Islamic State (which considers Shias as non-Muslims) to create sectarian tensions and threaten Iran's stability. Although the United Sates has vigorously denied any involvement, the heightened rhetoric within the region has created more confusion about the nature of the Islamic State and its goals and ultimately hindered regional and international efforts to counter the group. Compared to South and Central Asia, countries in Southeast Asia better coordinate in the fight against the Islamic State. Indeed, during the Maute Group's (an Islamic State affiliate in the Philippines) siege of Marawi City in Mindanao in 2017, the three Southeast Asian countries the group most threatened – Indonesia, Malaysia, and the Philippines – closely cooperated with one another to counter the group.

The Islamic State's future in Asia

Despite numerous challenges and obstacles, the Islamic State has shown a remarkable resilience in Asia. The group not only survived its decapitation, it continues to attract new members in various countries. Though the group's global appeal was waning after losing control of so much of its territory in Iraq and Syria, its narrative still resonates with various segments of the population in Asia and the idea of a global Islamic Caliphate has a certain popularity. Following setbacks by the Islamic State Central in Iraq and Syria in 2016 and 2017, the Islamic State's Khorasan province, headquartered in Afghanistan, emerged as the most powerful affiliated group outside the Middle East. Indeed, the Islamic State Khorasan has shown remarkable resilience despite major losses. Given the frequency of attacks the group claims and the number of casualties it has inflicted, especially in 2017 and 2018, the Afghanistan–Pakistan region became the group's main battlefront outside Iraq and Syria. Beyond this development, the Islamic State has a military presence and small sleeper cells in many other parts of Asia which, if activated, have the potential for creating major disturbances, especially in urban areas.

Unlike Marawi City, it is improbable that Islamic State militants in Asia will be able to replicate what the group achieved in Iraq and Syria. Nevertheless, the group will continue its efforts to expand with a strategy composed of five elements: (1) consolidate its presence in certain hot spots and the small pockets it already controls; (2) continue its recruiting efforts – both in the virtual and physical domains; (3) establish new safe havens to provide areas for respite, and to rearm and refit as well as host its fighters from other areas and countries; (4) provide networking opportunities for Islamic State factions in various parts of Asia to share expertise and resources (manpower, money, and materiel); and (5) establish sleeper cells in urban centers to conduct guerrilla attacks involving individual and group suicide attacks, targeted killings, and use of IEDs (improvised explosive devices). These groups may also engage in mobile warfare and war of position as the Islamic State pockets in Afghanistan and the Maute Group in Marawi City in the Philippines.

Since realities on the ground as well as the real and perceived grievances that drive alienation of individuals in the Islamic State affiliate groups in Asia differ in many respects from those in the Middle East, these groups also have different outlooks. Indeed, Islamic State affiliates in Asia have adapted to these local conditions and exploit the existing ideological and sectarian tensions and rivalries between different state and non-state actors. Further, the return of the Islamic State's foreign fighters from Iraq and Syria, who are generally well trained and highly radicalized, presents both a threat and a challenge. The returning foreign fighters strengthens the existing Islamic State affiliated groups in the region and any attempt at reintegration into their home societies remains a major challenge.

Conclusion

Since the establishment of its caliphate in June 2014, the Islamic State rapidly expanded from its base in Iraq and Syria to other parts of the Muslim world through a network of local militants who pledged their allegiance to Abu Bakar Al-Baghdadi. The establishment of the Islamic State Khorasan province in January 2015 marked the first time that the group had officially spread outside the Arab world. Khorasan comprises the largest Islamic State province (*wilayat*), both in terms of area and population, and includes Central Asia, South Asia, and parts of Southeast Asia.

While faced with numerous challenges, the Islamic State has managed to gain a foothold in a crowded militant market and changed the dynamics of militant landscape in South, Central, and Southeast Asia. Despite several setbacks and suffering losses of men and territory in various parts of Asia, the group has sought to remain relevant and to adapt to new circumstances, changing its plans and strategy by resorting to guerrilla tactics such as suicide attacks and targeted killings, but in some cases also retaining the ability to conduct mobile warfare and war of position as seen in the cases of Afghanistan and the Maute Group in the Philippines. Thus, the Islamic State in Asia will continue pursuing its objectives in various parts of Asia and inspire professional jihadists to join its quest for a global caliphate.

The Islamic State in Asia is the product and transformation of the decades-old unrest and conflict in the region. Insurgent groups thrive in conflict zones and unstable environments and exploit the real or perceived injustices among the people. Unless the states in Asia address the legitimate grievances of its populations and improve their living standards, groups like the Islamic State will find fertile ground to recruit and flourish.

Long-lasting rivalries and lack of mutual trust among several state actors, especially in South and Central Asia, are the most significant obstacles in countering the Islamic State in Asia. Instead of pursuing joint mechanisms to stabilize the wider region, including robust intelligence gathering and timely intelligence sharing, several state actors accuse each other of hidden agendas and supporting proxies. The Islamic State, like Al-Qaeda, is an international organization and countering such a group is beyond the capacity of a single state. Multilateral and coordinated approaches among regional countries are required to successfully respond to this threat.

Meanwhile, states, both individually and multi-laterally, must regularly assess the effectiveness of their present strategies and the attendant risks of executing these strategies. While state responses are generally centered around military operations and law enforcement activities, they tend to ignore the unintended consequences of an overly kinetic approach. Violations of the rule of law and international humanitarian rights as seen in the detention and torture of innocent citizens due to faulty intelligence and the damage to life and property of ordinary people in military operations lead to back lash.

Moreover, contradictory statements from government officials in many Asian countries indicate a lack of understanding of the scale and scope of the Islamic State within their borders. This has presented a major obstacle in formulating an adequate response at the national and regional levels. Rather than understanding the group's nature and addressing it during the early stages of its development, states in Asia have generally dismissed the existence of the Islamic State or underestimated its strength until the group manifested itself as a greater challenge.

The Islamic State's simple but powerful narrative poses a new challenge to peace and stability in Asia already afflicted with violent agitations and insecurity.

States within the region have largely ignored the narrative aspect and the socio-political context within which the Islamic State gained momentum. Although individual religious scholars and official *ulama* (religious scholars) councils in various Asian countries have tried to discredit the group's ideology on religious grounds, Muslim scholars and religious institutions in Asia do not have mechanisms at the regional level to coordinate effective responses and formulate measures for countering the narrative of violent extremism. Further, states lack efficient and effective de-radicalization programs for existing Islamic State prisoners and returnees, who will most likely continue to engage in radicalization and recruitment efforts, both in prison and in the society at large. Thus, a comprehensive counter-violent extremism strategy utilizing all instruments of national power that includes engagement with civil society will need to be worked out by the affected states to address the evolving threat.

Notes

1 "Islamic State Leader," *Site Intelligence Group*, July 1, 2014.
2 Hall, "The ISIS Map"; "ISIS Map of Areas"; "The New Face"; Hosken, *Empire of Fear*, x, 260.
3 Azami, "The Islamic State," 131–158.
4 Ibid.
5 Minault, *The Khilafat Movement*, 22–23.
6 Crilly, "Pakistani Terror Group"; "Ansar al-Tawhid."
7 "Shaam main Pakistani Taliban."
8 "IS Visits Militants"; "Lal Masjid Cleric"; " 'IS Flag' Near Islamabad."
9 "Pakistan say sau afraad."
10 "IS Emerging as a Threat."
11 "Islamic State a Serious Threat."
12 "Pakistan say dawlat-e islamiya."
13 "Nisar Rules Out,"; "No Daesh in Pakistan."
14 "Pakistan Launches Offensive."
15 " 'Daesh Recruits' Untraceable."
16 US Department of Defense, "Department of Defense Press Briefing"; "Atmar: 80 dar sad."
17 "Afghan Official: Chinese"; "Rais-e Sitad-e Artish"; Swami, "From Afghan Hideout"; Interviews with Afghan officials and Taliban sources who wanted to remain anonymous. September–October 2017.
18 Cunningham, "Islamic State Threatens."
19 "Syria Calling," 7.
20 "IMU Declares."
21 "Syria Calling," 1–3; "Some 7,000 nationals"; "The Lure of Islamic State."
22 Interview with a Chinese official who wanted to remain anonymous. October 2017.
23 "Islamic State 'Brand'."
24 "Jihadi Wave in India"; Burke, "Terror Threat to India."
25 Mishra, "IS Confirms Presence."
26 Janardhanan. "How Islamic State."
27 "Islamic State J&K"; "Pledge Allegiance to"; "Amaq Reports 2nd IS"; "Net Chats."
28 Interview with an Indian official who wanted to remain anonymous. October 2016.
29 "Muslim Organizations Launch."
30 Mishra, "IS Confirms Presence"; Hossain, "Who is Behind."
31 "Interview with the Amir."

32 "New Evidence Shows."
33 Lister, "ISIS Attack in Bangladesh."
34 Interview with a Bangladeshi official who wanted to remain anonymous. October 2017.
35 Borri, "The Maldives"; Burke, "Paradise Jihadis"; "Maldives Crisis."
36 The Abu Sayyaf, an Arabic combination meaning "Father of the Swordsman" was formed in 1993 by Abdurajak Janjalani, a native from Basilan, southern Philippines, who had met and befriended Afghan and Middle Eastern Jihadis including Osama bin Laden in Pakistan in late 1980s. See Banlaoi, *Al-Harakatul Islamiya*, 12–23.
37 "Philippine Conflict."
38 Rauhala, "Philippine Police Arrest."
39 "Philippines Unrest"; Weiss, "The Islamic State Grows"; "The Islamic State Group's"; "Islamic State 'Brand'."
40 "Indonesian Court Bans."
41 Ash-Shamālī, "Al-Qa'idah of Waziristan"; Ash-Shāmī, "The Qa'idah of Adh-Dhawahiri."
42 Ash-Shamālī, "Al-Qa'idah of Waziristan."
43 Ash-Shāmī, "The Qa'idah of Adh-Dhawahiri."
44 "A Fatwa for Khurasan," *Dabiq*, 18–24.
45 Azami, "The Islamic State"; Azami, "Why Taliban Special Forces"; "Islamic State Spokesman."
46 Mutmain, *Mullah Omar, Taliban*, 355–357; "Muhtaram Abu Bakar al-Baghdadi"; Azami, "Why Taliban Special Forces."
47 Central Intelligence Agency (CIA). "Statement by Director."
48 "Taliban Says Defeats."
49 Azami, "World Powers Jostle"; Azami, "Is Russia Arming."
50 Roggio and Joscelyn, "Central Asian Groups"; "The Islamic Movement of Uzbekistan"; Joscelyn, "Islamic Jihad Union."
51 Pakistani Taliban Movement (TTP). "TTP's *Sharai'* Stance."
52 Haqqani, "Prophecy & the Jihad."
53 Liow, "ISIS Goes to Asia."
54 "Senior Leader Breaks Ties"; "Daesh's Cruelty in Kot."
55 Soeriaatmadja, "Bashir Withdraws Support"; "The Islamic State Group's Influence."
56 Faiez, Rahim, "Afghan President"; "Ashraf Ghani: ISIL Defeated."
57 General John W. Nicholson, "The Situation in Afghanistan"
58 Gul, "US Military: Number of IS Members."
59 Habtoor, Abdul Hadi, "Mohammed Hanif Atmar."
60 "Pakistani Militias Fighting"; "Afghan Security Chief Seeks"; "Interview With Dr. Rangin."
61 "Is ISIS Setting"; "Pakistan Military Provides."
62 "India Building Contacts"; "Is Indo-Afghan Nexus"; "How RAW – ISIS."
63 "ISIS Training Militants"; Russian Foreign Ministry, "Comment by the Information"; "Russia Expresses Distrust."

Bibliography

"A Fatwa for Khurasan," *Dabiq*, no. 10 (July 2015), 18–24 available at https://archive. org/stream/Dabiq10_201509/Dabiq%2010#page/n17/mode/2up/search/umar.

"Afghan Official: Chinese, Uzbek IS Militants Killed in Raid," *The Hindu*, February 9, 2018 available at www.thehindu.com/news/international/afghan-official-chinese-uzbek-is-militants-killed-in-raid/article22700168.ece?utm_source=RSS_Feed&utm_medium=RSS&utm_campaign=RSS_Syndication.

"Afghan Security Chief Seeks to Unify Region Against Taliban," *Radio Free Europe/ Radio Liberty (RFE/RL)*, (September 4, 2016) available at https://gandhara.rferl.org/a/ afghanistan-nsa-hanif-atmar/27966541.html.

"Amaq Reports 2nd IS attack in Kashmir Valley, Targeting Indian Policeman," *Site Intelligence Group*, (February 26, 2018) available at https://ent.siteintelgroup.com/State ments/amaq-reports-2nd-is-attack-in-kashmir-valley-targeting-indian-policeman.html.

"Ansar al-Tawhid in the Land of Hind Pledges to IS, Repeats IS Spokesman's Call for Attacks," *SITE Intelligence*, (October 6, 2014) available at http://ent.siteintelgroup. com/Multimedia/ansar-al-tawhid-in-the-land-of-hind-pledges-to-is-repeats-is-spokesman-s-call-for-attacks.html.

"Ashraf Ghani: ISIL Defeated in Eastern Afghanistan," *Aljazeera*, (March 6, 2016) available at www.aljazeera.com/news/2016/03/isis-ashraf-ghani-defeated-eastern-afghanistan-160306093417163.html.

Ash-Shamālī, Abū Jarīr. "Al-Qa'idah of Waziristan: A Testimony From Within," *Dabiq*, no. 6 (December 2014) available at http://worldanalysis.net/14/wp-content/ uploads/2014/12/dabiq_6.pdf.

Ash-Shāmī, Abū Maysarah. "The Qa'idah of Adh-Dhawahiri, Al-Harari, and An-Nadhari and the Absent of Yemeni Wisdom," *Dabiq*, no. 6 (December 2014) available at http:// worldanalysis.net/14/wp-content/uploads/2014/12/dabiq_6.pdf.

"Atmar: 80 dar sad jangjoyaan-e daesh dar nangarhar Pakistani hastand" [Atmar: 80 Percent Daesh Fighters in Nangarhar are Pakistanis], *One TV*, (Dari), (October 27, 2015) available at www.1tvnews.af/fa/news/afghanistan/19189-2015-10-27-00-30-01.

Azami, Dawood. "Is Russia Arming the Afghan Taliban?" *BBC*, (April 2, 2018) available at www.bbc.co.uk/news/world-asia-41842285.

Azami, Dawood. "IS in Afghanistan: How Successful Has the Group Been?" *BBC*, (February 25, 2017) available at www.bbc.co.uk/news/world-asia-39031000.

Azami, Dawood. "The Islamic State in South and Central Asia," *Survival: Global Politics and Strategy* 58 (4) (2016), 131–158 available at http://dx.doi.org/10.1080/00396338.2 016.1207955.

Azami, Dawood. "Why Taliban Special Forces Are Fighting Islamic State," *BBC*, (December 18, 205) available at www.bbc.co.uk/news/world-asia-35123748.

Azami, Dawood. "World Powers Jostle in Afghanistan's New 'Great Game'," *BBC*, (January 12, 2017) available at www.bbc.co.uk/news/world-asia-38582323.

Banlaoi, Rommel. *Al-Harakatul Islamiyah: Essays on the Abu Sayyaf Group*. Quezon City: Philippine Institute for Political Violence and Terrorism Research (2008) available at http://pipvtr.com/pipvtr/files/Book_AHAI_Essays_on_ASG_Book_Banlaoi_2008.pdf.

Borri, Francesca. "The Maldives: Where Jihadists are Heroes," Norwegian Refugee Council, (March 14, 2017) available at www.nrc.no/perspectives/2017/where-jihadists-are-heroes/.

Burke, Jason. "Paradise Jihadis: Maldives Sees Surge in Young Muslims Leaving for Syria," *Guardian*, (February 26, 2015) available at www.theguardian.com/world/2015/ feb/26/paradise-jihadis-maldives-islamic-extremism-syria.

Burke, Jason. "Terror Threat to India Rising Again Six Years After Mumbai Attacks," *Guardian*, (November 26, 2014) available at www.theguardian.com/cities/2014/ nov/26/india-terror-threat-mumbai-attacks.

Central Intelligence Agency (CIA). "Statement by Director Brennan as Prepared for Delivery Before the Senate Select Committee on Intelligence," (June 16, 2016) available at www.cia.gov/news-information/speeches-testimony/2016-speeches-testimony/ statement-by-director-brennan-as-prepared-for-delivery-before-ssci.html.

Crilly, Rob. "Pakistani Terror Group Swears Allegiance to Islamic State," *Telegraph*, (July 9, 2014) available at www.telegraph.co.uk/news/worldnews/asia/pakistan/10955563/Pakistani-terror-group-swears-allegiance-to-Islamic-State.html.

Cunningham, Erin. "Islamic State Threatens More Bloodshed in Iran," *Washington Post*, (August 15, 2017) available at www.washingtonpost.com/world/middle_east/iran-was-at-the-forefront-of-the-fight-against-isis-now-it-has-to-face-the-militants-at-home/2017/08/14/e4fb735a-7dfe-11e7-b2b1-aeba62854dfa_story.html?utm_term=.9720905cbfe9.

"Daesh Commander Killed in Nangarhar Airstrike," *Tolo News*, (August 26, 2018) available at www.tolonews.com/afghanistan/daesh-commander-killed-nangarhar-airstrike.

"'Daesh Recruits' Untraceable After Release From Karachi Jail," *The News*, (September 28, 2017) available at www.thenews.com.pk/print/233035-Daesh-recruits-untraceable-after-release-from-Karachi-jail.

"Daesh's Cruelty in Kot District Unjustifiable: Muslim Dost," *Pajhwok*, (July 9, 2016) available at www.pajhwok.com/en/2016/07/09/daesh%E2%80%99s-cruelty-kot-district-unjustifiable-muslim-dost.

Faiez, Rahim. "Afghan President: Islamic State Being Wiped Out in Afghanistan." *Associated Press*, (March 7, 2016) available at www.military.com/daily-news/2016/03/07/afghan-president-islamic-state-being-wiped-out-in-afghanistan.html.

Gul, Ayaz. "US Military: Number of IS Members in Afghanistan Reduced to 700." *Voice of America (VOA)*, (March 1, 2017) available at www.voanews.com/a/afghanistan-islamic-state/3745401.html.

Habtoor, Abdul Hadi. "Mohammed Hanif Atmar: 55,000 Terrorists Present in Afghanistan," *Asharq Al-Awsat*, (February 23, 2018) available at https://aawsat.com/english/home/article/1184736/mohammed-hanif-atmar-55000-terrorists-present-afghanistan.

Hall, John. "The ISIS Map of the World," *Daily Mail*, (June 30, 2014) available at www.dailymail.co.uk/news/article-2674736/ISIS-militants-declare-formation-caliphate-Syria-Iraq-demand-Muslims-world-swear-allegiance.html#ixzz4riLoZ6Rm.

Haqqani, Husain. "Prophecy & the Jihad in the Indian Subcontinent," *Hudson Institute*, Washington, DC, (March 27, 2015) available at http://hudson.org/research/11167-prophecy-the-jihad-in-the-indian-subcontinent.

Hosken, Andrew. *Empire of Fear: Inside the Islamic State*. London: Oneworld Publications, 2015.

Hossain, Akbar. "Who is Behind the Bangladesh Killings?" *BBC*, (October 14, 2015) available at www.bbc.co.uk/news/world-asia-34517434.

"How RAW – ISIS is Spoiling Pakistan-Iran Relations," *Times of Islamabad*, (May 1, 2017) available at https://timesofislamabad.com/how-raw-isis-is-spoiling-pakistan-iran-relations/2017/05/01/.

"IMU Declares it is Now Part of The Islamic State," *Radio Free Europe/Radio Liberty (RFE/RL)*, (August 6, 2015) available at www.rferl.org/content/imu-islamic-state/27174567.html.

"India Building Contacts With ISIS, Al-Qaeda in Afghanistan," *The News*, Pakistan, (October 12, 2015) available at www.thenews.com.pk/print/15137-india-building-contacts-with-isis-al-qaeda-in-afghanistan.

"Indonesian Court Bans ISIL-linked Group Behind Deadly Attacks." *Aljazeera*, (July 31, 2018) available at www.aljazeera.com/news/asia-pacific/2018/07/indonesian-court-bans-isil-linked-group-deadly-attacks-180731055025110.html.

"Interview with Dr. Rangin Dadfar Spanta," *Afghanistan.ru*, (November 3, 2015) available at http://farsi.ru/doc/11547.html.

"Interview with the Amir of the Khilāfah's Soliders in Bangal Shaykh Abū Ibrāhīm al-Hanīf," *Dabiq*, no. 14 (April 2016) available at www.clarionproject.org/docs/Dabiq-Issue-14.pdf.

"'IS Flag' Near Islamabad Alarms Law Enforcers," *Dawn*, (September 25, 2017) available at www.dawn.com/news/1359907/is-flag-near-islamabad-alarms-law-enforcers.

"Is Indo-Afghan Nexus Pitching ISIS-K Against Pakistan?" *Daily Pakistan*, (June 24, 2017) available at https://en.dailypakistan.com.pk/pakistan/is-indo-afghan-nexus-pitching-isis-k-against-pakistan-mysterious-airdrops-to-militant-group-raises-eyebrows-in-islamabad/.

"Is ISIS Setting its Sights on India in Afghanistan?" *The Wire*, (December 8, 2016) available at https://thewire.in/85368/isis-india-afghanistan/.

"IS Emerging as a Threat, Warns IB Chief," *Dawn*, (February 11, 2016) available at www.dawn.com/news/1238771.

"IS Visits Militants in Baluchistan: Jundullah Spokesman," *Dawn*, (November 12, 2014) available at www.dawn.com/news/1143997.

"ISIS Map of Areas it Wants to Take Over by 2020 Includes India," *India Today*, (August 10, 2015) available at http://indiatoday.intoday.in/story/india-in-2020-if-isis-plans-succeed/1/457594.html.

"ISIS Training Militants From Russia in Afghanistan, 'US and UK Citizens Among Instructors'," *Russia Today (RT)*, (October 8, 2015) available at www.rt.com/news/317989-afghanistan-isis-train-russians/.

"Islamic State a Serious Threat to Pakistan, Foreign Secretary Admits," *Dawn*, (February 23, 2015) available at www.dawn.com/news/1165415/islamic-state-a-serious-threat-to-pakistan-foreign-secretary-admits.

"Islamic State 'Brand' Gains Ground Among Asian Militants," *Reuters*, (September 26, 2014) available at http://uk.reuters.com/article/2014/09/26/uk-southeast-asia-militants-idUKKCN0HL03120140926.

"Islamic State J&K Chief Among Four Terrorists Killed in Anantnag," *Times of India*, June 22, 2018. https://timesofindia.indiatimes.com/india/islamic-state-jk-chief-among-four-terrorists-killed-in-anantnag/articleshow/64695421.cms.

"Islamic State Leader Abu Bakr al-Baghdadi Encourages Emigration, Worldwide Action," *Site Intelligence Group*, (July 1, 2014) available at https://news.siteintelgroup.com/Jihadist-News/islamic-state-leader-abu-bakr-al-baghdadi-encourages-emigration-worldwide-action.html.

"Islamic State Spokesman Calls on Other Factions to 'Repent,' Urges Sectarian War," *The Long War Journal*, (June 23, 2015) available at www.longwarjournal.org/archives/2015/06/islamic-state-spokesman-calls-on-other-factions-to-repent.php.

Janardhanan, Vinod. "How Islamic State is Spreading Wings in Indian States, Slowly But Surely," *Hindustan Times*, (May 5, 2017) available at www.hindustantimes.com/india-news/how-islamic-state-is-spreading-wings-in-indian-states-slowly-but-surely/story-jdIwORErcAOJ1bBUcf1WAM.html.

"Jihadi Wave in India at a Height: Ex-NSA Narayanan," *Business Standard*, (September 23, 2014) available at www.business-standard.com/article/news-ians/jihadi-wave-in-india-at-a-height-ex-nsa-narayanan-114092301111_1.html.

Joscelyn, Thomas. "Islamic Jihad Union Participated in Siege of Kunduz." *Long War Journal*, (October 3, 2015) available at www.longwarjournal.org/archives/2015/10/islamic-jihad-union-participated-in-siege-of-kunduz.php.

"Lal Masjid Cleric Comes out to Support Islamic State," *The News*, (December 14, 2014) available at www.thenews.com.pk/archive/print/542487-lal-masjid-cleric-comes-out-to-support-islamic-state.

Liow, Joseph Chinyong. "ISIS Goes to Asia." *Foreign Affairs*, (September 19, 2014) available at www.foreignaffairs.com/articles/east-asia/2014-09-19/isis-goes-asia.

Lister, Tim. "ISIS Attack in Bangladesh Shows Broad Reach as 'Caliphate' Feels Pressure," *CNN*, (July 4, 2016) available at http://edition.cnn.com/2016/07/03/asia/bangladesh-isis-al-qaeda/index.html.

"Maldives Crisis: Opposition Leaders Meet US Officials, Ask Them to Act Urgently," *Hindustan Times*, (March 15, 2018) available at www.hindustantimes.com/world-news/maldives-crisis-opposition-leaders-meet-us-officials-ask-them-to-act-urgently/story-XlxGVScw4oYLnAvVRd5osO.html.

Minault, Gail. *The Khilafat Movement: Religious Symbolism and Political Mobilization in India*. New York: Columbia University Press, 1982.

Ministry of Defence of the Russian Federation. "Chief of the General Staff General of the Army Valery Gerasimov Took Part in the International Conference on Interaction of Defence Agencies on Security in Afghanistan and Central Asia in the New Environment," (October 8, 2015) available at http://eng.mil.ru/en/news_page/country/more.htm?id=12060200@egNews.

Mishra, Abhinandan. "IS Confirms Presence of Several Indian Fighters." *The Sunday Guardian*, (November 15, 2014) available at www.sunday-guardian.com/news/is-confirms-presence-of-several-indian-fighters.

"Muslim Organizations Launch Campaigns Against 'Un-Islamic' ISIS." *Times of India*, (September 29, 2015) available at http://timesofindia.indiatimes.com/india/Muslim-organizations-launch-campaigns-against-un-Islamic-ISIS/articleshow/49155295.cms.

Mutmain, Abdul Hai. *Mullah Omar, Taliban aw Afghanistan* [*Mullah Omar, Taliban and Afghanistan*]. (Pashto book), Kabul: Afghan Publishing Society, 2017.

"Muhtaram Abu Bakar al-Baghdadi ta da Islami Amaarat da Rahbari Shura lik" [Letter by the Leadership Council of the Islamic Emirate to Respected Abu Bakr al-Baghdadi]. *Afghan Taliban's website*, (June 16, 2015) available at http://alemara1.org/?p=17042.

"Net Chats Between Kashmir and Syria, Iraq on Security Radar," *The Deccan Herald*, (May 7, 2017) available at www.deccanherald.com/content/610319/net-chats-kashmir-syria-iraq.html.

"New Evidence Shows Deep Islamic State role in Bangladesh Massacre," *Reuters*, (December 1, 2016) available at www.reuters.com/article/us-bangladesh-islamicstate-insight/new-evidence-shows-deep-islamic-state-role-in-bangladesh-massacre-idUSKBN13P2WK.

Nicholson, General John W. "The Situation in Afghanistan." Testimony Before the Senate Committee on Armed Services, Washington, DC., (February 9, 2017) available at www.armed-services.senate.gov/imo/media/doc/Nicholson_02-09-17.pdf.

"Nisar Rules Out Presence of IS Militants in Pakistan," *Dawn*, (November 11, 2014) available at www.dawn.com/news/1143737.

"No Daesh in Pakistan, Says Interior Minister," *Dawn*, (February 13, 2016) available at www.dawn.com/news/1239280.

"Pakistan Launches Offensive Against IS Near Afghan Border," *BBC*, (July 16, 2017) available at www.bbc.co.uk/news/world-asia-40627093.

"Pakistan Military Provides Weapons, Training to ISIS in Afghanistan," *The Hindu*, (February 25, 2016) available at www.thehindu.com/news/international/pakistan-military-provides-weapons-training-to-isis-in-afghanistan/article8280787.ece.

"Pakistan say dawlat-e islamiya kay khatmay kaa kaam jaari" [Task of Elimination of Islamic State From Pakistan Continues], *BBC Urdu*, (September 1, 2016) available at www.bbc.com/urdu/pakistan/2016/09/160901_pakistan_islamic_state_zz.

"Pakistan say sau afraad Iraq jaa chokay hain" [100 People Have Gone to Iraq From Pakistan], *BBC Urdu*, (January 20, 2016) available at www.bbc.com/urdu/pakistan/2016/01/160104_punjab_islamicstate_as.

"Pakistani Militias Fighting Alongside ISIS Militants: Nangarhar NDS Chief," *Khama Press*, (June 16, 2017) available at www.khaama.com/pakistani-militias-fighting-alongside-isis-militants-nangarhar-nds-chief-02950.

"Philippine Conflict: Duterte Says Marawi is Militant-Free," *BBC*, (October 17, 2017) available at www.bbc.co.uk/news/world-asia-41647876.

"Philippines Unrest: Who are the Abu Sayyaf Group?" *BBC*, (June 14, 2016) available at www.bbc.co.uk/news/world-asia-36138554.

"Pledge Allegiance to Islamic State, Exhorts Video of 'Kashmiri Fighter'," *Times of India*, (December 26, 2017) available at https://timesofindia.indiatimes.com/india/pledge-allegiance-to-islamic-state-exhorts-video-of-kashmiri-fighter/articleshow/62245683.cms.

"Rais-e Sitad-e Artish-e Afghanistan: 2 hazar uzwe daesh kushtah shudah and" [Afghan Army Chief: 2 Thousand Daesh Members Killed], *BBC Persian*, (February 18, 2017) available at www.bbc.com/persian/afghanistan-39015171.

"Some 7,000 nationals of ex-Soviet Countries Fight for Islamic State – FSB Deputy Chief," *TASS*, (November 10, 2015) available at http://tass.ru/en/world/835147.

Swami, Praveen. "From Afghan Hideout, Kerala Jihad Leader Calls Faithful to Caliphate," *Indian Express*, (May 18, 2017) available at http://indianexpress.com/article/india/from-afghan-hideout-kerala-jihad-leader-calls-faithful-to-caliphate-4661362/.

Rauhala, Emily. "Philippine Police Arrest Man Suspected of Recruiting for the Islamic State," *Washington Post*, (February 19, 2018) available at www.washingtonpost.com/world/asia_pacific/philippine-police-arrest-suspected-islamic-state-recruiter-in-manila/2018/02/19/b9945e2e-1558-11e8-942d-16a950029788_story.html?utm_term=.d7bcea51e858.

Roggio, Bill and Thomas Joscelyn. "Central Asian Groups Split Over Leadership of Global Jihad." *Long War Journal*, (August 24, 2014) available at www.longwarjournal.org/archives/2015/08/central-asian-groups-split-over-leadership-of-global-jihad.php.

"Russia Expresses Distrust to US Over 'Helicopters Carrying Daesh' in Afghanistan," *Sputnik*, (September 21, 2017) available at https://sputniknews.com/middleeast/201709211057572844-russia-distrust-daesh-afghanistan-us-chopers/.

Russian Foreign Ministry. "Comment by the Information and Press Department on Repeated Accusations of Russian Support for the Taliban," (May 23, 2017) available at www.mid.ru/en/foreign_policy/news/-/asset_publisher/cKNonkJE02Bw/content/id/2764002.

Schmidt, Michael S. and Eric Schmidt. "U.S. Broadens Fight Against ISIS With Attacks in Afghanistan," *New York Times*, (January 31, 2016) available at www.nytimes.com/2016/02/01/us/politics/us-broadens-fight-against-isis-with-attacks-in-afghanistan.html.

"Senior Leader Breaks Ties With IS Khorasan Chief," *Pajhwok*, (October 19, 2015) available at www.pajhwok.com/en/2015/10/19/senior-leader-breaks-ties-khorasan-chief.

"Shaam main pakistani taliban kaa 'adah'" [Pakistani Taliban's 'Base' in Syria], *BBC Urdu*, (July 12, 2013) available at www.bbc.com/urdu/pakistan/2013/07/130712_pak_taliban_syria_zs.

Soeriaatmadja, Wahyudi. "Bashir Withdraws Support for ISIS," *The Straits Times*, (January 9, 2016) available at www.straitstimes.com/asia/bashir-withdraws-support-for-isis.

"Syria Calling: Radicalisation in Central Asia," *International Crisis Group (ICG)*, (January 20, 2015) available at www.crisisgroup.org/en/regions/asia/central-asia/b072-syria-calling-radicalisation-in-central-asia.aspx.

"Taliban Says Defeats Islamic State Fighters in North Afghanistan," *Reuters*, (August 1, 2018) available at https://uk.reuters.com/article/uk-afghanistan-islamic-state/taliban-says-defeats-islamic-state-fighters-in-north-afghanistan-idUKKBN1KM440.

"The Islamic Movement of Uzbekistan: An Evolving Threat," *Radio Free Europe/Radio Liberty (RFE/RL)*, (May 31, 2014) available at www.rferl.org/content/islamic-movement-uzbekistan-roundtable/25405614.html.

"The Islamic State Group's Influence in Indonesia," *BBC*, (July 20, 2016) available at www.bbc.co.uk/news/world-asia-35312624.

"The Lure of Islamic State for Central Asians," *RFE/RL*, (February 7, 2015) available at http://gandhara.rferl.org/content/central-asia-islamic-state/26835423.html.

"The New Face of Jihadism?" *Qantara*, (July 7, 2014) available at https://en.qantara.de/content/isis-leader-in-iraq-the-new-face-of-jihadism.

"TTP's *Sharai'* Stance Regarding The Caliphate Announced by Sheikh Abu Bakar Al-Baghdadi," *Pakistani Taliban's (TTP) Statement*, (May 2015) available at https://umar-media.files.wordpress.com/2015/06/to-download-the-file-in-urdu-click-here.pdf.

US Department of Defense. "Department of Defense Press Briefing by General Nichol-son," (September 23, 2016) available at www.defense.gov/News/Transcripts/Transcript-View/Article/954839/department-of-defense-press-briefing-by-general-nicholson-via-teleconference-fr/.

US Department of Defense. "Enhancing Security and Stability in Afghanistan," (June 2016) available at www.defense.gov/Portals/1/Documents/Enhancing_Security_and_Stability_in_Afghanistan-June_2016.pdf?source=GovDelivery.

US Department of State. "Foreign Terrorist Organization Designation of ISIL – Khorasan (ISIL-K)," (January 14, 2016) available at www.state.gov/j/ct/rls/other/des/266511.htm.

"U.S. Forces Target, Destroy ISIS-K Stronghold," *Resolute Support/NATO, Kabul*, (April 13, 2017) available at www.rs.nato.int/article/press-releases/u.s.-forces-target-destroy-isis-k-stronghold.html.

Weiss, Caleb. "The Islamic State Grows in the Philippines," *Long War Journal*, (June 24, 2016) available at www.longwarjournal.org/archives/2016/06/islamic-state-officially-creates-province-in-the-philippines.php.

2 The Philippines' counter-terror conundrum

Marawi and Duterte's battle against the Islamic State

Richard Javad Heydarian

Introduction

A year into office, Philippine President Rodrigo Duterte confronted an unexpected shock: the emergence of Islamic State (IS)-affiliated groups in his home island of Mindanao. The daring siege on the city of Marawi, the country's largest Muslim-majority city, by the Maute Group (MG) – a self-styled Jihadist group – sent shock waves across the region and beyond. The ensuing urban warfare saw a twentith-century conventional military struggling against twenty-first century insurgency. After weeks of heavy fighting, the largely destroyed Marawi resembled post-conflict Mosul and Aleppo in the Middle East. This inevitably raised major policy as well as political questions for the Philippine government and beyond.

First, the Battle of Marawi calls into question the competence of the government of the Philippines in anticipating, preventing, and countering the expansion of IS franchises in its troubled south. Second, it underscored the growing global footprint of IS in heavily populated, Muslim-majority areas in Asia, despite the extremist organization's seemingly irreversible military and strategic setbacks in western Iraq and eastern Syria. Third, it provoked profound anxiety among major Western powers, particularly the United States and Australia, which have feared the emergence of a *wilayat* (governorate) or so-called "distant caliphate" in Southeast Asia. Fourth, there are growing concerns over the peace process in Mindanao, which almost reached its apotheosis under the Benigno Aquino administration only to suffer a sudden breakdown in early 2015 after clashes between government forces and rebels in Mamasapano. Fifth, it also highlights failing state institutions as well as complex and combustible tribal-clan dynamics, which may have contributed to the emergence of the MG and its successful months-long siege on a large city like Marawi.[1]

Thus the crisis in southern Philippines is not only a reflection of the growing organizational reach of the IS but also the depth of democratic deficit and poor governance in Mindanao, which has provided a conducive atmosphere for radicalization and extremist mobilization. As early as 2015, Australia has been sounding alarm bells over the specter of IS in the Philippines. "We always consider the potential threat posed by radicalized Filipinos supporting the [IS]," a senior Australian police official told *Reuters* in 2015, underlining growing concerns in Canberra. "We are concerned with the risk of [IS] elements travelling to the country to promote violent extremism and, worse, to seek haven or use the country as a transit point in going to conflict zones."[2]

To explore these issues, this chapter methodologically discusses counterterrorism doctrine of the Philippines by looking at declassified government documents and making use of extensive interviews and discussions with senior members of the military and the National Security Council (NSC), media as well as diplomats. It also utilizes a comparative and trans-regional understanding of the evolution of IS and its expansion beyond its Middle Eastern heartland.

IS pivot to the Philippines

On May 23, 2017, a large contingent of violent radical Islamists launched a daring assault against key government offices in Marawi. The attack, which was planned months ahead,[3] came shortly after a failed raid on the suspected safe house of Isnilon Hapilon, the leader of Abu Sayyaf Group (ASG), who was also the IS-designated emir of jihadist fighters in Mindanao. The attack seemed particularly timely and opportunistic, since it took place precisely when almost the entirety of the defense leadership, including Defense Secretary Delfin Lorenzana, were thousands of miles away in Russia.

Initially, the Armed Forces of the Philippines (AFP) tried to downplay the whole incident. National Security Adviser Hermogenes Esperon claimed that the government was in "full control" of the city.[4] It did not take long, however, for the Duterte administration and the broader public to realize that the assault on the city was well planned and effectively executed. The attackers' blitzkrieg assault took the government forces by complete surprise. Within the first 24-hours, they rampaged across the city with impunity, setting fire to government buildings, attacking detention facilities and freeing prisoners, and planting the black IS flag at strategic locations within the city. They also took a Catholic bishop, Father Teresito Suganob, along with countless others, as hostages – later exploiting them as human shields and bargaining chips. These would give them significant leverage against government forces, which constantly faced the dilemma of inflicting civilian casualties during their bombardments of "militant" positions. After all, under a democratic republic, the government was under constant pressure to minimize casualties and maintain the support of the population. In the event, the latter were almost entirely displaced. As many as 400,000 individuals had to relocate to congested refugee camps on the outskirts of the city.

Three months into the siege, around 70 surviving militants managed to keep many as 7,000 troops at bay, but by late October 2017, after some five months, Philippine President Rodrigo Duterte triumphantly declared Marawi liberated from terrorists.[5] The statement came shortly after the AFP eliminated the attackers' leadership in the form of Isnilon Hapilon, the designated Emir of the prospective IS *wilayat*, and Omar Maute, whose brother, Abdullah was killed earlier in September by government troops. Dr. Mahmud bin Ahmad, the Malaysian conduit between the IS command in Raqqa and the Mindanao-based militants, was also among top leaders killed.

Ultimately, government forces managed to eliminate as many as 800 attackers, while saving as many as 1,700 hostages.[6] The number of civilian casualties was contested, with figures ranging from several dozen to as many as thousands.[7] There was no doubt, however, that the government's inaccurate and extensive use of air support in the first month of operations contributed not only to friendly-fire casualties and significant destruction of several neighborhoods, but likely also civilian casualties.[8]

Despite their eventual defeat, the IS-affiliates – as it eventually became clear they were – pulled off a major propaganda victory by holding off a large military, backed by foreign support ranging from the U.S. and Australia to China and Russia. The Maute Group (MG), which had assumed leadership of the effort, used its local and foreign fighters (mostly Indonesians) in sophisticated fashion, exploiting a network of underground tunnels to elude government forces, transport material, and refurbish its position. Months ahead of the attack, the group managed to preposition significant weaponry within the city and then, during the siege, making use of skilled marksmen at strategic locations. Other cutting-edge technology, such as Improvised Explosive Devices (IEDs), was integrated into the attack matrix.

Sensing an impending takeover of the city, Duterte declared Martial Law on May 24, not only in the epicenter of clashes but across the entire island of Mindanao. It later emerged that there were concerns that coordinated bombing and terror attacks were in the offing in other urban centers. MG in fact had targeted Duterte's home city of Davao the previous year (September 2, 2016). Officials were concerned that other IS-affiliated groups such as ASG, Bangsamoro Islamic Freedom Fighters (BIFF), Khilafa Islamiyah Mindanao (KIM), and Anshar Khalifa Philippines (AKP) could launch successive attacks against soft targets in major cities, from Basilan to Maguidanao to Zamboanga and Davao, either as a diversionary tactic – to weaken government offensive in Marawi – or to sow terror and secure a territorial footprint across Mindanao (see Figure 2.1).

The swift declaration of Martial Law proved controversial, especially among those who questioned Duterte's political intentions. Human rights groups and opposition legislators challenged the declaration as an affront to the country's democratic institutions, serving as a potential springboard for the creation of a full-fledged autocratic system.[9] Yet the Supreme Court eventually ruled in favor of the president, affirming his executive prerogative in deciding on matters of national security and the necessary scope of any Martial Law declaration.[10]

Touting a more independent foreign policy, the tough-talking Filipino president initially relied on the AFP to single-handedly stave off the IS threat in his backyard. Only a few weeks into the battle, though, it became clear that external help was necessary. In early June, the AFP and the Department of National Defense (DND) solicited direct American military assistance, a considerable turnabout from the president's consistent effort since his inauguration to claim he was in effect sending the U.S. packing. Subsequently, the Pentagon not only provided substantial individual weaponry but also deployed drones to generate

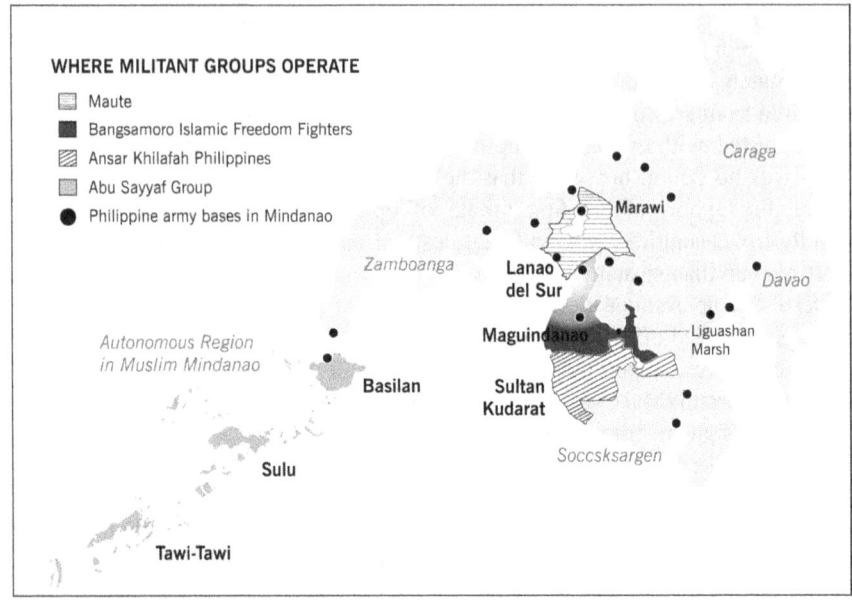

Figure 2.1 Areas of operation of IS-affiliated groups.

Source: Reuters, Philippine Army Recruitment Office.

real-time intelligence – as well as a Special Forces unit to train Filipino soldiers in modern urban warfare.[11] As Admiral Harry Harris, commander of the U.S. Pacific Command, told the author on the sidelines of the Shangri-La Dialogue in June 2017, "We are involved in activities in Mindanao to help the Armed Forces of the Philippines take the fight to [IS] in the Philippines."[12] Other Western leaders, particularly Malcolm Turnbull, also expressed all-out support for the Philippines amid the battle in Marawi, which dominated the discussion among regional defense officials in the Shangri-La Dialogue in Singapore in early June.[13]

When the U.S. Secretary of State Rex Tillerson visited the Philippines in early August, his discussions with Duterte primarily focused on counterterrorism cooperation in Mindanao. The Pentagon reportedly even considered direct drone attacks against IS-affiliated militants using MQ-9 Reaper drones.[14] In late August, Duterte also held an unusually open and friendly meeting with Nick Warner, the reclusive chief of Australian Secret Intelligence Service (ASIS). Counter-terrorism was also at the heart of Duterte's bilateral meeting with U.S. President Donald Trump at the sidelines of the Association of Southeast Asian Nations (ASEAN) Summit in early November 2017.[15] While Washington relishes a full-fledged treaty alliance with Manila, which allows for rotational American military presence on Philippine soil, Canberra, in turn, has a Status of Forces Agreement (SOF) with Manila, which underpins bilateral counter-terror cooperation and intelligence sharing.

Asian neighbors also pitched in. China offered a multi-million dollar defense aid package, providing assault weapons to the AFP. It has also proposed joint-military exercises as well as intelligence sharing with the Philippines. Russia proposed joint-naval exercises as well as intelligence sharing. Both China and Russia have been deeply worried about the prospect of Uighur and Chechen fighters joining the Battle of Marawi and gaining lethal experience that could be used at home.

Southeast Asian partners such as Malaysia and Indonesia have negotiated joint patrols with the Philippines in their highly porous tri-border, which covers the Sulu Sea (Philippines), the Celebes/Sulawesi Sea (Indonesia), and waters off the coast of the Malaysian state of Sabah – an area racked by piracy and terror activities of ASG. Aside from coordinated patrols on their overlapping maritime border, the three countries have also discussed intelligence-sharing and specific online and offline programs aimed at countering extremist ideology and mobilization.[16] Meanwhile, Singapore offered a C-130 aircraft for humanitarian aid transport, advanced urban warfare training in its facilities, as well as drones "to enhance the intelligence, surveillance and reconnaissance capabilities."[17] ASEAN moved toward consolidating intelligence-sharing and counter-terrorism cooperation among its member states, while Malaysia, Indonesia, and the Philippines are contemplating the prospect of sustained coordinated patrols in their maritime tri-border.

The emerging consensus among regional leaders is that the Philippines is the weak link in Southeast Asia, thus in need of maximum assistance lest the IS-affiliates manage to establish a *wilayat* in the region and inspire Marawi-like operations beyond Mindanao. The Marawi crisis and IS infiltration into the Catholic-majority nation marked a dangerous convergence of exogenous (i.e., IS pivot to Asia) as well as endogenous (e.g., Duterte's drug war, peace process breakdown, and systemic governance deficit) factors, which will be discussed in succeeding sections.

IS pivot to Asia

As late as the first quarter of 2015, some analysts pointed out that the main aim of IS was to establish a functioning and tangible Caliphate in the heartland of the Middle East. Thus, it was argued, IS differed from al-Qaeda, which has focused on targeting "far enemies" in the West rather than conquering actual territory and placing millions of people under its puritanical interpretation of *Sharia*.[18] As Graeme Wood argued, IS could not survive as an underground movement, since its dynamic of seeking pledges of allegiance and encouraging popular migration to its authentic Islamic state had tapped into a desire for just such an opportunity upon the part of a particular population (a number of whom actually desired martyrdom).[19] Yet, beginning in late 2014, perspicacious observers noticed a subtle but significant shift in the strategic disposition of IS brought by the growing shift in correlation of forces on the ground.

Throughout its initial phase of expansion, IS largely squared off against disorganized and demoralized Iraqi security forces, a heavily overstretched Syrian army, and poorly equipped Kurdish fighters. In the latter half of 2014, however, major regional and international powers directly stepped in, with Iran and the United States, together with Turkey and Persian Gulf allies, leading the way. What followed was coordinated and extensive aerial bombardment of IS troops and strategic assets across both Iraq and Syria. Tehran also deployed its Quds special forces to both theatres, providing tactical guidance to Iraqi and Syrian forces. Washington deployed special forces to Iraq and Syria, providing direct support to Kurdish fighters as well as Iraqi government troops. Ankara's concurrent decision to seal off its border with Syria represented another major blow, since it made it increasingly difficult for foreign fighters to migrate (*Hijra*) to the Caliphate in Raqqa. The final nail in the coffin was the decision of Russia to join the fray (September 2015) after the Syrian government in Damascus requested direct military assistance.

The direct entry of major powers heavily tipped the conventional warfare scale against IS. In its pronouncements, IS betrayed a growing sense of worry over its material survival in the Middle East and major setbacks it suffered in the conventional battlefields. As a result, the organization's leadership began to devise alternative strategies to ensure survival in the short-run as well as relevance in the global Jihadi landscape in the long-run. The shift in IS strategy was reflected in its main publication, *Dabiq* magazine, which began covering topics such as "From the Battles of Al-Ahzab to the War of Coalitions," "The Failed Crusade," and "Just Terror."[20]

IS turned its attention to targeting "far enemies" and expanding its global footprint by establishing new alliances with Jihadist groups in other Muslim-majority regions, a strategy that appeared not unlike that of al-Qaeda. Between October 2014 and November 2015, the group engaged in no less than 25 plots and attacks against Western citizens and interests, not to mention bombings in Turkey, Lebanon, Iraq, and across the Middle East and North Africa. It also aggressively pushed for expansion of its franchises in Sub-Saharan Africa (Nigeria), the Sinai, Libya, and Southeast Asia.

The new strategy had three key elements: first, expansion of so-called "lone wolf" operations against *kafir* (infidel) regimes, which entailed a heavy level of coordination with and financial and logistical support from the leadership in Raqqa; second, planning and executing high-profile "spectacular" terror attacks against soft targets and major capitals around the world; and third, establishment of *wilayats* in Muslim-majority regions of Africa and Asia.[21]

This reorientation would have major implications in the Philippines for at least four reasons. First, the southern island of Mindanao is home to millions of Sunni Muslims, a majority of whom live in abject poverty with minimal assistance from the national government. Throughout the past century, these Muslims in Mindanao have been subjected to systematic discrimination, large-scale land dispossession, and often maltreatment by the largely Christian military and security forces. This destructive dynamic reached its apogee during the Martial Law years under the Ferdinand Marcos dictatorship (1971–1986), which waged an intense military effort against Muslim separatist groups, namely the Moro

National Liberation Front (MNLF), which was then under the leadership of Nur Misuari, and later its breakaway and more Islamist-leaning faction, the Moro Islamic Liberation Front (MNLF), which was then under the leadership of Hashim Salamat. The result was significant civilian casualties and widespread popular displacement. Certain Muslim-majority provinces eventually gained a measure of political and socio-cultural self-determination under the Autonomous Regional of Muslim Mindanao (ARMM), which was established in 1990, but chronic corruption and capture of local government institutions by warlords and former rebels deprived communities of desperately needed public goods and services.[22] The dreary developmental landscape coupled with a dark and bloody history of conflict thus provided a conducive environment for radicalization, including among the educated and idealistic youth.

Second, Mindanao has extremely porous borders with the Muslim-majority nations of Indonesia and Malaysia, where regional Jihadist groups such as Jemaah Islamiyah (JI) have been well entrenched. The combustible combination of governance deficit and geographical accessibility facilitated the entry of transnational terrorist groups into Mindanao.

Third, these transnational terrorist groups were dominated by those from the Middle East. Throughout the 1990s, as al-Qaeda Central faced military pushback in Afghanistan, it gradually expanded into East Asia, particularly into Mindanao. Over time, together with JI, it established training camps and a thick network of alliances with local Jihadist groups as well as Moro Islamic Liberation Front (MILF). Thus, the prospect of a full-scale IS pivot to Mindanao, involving the deployment of Arab apparatchiks and fighters to southern Philippines became not altogether improbable.

Fourth, recent years have seen a troubling deadlock in peace negotiations between the government and the biggest rebel force in the country, MILF. After years of promising negotiations under the Benigno Aquino administration (2010–2016) had resulted in the signing of a Framework Agreement between the two parties in 2012, the peace process suffered a violent jolt in early 2015. The incident, better known as the Mamasapano Massacre, saw clashes between Philippine National Police' Special Action Force (SAF) and members of the MILF in the village of Mamasapano, Maguindanao.[23] The ensuing public backlash, particularly among the Catholic majority, undercut the Aquino administration's push for the passage of a Bangsamoro Basic Law (BBL), the foundational legal framework for establishment of a fully autonomous sub-state Bangsamoro entity encompassing all Muslim-majority provinces. In the face of public pressure, the Philippine Congress refused to endorse and approve the BBL before the end of Aquino's term.

With no clear prospect of full-autonomy, cynicism and agitation gained traction among the rank-and-file of the MILF, which warned of further defections within its ranks, especially among more hardline elements that insisted upon full independence under *Sharia* rather than autonomy within the Christian-majority Philippine state. In fact, the MILF detachment involved in the Mamasapano massacre was from the often controversial 105th Base Command, which was accused of cooperation with the IS-affiliated BIFF. Saudi-trained radical rebel Umbra Kato, a disgruntled former member of the MILF, formed the BIFF in the late 2000s with the aim of establishing a fully independent Islamic State in Mindanao.

The 2016 election of Rodrigo Duterte as the first president from Mindanao raised hopes of an end to the impasse. Yet Duterte, who boldly pushed for federalism and bravely advocated for the BBL during his presidential campaign, would end up spending much of his first year in office waging a bloody campaign against illegal drugs, which overstretched Philippine intelligence and security agencies. As one expert put it, Duterte's anti-drug campaign inadvertently led to "the weakening of the strategic capabilities of the state's security forces by way of the involvement and preoccupation of the military in a deep, multi-layered and intractable drug war."[24]

There is still debate as to whether the Marawi events represented a massive intelligence failure or/and failure to appreciate good intelligence.[25] What is clear is that the Duterte administration not only failed to pre-empt the siege on Marawi, later struggling to wrest control of the besieged town without inflicting massive destruction, but largely ignored the peace process with Muslim rebel groups. It was not until July, more than a month into the battle, that the president began to pay closer attention to peace negotiations with MILF and vowed to push for a revised BBL in the legislature, which is dominated by his allies.[26]

These four factors collectively turned Mindanao into a perfect destination for the broader global ambitions of IS. At first, it was not clear whether the operation in Marawi was a product of purely indigenous design among IS-affiliated Filipino *Jihadist* groups with political blessing from the IS command or, instead, directly aided by the leadership in Raqqa. In mid-2014, between July and August, a number of Mindanao-based Jihadist groups pledged *Bayah* (allegiance) to the IS leadership. The emerging coalition was composed of the ASG, MG, BIFF, and lesser-known groups such as KIM and Anshar Khalifa Philippines (AKP). They all aimed to leverage the ideological appeal and organizational resources of the transnational terror group in order to establish an Islamic, *Sharia*-based society in Mindanao. Though a relative newcomer, MG, due to the organizational and military acumen of the Maute brothers (Abdullah and Omarkhayam), quickly rose to become a leading force among the IS-affiliated groups in Mindanao.

There was also an element of geography, since the Maute family were a force in Lanao Del Sur, where Marawi – the ultimate prize for the IS affiliates in the Philippines – is located. Beginning in 2016, the IS leadership provided direct and full recognition of the Philippine Jihadist groups as the harbingers of a *Daulah Islamiyah Wilayatul Mashriq* (Islamic State – Eastern Region). ASG leader Isnilon Hapilon was given the title *Emir* of a prospective *wilayat* in Southeast Asia.

What stood between the Filipino jihadists and the *wilayat* in the Orient, however, was the hard task of conquering actual territory, with significant resources and population. The Islamic State of Marawi proved a perfect target.

Over the years, Hapilon deftly used the eschatological and organizational appeal of IS to unite and invigorate disparate Jihadist groups spread across maritime mainland Mindanao and hailing from divergent ethnic-tribal backgrounds.[27] In short, it was IS that allowed the ASG leader to transcend, at least for the moment, the complex ethnic, tribal strategic and tactical faultlines that had long divided Islamist insurgents and rebels. Moreover, by creating a united force, they convinced the IS leadership of their reliability and strength.

As IS suffered successive military setbacks in the heartland, losing Mosul and on the verge of exiting Raqqa, it explored alternative havens in the Far East. The deadlock in peace negotiations between the government and MILF provided a perfect opening for the IS-affiliated groups to seize the initiative and overwhelm the heavily distracted Duterte administration. Through its Indonesian and Malaysian fighters, the IS leadership in Raqqa not only gave the green light for the Marawi operation but also provided direct financial and logistical assistance. As one authoritative report explains:

> Marawi operations received direct funding from ISIS central and reveal a chain of command that runs from Syria through the Philippines to Indonesia and the rest of Southeast Asia. ISIS central seems to have been represented by Khatibah Nusantara, the fighting unit led by the Indonesian named Bahrumsyah and his associate, Abu Walid. Khatibah Nusantara in turn sent funding through Dr Mahmud Ahmad, a Malaysian who sits in the inner circle of the Marawi command structure. Dr Mahmud controlled recruitment as well as financing and has been the contact person for any foreigner wanting to join the pro-ISIS coalition in the Philippines. Tactical decisions on the ground are being made by the Philippine ISIS commanders themselves, but the Syria-based Southeast Asians could have a say in setting strategy for region when the siege is over.[28]

Philippines counter-terror infrastructure

Thus, Philippines became the new front in the global struggle against IS. The crisis in Marawi was but the latest round in a long-running cycle of conflict and violence not only in Mindanao but also across the country's restive rural areas. Since its inception, the Philippine state has had to confront two major sources of insurgency, both using terrorism as but one method: the Islamist-Moro rebels, primarily in Mindanao, and the communist rebels, who have been active across the country but mostly concentrated, especially in recent decades, in poverty-stricken rural areas. MNLF, MILF, and the Communist Party of the Philippines (CPP), together with its military wing, the New People's Army (NPA), were eventually recognized as legitimate revolutionary groups. In response, the Philippine government engaged in peace negotiations with all three, anchored by the ultimate objective of finding a mutually satisfactory political compromise. The ensuing Disarmament, Demobilization, and Reintegration (DDR) of these rebel-revolutionary groups would, the government hoped, gradually transform them into constitutive elements of the Philippine state and political leadership. This has partially happened in the case of the MNLF, the leadership of which was incorporated into the governance structure of the ARMM.

The creation of a Bangsamoro nation is supposed to bring MILF into the fold, as well. As for the CPP-NPA, the government hopes to strike a final agreement that will incorporate the legitimate grievances of the group (particularly on questions of land reform and social justice) into the operating principles and policies of the Philippine state. So far, Manila has had only partial success in the case of MNLF; it has yet to overcome legal hurdles (i.e., passage of BBL) for a final peace agreement with the MILF; and has struggled to prevent negotiations with CPP-NPA from collapsing. Now, the specter of IS terrorism has risen in Mindanao.

Defining terrorism is often an intellectually fraught and politically sensitive undertaking. It presents a policy conundrum to national governments as to characterization of specific groups which challenge the legitimacy of the state and its monopoly on the use of force. After all, categorizations have direct implications for policy, since if a specific group is deemed as revolutionary, then there is room for political compromise, but if it is, alternatively, classified as a purely terrorist organization, then it has to be degraded, contained, and, if possible, eliminated. In practice, however, certain groups move from one category to the other and swing back and forth depending on their disposition toward the state, their strategic calculus, and the ideological dialectics of the leadership. This has been the case for the MNLF and the MILF, though the CPP-NPA is still formally classified as a terrorist organization by Washington[29] even as it is in practice (e.g., by special forces) treated as what it is, an insurgency.

The government defines terrorism as "the premeditated use or threatened use of violence or means of destruction perpetrated against innocent civilians or non-combatants, or against civilian and government properties, usually intended to influence an audience."[30] This agrees with other definitions in this volume. Terrorists may resort to varying methods in achieving their objectives, ranging from piracy, assassination, to hostage taking, and indiscriminate use of violence against civilians.[31] In response, the Philippine government has adopted the "Fourteen Pillars of Policy and Action Against Terrorism" that comprise the essence of the country's "CT/CVE" doctrine,[32] further operationalized under Memorandum Order Number 37 (Article I), signed October 12, 2001. It has the following specific measures to combat terrorism:

> Join the international counter terrorist coalition and work with the United Nations; work closely with the United States on intelligence and security matters concerning terrorism; make available Philippine airspace and facilities if the latter are required as transit or staging point; contribute logistical support in the form of food supplies, medicine and medical personnel; subject to the concurrence of the Philippine Congress, provide combat troops if there is an international call for such troops; and prevent the flow of funds to terrorist groups in accordance with the Anti-Money Laundering Act of 2001 (Philippine Republic Act No. 9160) and other laws.[33]

The Cabinet Oversight Committee on Internal Security (COC–IS) is in charge of supervising the implementation of the government's response to terrorism. More specifically, the COC–IS is in charge of managing crises, which meet the threshold of national security threats. It is also in charge of the National Peace and Order Council (NPOS), with the Secretary of the Interior and Local Government (SILG) and the Secretary of National Defense (SND) as co-chairs. For intelligence gathering, the AFP, PNP, the National Bureau of Investigation (NBI), and the National Intelligence Coordinating Agency (NICA) are key elements. The NSC, as the principal adviser to the president in matters of security and intelligence, assists him in crafting national security policy and assessing its implementation. It sources its intelligence from AFP, PNP, NBI, and NICA. The NBI, NICA, and AFP conduct regular meetings, at both national and regional levels, under the Regional Intelligence Coordinating Committee (RICC) to coordinate/triangulate/share intelligence on a need-to-know basis.

In order to translate intelligence into concrete policies, the NSC consults the NICA, which represents the main intelligence agencies of the government. NICA oversees the operations of the National Intelligence Committee (NIC) and the Regional Intelligence Committee (RIC). Meanwhile, the National Intelligence Board (NIB) is the main forum for intelligence exchange among high-ranking officials from various lead departments. It is also tasked with providing the NICA Director-General with independent advice on the effectiveness of the Intelligence Community (IC). It also facilitates coordination and integration of intelligence from local and international operations. In order to ensure effective intelligence gathering, the government established the Special Monitoring Committee (SMC) in 2002 for the purpose of counterintelligence operations. The SMC plays a key role in monitoring threats to high-ranking officials (i.e., assassination plots) as well as the general public (i.e., terrorist plots).

However, it was not until the Gloria Arroyo that the country began to develop a robust and coherent CTE/CVE apparatus and doctrine, especially as Washington stepped up its Global War on Terror (GWOT) campaign in the aftermath of the 9/11 attacks by the Al-Qaeda (AQ). Previously, counterinsurgency doctrine had served the purpose for meeting internal threats. Under Executive Order Number 21, Series 2001, known as the "National Internal Security Plan," the government forwarded a Strategy of Holistic Approach (SHA) to address both insurgency (rebellion) and terrorism, particularly in Mindanao. Not surprisingly, given the country's long irregular warfare experience, terrorism and insurgency were assessed to be interrelated elements, which had non-military genealogy, thus the necessity for a SHA, which employs politico-diplomatic, socio-economic, and military approaches to address the roots of insurgency and terrorism. This is but the historically effective "left hand/right hand" approach that has featured in Filipino documents since the immediate post-Marcos years.[34] As a result, the Government of the Philippines' (GOP) CTE/CVE strategy involves all key national departments to address drivers of insurgency and terrorism. The SHA relies on both short-term CTE response, mainly led by the PNP and AFP, but also more long-term policies to mitigate and eliminate conditions that trigger armed rebellion and enable extremist ideology to morph into organized acts of terror (see Figure 2.2).

Contrary to almost all its neighbors, the CTE/CVE strategy of the Philippines has been heavily dependent on external guidance and assistance, as well as close and institutionalized cooperation. This has been provided by the U.S. through "Operation Enduring Freedom – Philippines." Originally implemented during the Arroyo administration, it incorporated best practices from previous administrations and around the world. As noted above, it built on past efforts, which also relied on extensive American assistance: the "All Out Friendship or All Out Force" campaign under defense minister and later president Ramon Magsaysay (1953–1957), who partnered with the extraordinary Lansdale/Bohannan team, military officers seconded to the CIA at the height of Cold War, as well as a fair-sized military advisory effort; the *Oplan Katatagan* campaign of President Ferdinand Marcos (1965–1986), which proved the bloodiest with questionable efficacy; and the *Lambat-Bitag* campaign of President Corazon Aquino (1986–1992), which drew its animating essence from the extraordinary career of former Philippine Constabulary (PC)-turned NPA guerrilla-turned APF chief of intelligence Victor Corpus.[35]

The incorporation of the Philippines into the Operation Enduring Freedom matrix saw the U.S. Special Operations Command Pacific (SOCPAC) playing a pivotal role in guiding and assisting the Arroyo administration's CT operations against groups such as JI, ASG (previously known as Al Harakatul Islamiyah), and BIFF, as well as lesser known threats such as the Pentagon Gang and the Rajah Solaiman Movement (RSM), now renamed Syuful Khilafa Fi Luzon. By 2003, Washington acknowledged joint CT operations against ASG, which saw

Figure 2.2 Philippine Strategy of Holistic Approach (SHA).
Source: Pena 2007.

the participation of several hundred U.S. special operations personnel deployed in support of their Filipino counterparts. Under the Joint Special Operations Task Force-Philippines, as many as 750 American military personnel were stationed in Mindanao, albeit rotationally, because the Philippine constitution prohibited direct foreign participation in combat (a legacy of left-wing influence in the post-Marcos years). At one point, the effort was described by the *New York Times* as the "the largest single deployment of American military might outside Afghanistan to fight terrorists since the Sept. 11 attack."[36] More recently, it has officially ended.

Due to its extensive terrorism and criminal activities (especially kidnap-for-ransom), ASG proved the key target. It had expanded from a few hundred fighters into a force estimated at several thousand at its peak. Its violent radical Islamism was at times overshadowed by its criminality, but its demonstrated international links with global terror drove the combined response. Following a spate of kidnapping prior to 9/11, including in a luxury resort in Sipadan, Malaysia in 2000, ASG operations moved to large-scale terrorist attacks, including bombing of a super ferry in February 2004 and the Valentine's Day bombings in three cities in February 2005.

Philippine–American CT cooperation, though labeled CT, operated from a counterinsurgency perspective as taught in the U.S. and Filipino military schools and reflected in written doctrine and practices. As such, it fielded not only a developmental component but a robust element of high-visibility targeting. Overall, the campaign proved highly successful, decimating ASG, including the elimination of much of its leadership, and pushing other less known groups such as RSM to the brink of extinction. JI was also largely contained in Mindanao, while BIFF became a marginal force located in the MILF-dominated Maguidanao. Meanwhile, the Arroyo administration also managed to severely weaken the CPP-NPA – which had already been decimated by the late 1980s – while opening negotiations with MILF. These culminated in a preliminary peace agreement, the Memorandum of Agreement on Ancestral Domain (MOA-AD), which was later struck down by the Supreme Court as unconstitutional. The Aquino administration (2010–2016), headed by the previous Aquino president's son, built on the CT successes and peace initiatives of its predecessor.

Looming challenges

The year 2014 represented the high water mark in joint Philippine–US CT cooperation, especially as, first, the government and the MILF pushed ahead with finalizing their peace agreement, and, second, groups such as ASG were greatly reduced. This, at least, was the assessment of Washington, which began to draw down its military presence in the country. There was also an assumption that the AFP had developed sufficient expertise to address residual terrorist threats largely on its own. What both failed to appreciate, though, was that the deadlock in the Manila–MILF peace negotiations (occasioned by the previously discussed Mamasapano massacre) would provide a perfect opportunity for

Jihadist groups, including the upstart MG, to seize the initiative and claim the narrative high ground by launching (ultimately) their daring assault on Marawi.

There had been no shortage of warnings.[37] In March 2016, the MILF chief, Murad Ibrahim, openly warned about growing radicalization among disaffected communities, including the MILF's own rank-and-file, due to the non-passage of the BBL and uncertain trajectory of the peace negotiations. He warned that radical Jihadist groups aligned with IS were now able to "capitalize on this because the (frustration) of the people in the area is now very strong" and raised the alarm over "some efforts of penetration" by IS in Mindanao.[38]

The greater threat has been further fragmentation of MILF, whose ranks are rent by fissures and filled with estranged hardliners, who often grumble about the peace negotiations with the infidel government. In recent years, as many as 300 MLF-affiliated fighters have reportedly defected to MG.[39] In 2010, a similar number of fighters had defected to form the more hardline BIFF. Further manpower hemorrhage and defections from MILF ranks cannot be ruled out unless the peace process is placed back on track. It is imperative for the Duterte administration to certify the revised version of the BBL as priority legislation and get it passed as soon as possible in order to, crucially, separate it from the parallel process of Congressional deliberation and negotiation over a prospective charter change – establishing a federal-parliamentary system – which is a complex undertaking that may take years. For its part, to prove its commitment to the goal of peaceful co-existence with the Philippine state, MILF has stepped up its coordinated attacks with the AFP against IS-affiliated elements such as BIIF, a development that has been much encouraged by Duterte, who has called for a common front against IS. After all, the IS-affiliated groups subscribe to a *Takfiri* tradition of branding more moderate rebel groups such as the MILF leadership as apostates, who should be killed in a holy war.[40]

But the government will have to go beyond just CT, CVE, and negotiations to ensuring that it provides sufficient political and fiscal support to the emerging Bangsamoro sub-state entity, while pushing back against corruption and criminal activates that were likely drivers behind the Marawi episode. For some, MG's plan to seize the city was not only an act of terror but part of a full-scale social rebellion against the discredited political order in the area. MG, for instance, would not have been able to build underground tunnels and preposition equipment across the city absent some degree of public support and sympathy, not so much on grounds of ideology as opposed to dissatisfaction with the status quo.[41]

The other big challenge is the reconstruction of heavily devastated Marawi. The Duterte administration, which has promised to make the city "beautiful again," initially pledged close to $400 million dollars to this end, though the cost could run to the billions. Of equal importance, though, is making sure reconstruction efforts are not hobbled by corruption and the government's limited bureaucratic capacity.[42] The proposed reconstruction blueprint of the city by a Chinese-led consortium has provoked outrage among some residents. Given the depth of unemployment and poverty among the displaced citizens, there are concerns over an influx of foreign labor and business entities to the exclusion of locals, not to mention, the broader anxieties over bidding competitiveness, good governance, and consultation, as well as collaboration with local stakeholders in the conceptualization and execution of the project.[43]

It is also crucial for the Duterte administration to continue soliciting maximum military support from all strategic partners, both Western and Eastern, in order to stave off the short-to-medium-term CT challenge in Marawi and other areas where IS-affiliated groups are active, especially in the Sulu Sea area, a bastion of ASG. Coordinated and sustained patrols and joint naval exercises in the area are crucial to eliminating IS access to Mindanao and its ability to financially and logistically support its designated proxies in the area. Thorough and institutionalized intelligence sharing among key law enforcement and military sectors of Malaysia, Indonesia, and the Philippines is hence vital.

Moreover, the Duterte administration and its legislative allies are reviewing the possibility of strengthening anti-terror legislation, namely the Human Security Act of 2007, in order to ensure law enforcement agencies have more leeway to nip terrorist threats in the bud. There is a growing belief that one reason neighboring countries such as Singapore and Malaysia have been more effective in their CT operations is because of their tough Internal Security Act (ISA) legislations.[44] If the Philippines fails to contain the IS threat in Mindanao, it could face the nightmare scenario of large-scale migration of Jihadists from not only Southeast Asia[45] but also from the Caucasus, Central Asia, and the Arab world.[46] Recruitment has apparently not been a concern for Jihadist groups based in Mindanao. Given its accessible geography and enduring status as the cauldron of ethnic-religious discontent in Southeast Asia, Mindanao has been a top destination for regional and global Jihadi groups since the 1980s.

Widespread poverty, combined with religiously driven grievance and weak state institutions that fail to provide basic law and order, has also been a dependable driver of extremist recruitment among local Jihadist groups spread across the island. In short, Mindanao remains a security vacuum that now threatens to be filled by transnational Jihadist groups in tandem with enduring indigenous radical movements.[47] More worrying, as prominent terrorism expert Sydney Jones put it, the siege of Marawi made the southern Philippines the "new sexy destination" for global Jihad, especially in light of the precipitous decline of IS in the Middle East.[48]

Currently, the AFP is overstretched, undertaking the triple role of facilitating the return of Marawi residents to the city, clearing the battle zone of booby traps and IEDs, and neutralizing remaining extremist elements, as well as fighting against IS ideology.[49] Fortunately, the immediate likelihood of another Marawi-like siege in the country is low. Amin Baco, a Malaysian radical jihadist and protégé of Dr. Mahmud bin Ahmad, replaced Hapilon as the new emir of IS in Southeast Asia.[50] He was killed not long after and replaced by a purported relative of Maute brothers, Abu Dar, who has had military training in Afghanistan, with long experience of smuggling foreign fighters into Mindanao over the years. He reportedly escaped during the final hours of the Marawi siege in October 2017, taking with him around $10 million in loot and as many as 300 fighters.

Philippine authorities have raised concerns about the reactivation of the IS cells under Abu Dar, an ethnic Maranao with extensive kinship ties and operational experience in the Greater Lanao area.[51] Notwithstanding such concerns, it remains to be seen if the new leadership can bring together the unique set of skills and resources that their predecessors managed to mobilize. Hapilon was an astute political leader, who successfully brought together jihadists from rival ethnic groups (Tausug, Maranao, and Maguindanao) under one banner, while the Maute brothers offered cutting-edge insurgency training and resources. That proved a lethal combination.

The threat under Abu Dar's leadership could come through so-called "spectacular attacks" that focus on soft targets in urban areas in Mindanao and beyond; maritime piracy and terrorism via ASG group in porous borders of Malaysia, Indonesia, and the Philippines; and occasional harassment of the security forces as well as recruitment efforts among disenchanted internally displaced people in Marawi and surrounding regions. Thus the state will have to double-down on its intelligence-gathering and CVE operations, while cracking down on sleeper-cells of IS-affiliated elements across Mindanao and beyond. There is no room for complacency.

Notes

1 Banlaoi, *The Maute Group.*
2 ABC News, *Australia and Philippines Vow.*
3 Liang, *Video Shows Militants.*
4 ABS CBN News, *Maute, puwersa ng gobyerno.*
5 Tubeza, *IS Terrorists Guarding*; Hincks, "Philippine President Rodrigo Duterte Declares Marawi 'Liberated' From ISIS Terrorists."
6 Heydarian, *Beyond Battle of Marawi*
7 Mangosing, *Lorenzana on 2,000 Civilian.*
8 Aljazeera English, *Philippines: 11 Soldiers Killed.*
9 Lopez, *Opposition Congressmen Challenge.*
10 Martial Law, for instance, suspends the Writ of Habeas Corpus for suspected terrorists, specifically those who are "judicially charged for rebellion or offenses," providing the government additional wiggle room to pre-empt, apprehend and chase down suspected terrorists outside. Yet, in the same breath, it also provides room for abuse and human rights violations, although the 1987 constitution has myriad safeguards against large-scale human rights violations in areas covered by Martial Law. Citizens can, for instance, challenge the validity of the Martial Law at the Supreme Court, while the Congress has power over the declaration's extension beyond 60 days. Not to mention, the normal operations of civilian courts isn't disrupted, while with the exception of those charged with rebellion and treason, the writ of habeas of everyone else remains intact.
11 Heydarian, *Mindanao Crisis Stops Duterte.*
12 Exchanges with the author, June 4, Shangri La Hotel, Singapore. President Duterte, however, implied that he didn't seek American assistance personally, but only retroactively acquiesced to his general's decision to do so.
13 Partly based on exchanges with the author, June 2, Shangri La Hotel, Singapore, and exchanges with Brig. General Rolando Bautista March 6, 2018, Embassy of the Republic of the Philippines, Australia.
14 Kube, *U.S. May Begin Airstrikes.*

15 SBS News, *Turnbull and Duterte Talk.*

16 Cheng, *Philippines, Indonesia, Malaysia.*

17 Dancel, *Singapore Offers Drones.*

18 Stern and Berger, *ISIS: The State of Terror.*

19 Wood, *What ISIS Really Wants.*

20 Ashour, *After Paris.*

21 Byman, *ISIS Goes Global.*

22 Abineles and Amoroso, *State and Society.*

23 Up to 44 members of the Special Action Force (SAF) of the Philippine National Police (PNP), purportedly assisted by United States Army Special Forces, were killed at Tukanalipao, Mamasapano, Maguindanao on January 25, 2015. They were on a mission to kill or capture Malaysian bomb-maker Zulkifli Abdhir, who had taken refuge with BIIF fighters, who, in turn, were allegedly protected by members of MILF. The MILF leadership, which was technically in control of the area under a ceasefire agreement with the AFP and government, was not informed about the operation, which it condemned as a violation of the provisions of the Framework Agreement; neither was the AFP leadership informed. A video of the massacre of SAF members by MILF troops was later uploaded on the internet, provoking massive uproar among the public and members of the political establishment in Manila. The incident almost ended Aquino's term, cost him much political capital, and was later used by the Duterte administration to intimidate the former president by raising the prospect of prosecution and imprisonment.

24 Liow and Ghandour, *The Warning from Marawi for Regional Security.*

25 Discussions with senior Philippine army officers (including intelligence officers) in charge of Marawi in March 2018, who spoke to the author on condition of anonymity, suggest that it was more a failure by the political leadership to appreciate available intelligence rather than failure of availability of intelligence that was at fault.

26 Gita, *Duterte Reiterates Vow.*

27 Ibid.

28 Institute for Policy Analysis of Conflict, *Marawi.*

29 Galag, *CPP-NPA.*

30 See Memorandum Order No. 121, s. 2000, available at www.officialgazette.gov.ph/2000/10/31/memorandum-order-no-121-s-2000/.

31 More specifically, the government looks at:

> a) Hijacking or sabotage of an aircraft, vessel or vehicle; b) Kidnapping, detaining without consent, of a person or persons; c) Use of any biological and/or chemical agent, or radioactive material, or nuclear devise, explosive, firearm or other weapon, with the intent to endanger, directly or indirectly, the safety of one or more individuals, or to cause great damage to property; d) Cyber-terrorism which includes the unauthorized access to, destruction or disruption of government data and finally; e) Act of assisting terrorists in any way in the commission of their crime, as an accessory or accessories.

See Primer on the National Plan to Address Terrorism and Its Consequences (2002), 2.

32 In broadly abstract terms, they are: supervision and implementation of policies and actions of the government against terrorism; intelligence coordination; internal focus against terrorism; accountability of public and private corporations and personalities; synchronizing internal efforts with global outlook; appropriate legal measures; promotion of Christian and Muslim solidarity; vigilance against the movement of terrorists and their supporters, equipment, weapons and funds; contingency plans; comprehensive security plans for critical infrastructure; support of overseas Filipino workers; modernization of the AFP and the Philippine National Police (PNP); media support; political, social and economic measures.

33 See Memorandum Order Number 27, available at www.officialgazette.gov. ph/2001/10/12/memorandum-order-no-37-s-2001/.
34 See especially, Thomas A. Marks, *Maoist Insurgency Since Vietnam*.
35 Ibid., 134–137.
36 Schmitt, *A Nation Challenged*.
37 Including from the author, who gave a briefing to several presidential cabinet members of the Aquino administration on the rising threat of IS infiltration in Mindanao in October 2015.
38 Philippine Daily Inquirer, *MILF Warns IS*.
39 Banlaoi, *The Maute Group*.
40 Liow, *ISIS Goes to Asia*.
41 Simons, *Pagdaro Sa Kalinaw*.
42 Tara, *Gov't Bares Funding*.
43 See, for instance, Solomon and Villamor, *Filipinos Get a Glimpse*.
44 Romero, *Harsher Anti-terror Law Pushed*.
45 Singh and Bin Jani, *The Marawi Narrative*.
46 Partly based on discussions with senior Philippine army officials in early March 2018.
47 See for instance, Jayakumar, *The Islamic State Looks East*.
48 Agence France Presse, *Brothers Who Brought Death*.
49 Acosta, *AFP's 'war' in Marawi*.
50 Singh, *Southeast Asian Jihadi*.
51 Reuters, *Looted Cash, Gold*.

Bibliography

ABC News. "Australia and Philippines Vow Cooperation Against Islamic State, Boost Surveillance of Suspected South-East Asian Militants," *ABC News*, (December 15, 2016) available at www.abc.net.au/news/2015-12-11/australia-philippines-vow-cooperation-against-islamic-state/7020372 (accessed January 1, 2018).

ABS CBN News. "Maute, puwersa ng gobyerno, nagpalitan ng putok sa Marawi," *ABS CBN News*, (December 1, 2017) available at http://news.abs-cbn.com/news/05/23/17/maute-puwersa-ng-gobyerno-nagpalitan-ng-putok-sa-marawi (accessed January 1, 2018).

Acosta, Rene. 2018. "AFP's 'War' in Marawi Not Totally Over," *Business Mirror*, (January 21, 2018) available at https://businessmirror.com.ph/afps-war-in-marawi-not-totally-over/(accessed January 1, 2018).

Abineles, Patricio and Donna Amoroso. *State and Society in the Philippines*. Rowman & Littlefield Publishers: Maryland, 2005.

Agence France Presse. "Brothers Who Brought Death and Ruin to Marawi City," *Agence France Presse*, (June 18, 2018) available at http://newsinfo.inquirer.net/906559/brothers-who-brought-death-and-ruin-to-marawi-city.

Aljazeera English. "Philippines: 11 Soldiers Killed in Misdirected Air Rail." *Aljazeera English*, (June 2, 2017) available at www.aljazeera.com/news/2017/06/philippines-10-soldiers-killed-friendly-fire-170601031851380.html (accessed January 1, 2018).

Ashour, Omar. "After Paris: ISIL's Strategy Against the 'Far Enemy'," *Aljazeera English*, (November 24, 2015) available at www.aljazeera.com/indepth/opinion/2015/11/paris-isil-strategy-enemy-151124074915465.html (accessed January 1, 2018).

Banlaoi, Rommel. "The Maute Group and Rise of Family Terrorism," *Rappler*, (June 15, 2017) available at www.rappler.com/thought-leaders/173037-maute-group-rise-family-terrorism (accessed January 1, 2018).

Byman, Daniel. "ISIS Goes Global," *Foreign Affairs*, available at www.foreignaffairs.com/articles/middle-east/isis-goes-global (accessed January 1, 2018).

Cheng, Willard. "Philippines, Indonesia, Malaysia Boost Joint Efforts vs Terrorism," *ABSCBN News*, (June 22, 2017) available at http://news.abs-cbn.com/news/06/22/17/philippines-indonesia-malaysia-boost-joint-efforts-vs-terrorism (accessed January 1, 2018).

Dancel, Raul. "Singapore Offers Drones, Urban Warfare Training Grounds, Aid to Help Philippines Fight Militants in Marawi," *The Straits Times*, (July 19, 2017) available at www.straitstimes.com/asia/se-asia/singapore-offers-saf-assistance-to-philippines-in-fight-against-terrorism (accessed January 1, 2018).

Galag, Ron. "CPP-NPA is Still a Terrorist Group, U.S. Says," *ABSCBN*, (February 3, 2017) available at http://news.abs-cbn.com/news/02/02/17/cpp-npa-is-still-a-terrorist-group-us-says (accessed January 1, 2018).

Gita, Ruth. "Duterte Reiterates Vow for Immediate Passage of BBL," *Sun Star*, (July 24, 2017) available at www.sunstar.com.ph/manila/local-news/2017/07/24/duterte-reiterates-vow-immediate-passage-bbl-554679 (accessed January 1, 2018).

Heydarian, Richard. "Mindanao Crisis Stops Duterte from Cutting the US Cord," *Nikkei Asian Review*, (July 13, 2017) available at https://asia.nikkei.com/magazine/20170713/Viewpoints/Richard-Heydarian-Mindanao-crisis-stops-Duterte-from-cutting-the-US-cord (accessed January 1, 2018).

Heydarian, Richard. "Beyond Battle of Marawi: Duterte Needs Both U.S. and China," *US China Focus*, (November 1, 2017) available at "www.chinausfocus.com/peace-security/beyond-battle-of-marawi-duterte-needs-both-us-and-china (accessed January 1, 2018).

Hincks, Joseph. "Philippine President Rodrigo Duterte Declares Marawi 'Liberated' From ISIS Terrorists," (October 17, 2017) available at http://time.com/4985282/marawi-city-liberated-isis-duterte/ (accessed July 1, 2018).

Institute for Policy Analysis of Conflict "Marawi, The 'East Asia Wilayah' and Indonesia," IPAC Report, 38, (July, 2017), 1.

Jayakumar, Sashi. "The Islamic State Looks East: The Growing Threat in Southeast Asia." Combating Terrorism Center at West Point, (February, 2017) available at https://ctc.usma.edu/the-islamic-state-looks-east-the-growing-threat-in-southeast-asia/.

Kube, Cortney. "U.S. May Begin Airstrikes Against ISIS in Philippines," (August 7, 2017) available at www.nbcnews.com/news/world/u-s-may-begin-airstrikes-against-isis-philippines-n7902 (accessed January 1, 2018).

Liang, Annabelle. "AP Exclusive: Video Shows Militants in Philippine Siege Plot," (2018) available at https://apnews.com/8f6649e404964ab4a2d280537c86c83f (accessed January 1, 2018).

Liow, Joseph Chinyong. "ISIS Goes to Asia," *Foreign Affairs*, (March 13, 2018) available at www.foreignaffairs.com/articles/east-asia/2014-09-19/isis-goes-asia (accessed January 1, 2018).

Liow, Joseph Chinyong and Salem Ghandour. "The Warning from Marawi for Regional Security," *The Straits Times*, (June 12, 2017) available at www.straitstimes.com/opinion/the-warning-from-marawi-for-regional-security (accessed January 1, 2018).

Lopez, Virgil. "Opposition Congressmen Challenge Martial Law Declaration in Mindanao Before SC." *GMA News*, (June 5, 2017) available at www.gmanetwork.com/news/news/nation/613273/opposition-congressmen-challenge-martial-law-declaration-in-mindanao-before-sc/story/ (accessed January 1, 2018).

Mangosing, Frances. "Lorenzana on 2,000 Civilian Deaths in Marawi: Avoid Sharing Unverified Data," *Philippine Daily Inquirer*, (July 11, 2017) available at http://news-info.inquirer.net/912785/lorenzana-on-2000-civilian-deaths-in-marawi-avoid-sharing-unverified-data (accessed January 1, 2018)

Marks, Thomas A. *Maoist Insurgency Since Vietnam.* Frank Cass & Company: London, 1996.

Pena, Leonardo, "Finding the Missing Link to a Successful Philippine Counterinsurgency Strategy." Masters Thesis. California: Naval Postgraduate School. 2007.

Philippine Daily Inquirer. "MILF Warns IS May Gain from BBL Delay," *Philippine Daily Inquirer*, (March 7, 2016) available at http://globalnation.inquirer.net/137460/milf-warns-is-may-gain-from-bbl-delay (accessed January 1, 2018).

Reuters. "Looted Cash, Gold from Overrun Town Helps Islamic State Recruit in Philippines," *Japan Times*, (January 23, 2018) available at www.japantimes.co.jp/news/2018/01/23/asia-pacific/looted-cash-gold-overrun-town-helps-islamic-state-recruit-philippines/#.WwSms62B10s.

Romero, Paolo. "Harsher Anti-terror Law Pushed," *Philippine Star*, (July 23, 2017) available at www.philstar.com/headlines/2017/07/24/1721007/harsher-anti-terror-law-pushed (accessed January 1, 2018).

SBS News. "Turnbull and Duterte Talk Counter-terrorism and 'War on Drugs' in First Ever Bilateral Meeting," *SBS News*, (November 14, 2017) available at www.sbs.com.au/news/turnbull-and-duterte-talk-counter-terrorism-and-war-on-drugs-in-first-ever-bilateral-meeting (accessed January 1, 2018).

Schmitt, Eric. "A Nation Challenged: Pacific Terror; U.S. and Philippines Setting Up Joint Operations to Fight Terror," *New York Times*, (January 16, 2002) available at www.nytimes.com/2002/01/16/world/nation-challenged-pacific-terror-us-philippines-setting-up-joint-operations.html?mcubz=0 (accessed January 1, 2018).

Simons, Jeremy. "Pagdaro Sa Kalinaw: Dureza's Betrayal and Duterte's Hypocrisy in Marawi," *MindaNews*, (September 8, 2017) available at www.mindanews.com/mindaviews/2017/09/pagdaro-sa-kalinaw-durezas-betrayal-and-dutertes-hypocrisy-in-marawi/ (accessed January 1, 2018).

Singh, Bilveer. 2018. "CO18007 | Southeast Asian Jihadi Leaders in the Post-Marawi Era," RSIS Commentaries, (January 16, 2018) available at www.rsis.edu.sg/rsis-publication/cens/co18007-southeast-asian-jihadi-leaders-in-the-post-marawi-era/#.WnIu9a2B2jQ (accessed January 1, 2018).

Singh, Jasminder and Muhammad Haziq Bin Jani. "The Marawi Narrative: 'Inside the Caliphate'," *RSIS Commentaries*, (September 19, 2017) available at www.rsis.edu.sg/wp-content/uploads/2017/09/CO17169.pdf?utm_source=getresponse&utm_medium=email&utm_campaign=rsis_publications&utm_content=RSIS+Commentary+169%2F2017+The+Marawi+Narrative%3A+"Inside+the+Caliphate"+by+Jasminder+Singh+and+Muhammad+Haziq+Bin+Jani+ (accessed January 1, 2018).

Solomon, Ben and Felipe Villamor. "Filipinos Get a Glimpse of Their Ruined City. The Chinese Get the Contract," *New York Times*. (April 10, 2018) available at www.nytimes.com/2018/04/10/world/asia/marawi-duterte-china-rebuilding.html (accessed July 1, 2018).

Stern, Jessica and J. M. Berger. *ISIS: The State of Terror.* New York, NY: Ecco, 2016.

Tara, Lin. "Gov't Bares Funding Source for Marawi 'Multi-year' Rehabilitation," *CNN Philippines*, (July 15, 2017) available at http://cnnphilippines.com/news/2017/07/15/Govt-bares-funding-source-for-Marawi-multi-year-rehabilitation.html (accessed January 1, 2018).

Tubeza, Philip. "IS Terrorists Guarding P1.4-B Loot in Marawi," (July 22, 2017) available at http://newsinfo.inquirer.net/915973/is-terrorists-guarding-p1-4-b-loot-in-marawi (accessed July 27, 2017).

Wood, Graeme. "What ISIS Really Wants," *The Atlantic*, (March 2015) available at www.theatlantic.com/magazine/archive/2015/03/what-isis-really-wants/384980/ (accessed January 1, 2018).

3 The evolution of violent extremism and state response in Indonesia

Geoffrey Macdonald, Rhonda Mays and Luke Waggoner

Introduction

In Indonesia, the nature of violent extremism and the characteristics of the state's response have evolved over time. Nevertheless, both have been deeply rooted in the contested identity of the Indonesian state. Indonesia's nationalist movement, which prevailed over Dutch colonialism in 1945, was divided primarily along devotional lines: Muslim secularists versus political Islamists. The secular nationalists, led by Sukarno, ultimately prevailed, but the country never fully unified around a secular identity. This disconnect has continued to plague the Indonesian state, and officials have vigilantly applied force in their response to political violence. Indonesia's fractured nationalism set the stage for two distinct periods of violent extremism and state response in Indonesia: the rise of Jemaah Islamiyah and other terrorist organizations (1999–2009), and the ascendency of decentralized terrorism inspired by the Islamic State (IS) (2016 to the present). During these periods, both the nature of violent extremism and the state's response shifted in important ways. Specifically, violent extremism has become increasingly global in nature, embracing international jihadism. In response, the government has used hard and soft power approaches and has leveraged regional and international assistance. Indonesian counterterrorism policy has been both praised and criticized by domestic and international observers.

This chapter is structured as follows: Section II details the weak development of Indonesian nationalism, early challenges to the state, and the government's use of disproportionate force. Section III outlines the nature of violent extremism, its key drivers, and the state's response from 1999–2009 and 2016–present. This section identifies the similarities and differences in the characteristics of violent extremism and government policy across these two periods. Section IV discusses how violent extremism and counterterrorism policies in Indonesia are expected to evolve, and offers recommendations for a more holistic strategy that combines military action, intelligence gathering, de-radicalization policies, and the application of more responsive and inclusive governance practices.

Indonesia's fractured nationalism and the persistence of Islamic opposition

At first glance, it would be understandable to characterize violent extremism in Indonesia as a modern phenomenon born out of global extremist organizations seeking to expand their struggles across the globe. Proponents of this argument would cite the fact that violent extremist organizations in Indonesia have only been able to execute large-scale attacks since the early 2000s – successes that were largely enabled by support and inspiration from Al-Qaeda and IS. However, the seeds of violent extremism in Indonesia can be traced back to contestations and societal influences that pre-date Indonesia's existence as a nation-state. The country's contested definitions of nationhood, its relationship with political violence, and the state's legacy of ineffectively dealing with those two issues are evident in the development and rhetoric of contemporary violent extremist organizations in Indonesia.

With approximately 17,500 islands, more than 300 languages, and variety of religions, Indonesia is an unlikely nation-state. As the idea of the Indonesian nation emerged in the early twentieth century, debates arose over the characteristics of the nation and who belonged within its boundaries. One of the most contentious disputes arose over whether the new nation-state would be secular or Islamic. In June 1945, as the victory of the Allied powers ended Japanese rule in Indonesia, debates over the character of an independent Indonesian nation intensified. Indonesia's secular nationalist leader Sukarno spoke of five principles of nationalism known as *Pancasila* that would define the character of the new Indonesian nation-state. These principles were the belief in one god, a just and civilized humanity, the unity of Indonesia, deliberative democracy, and social justice for all people. The first principle, belief in one God, recognized the inherent religiosity of the Indonesian people, while acknowledging the religious diversity of the vast archipelago by not forcing the entire population to accept Islam and a Muslim God.

Whereas Sukarno's cadre imagined a modern, secular state, political Islamist leaders like Kartosuwirjo and his Darul Islam (DI) movement, wanted the constitution to require a Muslim head of state and for Islamic law to govern Muslim citizens. Yet when the 1945 constitution was ultimately approved, it referenced only *Pancasila*, and rejected provisions enshrining the superior role of Islam in the government and laws of the country. This alienated Islamist nationalists and set the stage for an enduring battle over Indonesia's religious character that continues to this day.

Layered beneath this rhetorical debate over the Islamic character of the state was burgeoning political violence. The nation-state of Indonesia was born out of conflict, which strongly influenced society's understanding of and attitudes toward violence. Like many of its neighbors, World War II enabled the Indonesians to cast off the yoke of nearly 350 years of colonial rule. However, Indonesia's path to independence was not linear. Sukarno and Mohammad Hatta, two of Indonesia's founding fathers, declared independence on August 17, 1945, just days after Japan's final defeat, but the country did not achieve full independence until 1949. As the debate raged over the Islamic character of the constitution, the Indonesian armed forces fought a bloody war against the Dutch to establish Indonesia as a free nation-state. This violent birth established a place for legitimized violence in the national ethos, a characteristic that would be reinforced through a series of violent conflicts and episodes beginning with the Darul Islam revolt immediately following independence.

Increasingly frustrated with the secular nationalists' unwillingness to grant Islam special status in the new republic, Darul Islam waged an insurgency from 1949 to 1962 to establish and expand the Islamic State of Indonesia based in West Java. Although ultimately crushed by the republican government, the Darul Islam movement demonstrated the intense and persistent conflict over the role of Islam in defining the Indonesian nation. The revolt also helped to solidify acceptance of violence as a legitimate expression of grievances against the state.

Darul Islam's anti-state violence was followed by the Free Papua Movement (Organisasi Papua Merdeka or OPM) beginning in 1963 and again in 1976 by the Free Aceh Movement (Gerakan Aceh Merdeka, or GAM). Both OPM and GAM used violence to express grievances against the Indonesian central government for its perceived exploitations of both Papuan and Acehnese people, as well as the natural resources found in the provinces. The rhetoric and tactics of these movements are reflected in the grievances and actions of modern violent extremist organizations in the country. These grievances range from the belief that the government is not sufficiently Islamic and is not protecting the special rights of Muslim citizens to outright rejection of the Indonesian state's legitimacy and authority.

Since independence, the Indonesian state has been singularly focused on preserving its existence in the face of religious and ethno-linguistic challenges. Rather than embrace its diversity, the state has often shrunk political space to control its multi-religious population and suppress debates over religion, citizenship, and the role and character of the state. Throughout its history, when contestations have arisen, the state has invoked *Pancasila* to shut down criticism and serve as the ultimate arbiter of legitimacy. *Pancasila* has come to define Indonesians' religiosity and expresses the nature of the Indonesian state. Acceptance of *Pancasila* determines who is and is not a member of the Indonesian nation. However, *Pancasila* has never fully encapsulated the heterogeneous identities and aspirations of Indonesia's citizens. Among those disenfranchised by *Pancasila*'s generic spirituality are political Islamists who believe the Indonesian state should reflect the religion of the majority of the country's citizens. This disenfranchisement was a driving force in the Darul Islam revolt, and it remains a motivating factor for current extremists who do not see themselves or their beliefs reflected in the definitions and characteristics of the Indonesian nation.

The evolving threat from Islamic extremism in Indonesia

Indonesia's fractured nationalism and contested identity precipitated significant subnational violence between Christian and Muslim communities and different ethnic groups – a situation that intensified after the fall of Suharto in 1998. This violence was localized, often revenge-driven, and lacked a unifying ideology. Simultaneously, violent extremist groups emerged, using both conventional and unconventional tactics in pursuit of a unified and radical political agenda: to transform Indonesia into an Islamic theocracy. This ideology distinguished these groups from other violent actors in Indonesia and closely linked them to the country's political Islamist tradition.

While the Indonesian state effectively blunted Darul Islam in the 1960s, the religious ideology that underpinned the movement remained, along with many of its leaders. After the end of Suharto's authoritarian regime, Indonesia's repressed Islamism metastasized into violent extremism, embracing interreligious conflict and terrorist tactics. Since 1998, there have been two primary periods of violent extremism, which illuminate the evolving nature of both the threat and the state's response: the ten-year period from 1999–2009, which saw the rise of Jemaah Islamiyah; and 2016 to the present, with the rising influence of the Islamic State (IS). While violent Islamic extremism is not the only type of political violence in Indonesia, it is unique in its ideology, tactics, deadliness, and persistence, and has elicited a direct response from the state.

1999–2009: Jemaah Islamiyah and the first phase of Islamic extremism

The turn of the twenty-first century brought a new variant of an old problem to Indonesia. Jemaah Islamiyah combined the political Islamism of Darul Islam with terrorist tactics. The drivers of this new violence were multifaceted, motivated by long held grievances, new anxieties, and foreign inspiration. Indonesia's newly democratic government devised a multipronged counter-terrorism policy to combat this threat.

Nature of the threat

In April 1999, a bomb exploded on the first floor of Jakarta's Istiqlal mosque. Although there were no fatalities, the attack marked the beginning of a distinct shift in the ethnic and religious conflict that had divided Indonesia since independence. Whereas violence in the previous decade tended to be sporadic and localized, this new brand of Islamic extremism was premediated, national in character, and characterized by a distinct Islamist ideology.[1] Two main strands of Islamic extremism emerged in the early 2000s: paramilitary groups and terrorist groups. While the former was more regionally localized than the latter, both strands embodied a new form of militarized Islam.

The first strand of Islamic extremism, which featured armed paramilitary groups, was Islamist in orientation, but lacked the premediated planning and totalistic Islamist ideology that defined the second strand. Ambon, the capital of the Maluku province, was a focal point of paramilitary Islamism. From 1999 to early 2002, intercommunal conflict between Muslims and Christians led to interventions by paramilitary Islamist organizations. In January 1999, a small personal conflict between Christian and Muslim residents spiraled into widespread reciprocal religious violence.[2] Christian and Muslim rioters destroyed homes and places of worship and murdered those of the opposite faith. The escalating violence inspired the creation of Lashkar Jihad (LJ), an Islamist paramilitary group organized to defend Muslims in Maluku. Thousands of LJ fighters were trained in camps on neighboring islands before leading coordinated attacks on Christian villages. While LJ's intervention in Ambon had an ostensibly narrow goal, the organization was broadly dedicated to the introduction of Sharia law in Indonesia.[3] A similar form of paramilitary Islamism also ravaged the Central Sulawesi city of Poso.[4]

The second strand of Islamic extremism in this period was a bombing campaign that fully embodied a national Islamist agenda. Rather than conducting an organized insurgency confined to a particular area, from 1999 to 2009 terrorist organizations carried out a series of bombings designed to pressure the state and citizens to embrace political Islamism. The Istiqlal mosque bombing was the first major attack of this period. Scholar John Sides maintains that the mosque was targeted in 1999 because it represented to extremists "the accommodation between Islam and secular nationalist forces in Indonesia" and was therefore "a sacrilegious symbol of Islam's subordination to secular state power."[5] Over the next decade, at least 20 high profile terrorist attacks were carried out, killing hundreds of civilians. Among the most prominent were the 2002 and 2005 Bali bombings, which together killed more than 200, and the nearly simultaneous suicide bombings carried out at the Marriott and Ritz-Carlton hotels in Jakarta in 2009.

The primary perpetrator of these attacks was Jemaah Islamiyah (JI), a transnational Islamic extremist group operating in several Southeast Asian nations. JI was little known until the 2002 Bali bombings, which received significant international attention. Despite its recent rise, JI and other Islamic extremist groups such as Lashkar Jihad are part of a longstanding intellectual rebellion against Indonesia's secular state.

Drivers of extremism

The rise of violent extremism in Indonesia in 1999 and throughout the first decade of the 2000s was due to several underlying and proximate causes, including the historical divide over the Islamic character of the state, foreign religious influence, the recrudescence of secular democracy as opposed to political Islam, and the inspiration of foreign global jihadist groups.

The broadest underlying driver of violent extremism in Indonesia is the bitter divide over the character of the Indonesian state. Indonesia's secular nationalists subjugated religious alternatives and suppressed ethnic and communal violence under Sukarno and Suharto. Since the return of democracy in 1998, the role of religion in lawmaking and politics was opened up for debate among portions of the population and political elite that continue to favor Islamic law. At the violent fringe of this debate in the early 2000s was Jemaah Islamiyah. JI and other Islamist groups are heir to a political tradition in Indonesia that inspires both mainstream Islamist parties and radical terrorist groups. Thus, what was new in Indonesia after 1998 was not Islamic extremism itself, but the tactics deployed in its name.

Foreign influence is an additional underlying cause of the rise of extremism at the turn of the twenty-first century. There is a direct connection between JI in 2000s, Darul Islam in the 1950s and 1960s, and early 1900s Islamic education networks populated by Indonesians of Arab descent and funded and influenced by Middle Eastern governments and religious scholars. At the time, organizations within this education network promulgated "the strictest interpretations and applications of Islam" and cultivated a sense of separateness and linkage to the Middle East.[6] The leaders of these organizations were strongly opposed to secular constitutional democracy. Decades later, foreign powers have continued to perpetuate this Islamism. Saudi Arabia has directly funded Islamic educational institutes in Indonesia that foster strict religious conservativism: for example, the Saudi charity Al Haramain is alleged to have provided covert funding to JI.[7] Members of Darul Islam and JI members received training and combat experience during the 1980s in Afghanistan.[8] The rise of extremism in Indonesia after 1999 therefore had close connection to past and ongoing foreign influence in the country.

These underlying causes of Islamic extremism – the contested identity of the state and foreign influence – cannot explain why terrorist attacks began rising in 1999 as opposed to years earlier. The re-constitution of a secular, liberal, and democratic Indonesia after years of authoritarianism was deeply disillusioning to Islamists. Starting in 1999, the weakness of Islamist political parties in early post-Suharto elections suggested little public appetite for religious governance, in what John Sidel describes as "the eclipse and evisceration of the Islamist project in the country, in a rather sudden and dramatic reversal of fortunes."[9] Furthermore, the political ascendance of the secular-nationalist politician Megawati Sukarnoputri and her secular-nationalist party, the PDIP, likely pushed Islamists to more extreme tactics to pursue their goals. In sum, the exclusivity of Indonesia's *Pancasila*-based secular democracy has alienated and empowered radical elements of society.

The influence of foreign terrorist organizations served as a second proximate cause of extremism's rise in Indonesia. The September 11, 2001 terrorist attacks in the United States became a major source of inspiration to struggling Islamists. The end of the Cold War had sparked unprecedented democratization around the globe. This growth of liberal market democracy was antithetical to Islamism's all-encompassing and undemocratic tenets, and al-Qaeda's strike against the United States demonstrated a strategy and tactic of resistance. There is evidence that al Qaeda's attacks had a "contagion effect" for other groups: one attack inspires more.[10] As terrorism scholar Marc Sageman writes:

> The present threat has evolved from a structured group of al Qaeda masterminds, controlling vast resources and issuing commands, to a multitude of informal local groups trying to emulate their predecessors by conceiving

and executing operations from the bottom up. These "homegrown" wannabes form a scattered global network, a leaderless jihad.[11]

Though it is difficult to substantiate what inspiration JI drew from al-Qaeda and other transnational terrorist groups, the nature of its attacks and extremity of its beliefs increasingly mimicked other Islamic extremism movements around the world.

State response

Shortly after the Bali bombings in 2002, the Indonesian government initiated a counterterrorism policy that had two core components: intelligence and law enforcement, and rehabilitation and deradicalization. The government created an elite police unit known as Detachment 88, and later the National Counter-Terrorism Agency, to coordinate counterterrorism efforts, and introduced rehabilitation and de-radicalization programs. Although the results of the latter programs have been mixed, the government has successfully reintegrated many former terrorists.[12] Today, Indonesia continues to pursue both these soft and hard-power approaches, which have – despite some deserved criticism – been lauded by many foreign governments and international observers.

From the onset of the problem, the Indonesian government primarily addressed the threat of terrorism through the lens of law enforcement.[13] In the immediate aftermath of the Bali bombings, the government of Megawati Sukarnoputri passed an interim Anti-Terrorism Law of 2002 (later confirmed in 2003). This legislation was designed to supplement existing criminal codes with legal mechanisms to prosecute and convict suspected terrorists quickly. However, this law has been criticized for applying an overly broad definition of terrorism that critics warn could include criminalization of various acts of dissent.[14] In 2010, the government passed Presidential Regulation No. 46, which established the National Counter-Terrorism Agency (BNPT) to oversee all counterterrorism agencies and units in Indonesia.[15] This counterterrorism framework has been generally praised, particularly in comparison to regional neighbors. In a comparative study of counterterrorism approaches, researchers at West Point's Combatting Terrorism Center concluded, "The Indonesian law enforcement-based CT approach has been more effective than the military-based Philippine CT approach."[16]

Detachment 88, an elite special forces unit, was formed in 2003 to lead these counter-terrorism efforts in Indonesia. Over the ensuing decade, Detachment 88 and other counter-terrorism units arrested[17] and convicted[18] hundreds of Jihadists. However, international human rights groups and researchers have alleged abuses by the army and police. According to John Sidel, "well-publicized incidents of torture and summary executions of Lashkar Jihad members" took place in Maluku in the 2000s. Citing a long history of abuse, Amnesty International claimed in 2016 that "torture is rife in Indonesia."[19] The Indonesian National Human Rights Commission has identified 121 terrorism suspects who have died in custody since 2007.[20] Even while acknowledging past abuses, the government denies current allegations of torturing suspects. "Now we have a more demo-cratic country – with open investigation and no torture any longer," said a senior government official in 2008.[21]

Indonesia's counterterrorism policy also includes de-radicalization programs. The "De-radicalization Blueprint," published by the National Counterterrorism Agency in 2013, formalized a longstanding policy to re-socialize terrorists, and includes vocational training, counseling, and discussion and dialogue.[22] It also offers assistance to communities willing to reintegrate former terrorists.[23] However, the program has struggled with recidivism[24] and some unclear results. A study of Indonesia's rehabilitation program, which included focus group dis-cussions with program participants, highlighted a high degree of skepticism among participants. According to one participant, "The government tries to change us, change our spirit of jihad, using the de-radicalization program, they actually don't understand us and how to improve us." Others complained that the government's uniform approach to de-radicalization failed to account for different personalities and backgrounds of prisoners.[25]

2016–present: ISIS and the second phase of Islamic extremism

In 2016, a new brand of violent extremism emerged in Indonesia. With JI largely defeated and disbanded, terrorism inspired and directed by the so-called Islamic State (IS) escalated the frequency of attacks around the country, and ushered in new targets and methods of terrorism. Local extremist groups in Indonesia have re-oriented their attacks to fit the global objectives of the Middle East-based, IS leadership. This new era of violent extremism is challenging the Indonesian state to balance security objectives with human rights while cooperating with regional and international partners.

Nature of the threat

On Sunday morning, May 13, 2018 a family of six split into three groups – a mother and her two young daughters, two teenage boys, and the father – and blew themselves up in three different churches in the Javanese city of Surabaya. The coordinated blasts killed seven and injured dozens. The church bombings were unique in Indonesia not only because an entire family of six carried them out, but also because the family is among 500 other Indonesians that recently returned from Syria. In the four years since IS declared itself to be the new Islamic caliphate, IS has directed and inspired attacks all over the world, includ-ing London, Brussels, Paris, and Orlando, and has shown its capacity to provide the resources and direction for successful acts of terror throughout the West. The May 2018 church bombings demonstrated IS' new means of exporting terrorism, and with it, have ushered in the newest phase of international jihad.

The church bombings highlighted the Islamic State's new plan of redirecting its fighters who have returned home to commit acts of terror in their own neigh-borhoods. This new offensive is supplementing its previously established use of propaganda and proxy fighters to expand its reach beyond the Middle East and North Africa. Unlike al Qaeda, IS is not devoting significant resources to setting up cells all over the world – preferring instead to expand through existing radical movements. JI's downfall at the end of the 2000s left its followers scattered, which prevented the formation of an al Qaeda-type terror cell. However, this crop of "homeless radicals" were desperate to join and fight for an organized jihadi network. IS recognized this and has exploited these radicals to carry out attacks in Indonesia.

While IS has created a Southeast Asian military contingent based in Raqqa – Katibah Nusantara – the group relies on preexisting networks of radical individuals in target countries to carry out attacks.[26] In contrast to al Qaeda's long-term and planned approach, IS has aggressively and hastily sought to control physical territory in Iraq and Syria and to provide support and inspiration for would-be combatants in societies controlled by non-Muslims or apostates. IS seeks to empower and, when possible, equip existing radical groups and militants to fight on its behalf, many of whom have been fighting other conflicts for years. By doing this, IS has set up ideological affiliates all over the world – as with the Jamaah Ansharut Daulah (JAD) terrorists who carried out the 2016 Jakarta attack.

In Indonesia, the IS–affiliates include (JAD), a network of several small radical factions that coalesced and pledged allegiance to Abu Bakr al Baghdadi – the leader of IS – in 2015, months before the first IS-directed attack in Indonesia.[27] The group was formed from members linked to Hizb at-Tahrir and Timorese radical groups such as Mujahidin Indonesia Timor. With its core opposition to a "secularizing" and "marginalizing" state, JAD fit the profile of the type of groups IS recruits to carry out terrorist operations around the world.[28]

IS has made a concerted effort to promote its brand in Southeast Asia by building on JI's legacy. For example, Katibah Nusantara is made up of battle-tested jihadi fighters who "fought in Afghanistan in the 1980s [and] formed the backbone of the Jemaah Islamiyah in 1990s and the first decade of 2000."[29] The military successes achieved by this unit have featured prominently in IS propaganda, and the unit has led the effort to translate and promote IS materials for a Southeast Asian audience. According to the *New York Times*, Katibah Nusantara's prominence could indicate IS' intention to continue to recruit and launch attacks in Southeast Asia.[30] This regional contingent is thought to be led by Syria-based Bahrun Naim, who is also believed to be leading IS' recruitment efforts in Indonesia. Naim advises IS and IS affiliates on bomb-making and develops and disseminates plots for attacks – including the January 2016 Jakarta attack, which utilized Naim's network of violent radicals in Indonesia.[31] Naim was also behind the attempted suicide bombing of the presidential palace in late 2016, which would have marked the first incident involving a female suicide bomber in Indonesia.

The May 2018 church bombing is not the first attack for which IS has used proxy fighters. On January 14, 2016, a series of bomb blasts and gunfire reverberated through Central Jakarta in close proximity to several embassies and the

United Nations headquarters. The following year, on May 24, 2017, a series of bombings hit the Kampung Melayu bus terminal in East Jakarta. Four people were killed in the January 2016 attack and another ten people were killed, including three police and the two attackers, in the May 2017 bombings. The bus terminal attack was carried out by Mudiriyah Bandung Raya, an organization with links to JAD, Bahrun Naim, and IS. The attack targeted the police, representing the country's single most deadly attack on law enforcement, and showcased IS' ability to coopt local terrorist groups. It also indicated a shift in tactics for local groups: whereas earlier attacks largely targeted tourists and foreigners, under IS' inspiration and direction, Indonesian extremists have begun to target the state itself.

Drivers of extremism

In the age of IS-inspired and directed attacks, Indonesia's radical Islamists have channeled decades-old grievances into an IS narrative of bringing about the global caliphate. Despite this new age of terrorism in the archipelago, the underlying causes that are motivating Indonesian Islamists to commit acts of terror are largely unchanged since the age of JI in the late 1990s and 2000s. However, the prerequisite for receiving IS support and acknowledgment requires an additional objective: belief in and promotion of the global caliphate as imagined by IS. The IS doctrine of establishing a global caliphate has linked otherwise disparate radical groups in Indonesia with a worldwide network of groups and individuals waging the same war, and is the primary distinguishing factor of this new incarnation of violent extremism in Indonesia.

Yet despite the requirement to swear allegiance to IS and its global ideology, most Indonesian radical groups remain motivated by the same grievances that drove terrorist attacks in the 1990s and 2000s. Like the terrorists behind the Bali bombings, the statements of modern-day Indonesian jihadists identify the secularizing state as motivation for committing acts of violent extremism. IS has guided groups such as JAD and Mudiriyah Bandung Raya to act on these grievances by actually targeting agents of the state, as seen in the May 2017 bombings targeting police in retribution for their role in foiling terrorist attacks.

IS's characterization of Indonesian state officials as "infidels" resonates with many conservative Muslim Indonesians who have felt marginalized by the secular government. The continued liberalization of Indonesia's policies since the reintroduction of democracy in 1998 have intensified this feeling of alienation among Islamists. The recent blasphemy trial of the governor of Jakarta, Basuki Tjahaja Purnama (known locally as Ahok) underscored these tensions. In September 2016, Governor Ahok, a Christian, said that the Quran does not contain a prohibition against Muslims being led by non-Muslims. After a doctored video of the quote went viral, and the governor was charged with blasphemy and taken to court. Following the charge, thousands of Indonesians took to the streets to protest or support Ahok. For many Islamist groups, having a Christian leader of the Indonesia's largest city represents a major blow in a larger identity war. Thus, Ahok's statement reaffirmed the fears of many radical Muslims that their social and national identity as Muslim was being stripped away. Those in support of Ahok rallied against religious limitations on free speech and the undemocratic implications of such a verdict. In this sense, the trial has served as a microcosm of the tensions between the secular and conservative segments of Indonesian society.[32]

State response

The Indonesian state's response to this new phase of violent extremism, which has largely continued the policies devised in the early 2000s, has been effective in breaking up existing terrorist organizations. Only five terrorist attacks were reported in Indonesia in 2016, resulting in a total of five casualties.[33] Scores of attacks have been prevented, including the attempted bombing of the presidential palace in December 2016, and 150 arrests on terrorism charges were made in 2016. Indonesia security forces have killed several other high-profile terrorists in raids, including Santoso, the leader of the Mujahidin Indonesia Timor – one of the groups who joined JAD – in Central Sulawesi.[34]

The state response to IS-inspired and directed terrorism has been bolstered by political will, strong counterterrorism laws, an effective security and intelligence apparatus, and a continuing focus on de-radicalization laws and programs. However, human rights advocates remain critical of some of these policies. Human Rights Watch (HRW) has warned the proposed changes to the country's terrorism laws are, "overbroad, vague, and would unjustifiably restrict basic rights."[35] HRW has noted its concern over proposals to expand the definition of "law-violating actions" to include "speech, writings, pictures, symbols or bodily gestures, through electronic or non-electronic facilities, causing other people to be fearful in order to fetter an individual/the society's freedom."[36] The state must try to balance an aggressive legal response to prevent violent extremism while maintaining the national rights set forth in the Indonesian constitution.

One way in which the Indonesian government is attempting to diversify its response to violent extremism is by employing the assistance of other countries that are facing the same threat from IS-inspired and directed violent extremism. As the threat grows throughout the region, there has been increased cooperation between Southeast Asian countries. Perhaps most notably, Indonesia entered into a cooperative agreement with Malaysia and the Philippines to form the Trilateral Maritime Patrol Indomalphi in June 2017. The patrol focuses on disrupting the maritime routes of militants and arms among the three countries' archipelagos, which have been historically vulnerable to terrorist groups seeking safe havens due to the lack of cooperation in conducting patrols and the difficulty of monitoring such a vast area. This agreement highlights not only the severity of the problem of violent extremism in Southeast Asia, but also the seriousness with which the leadership of these three countries are treating the problem.

The future of countering violent extremism in Indonesia

While methods and alliances may change, the underlying motivations of most Islamist terrorists in Indonesia have remained rooted in opposition to secular nationalism as a core identity of the Indonesian state. The influence of IS may have expanded the reach and capacity of Indonesia's violent extremists, but the localized grievances that motivate terrorism by domestic actors persist.

IS' leadership and direction has introduced a new form of transnational violent extremism in Indonesia. From logistical and financial support to the targets and terrorists themselves, Indonesia today is faced with a *global* violent extremist challenge. This dynamic has pushed the state's leadership to seek assistance beyond its own borders. The naval patrol partnership with Malaysia and the Philippines has created a more comprehensive and responsive strategy to stem IS' influence in Southeast Asia. However, while the Philippines, Malaysia, and Indonesia have allowed each other to conduct "hot pursuits" of suspected terrorists in their respective littoral territories, there has been little cooperation on land-based offensives or intelligence sharing.

IS and its affiliates use the migration routes between Indonesia, Malaysia, and the Philippines to traffic arms, money, and fighters. The "Indomalphi" alliance will help stem this flow, but the lack of comprehensive cooperation will likely hinder efforts to address the departure and arrival of terrorist resources in all three countries. The recently proposed counter-IS taskforce suggests that the three regional partners might soon establish a more thorough alliance against IS and its extensions.[37] The proposed agreement would not only establish a trilateral taskforce, but include provisions that would open the Philippines' borders to Indonesian and Malaysian security forces as they hunt terrorists. Indonesia is also reliant on military and intelligence support from the international community. The United States and other international actors have a long history of military cooperation that will likely continue to support this new phase of countering violent extremism in Indonesia.

While hard power is an essential element in countering violent extremism, the Indonesian government has often undervalued the role of citizen-centered, democratic governance. Citizen marginalization, which is a decades-old driver of political violence and extremism in Indonesia, can only be effectively countered through more responsive national and local governance and other "soft power" initiatives. More direct communication between local government officials and the citizenry, local efforts to increase transparency in procurement and government services, and prioritizing quality, reliable service delivery are key elements that should be featured in the government's soft-power approach to addressing grievances felt by at-risk populations, and would better address unique vulnerabilities on the community level.

From de-radicalization programs in prisons to support for civil society-led violent extremism prevention programs, the United States, Australia, Canada, the United Kingdom, and other foreign governments are contributing large amounts of money to support Indonesia's efforts to counter violent extremism through nonmilitary means. The international community must continue to support these and other efforts to address governance that contribute to greater vulnerability to violent extremism.

As IS continues to exploit old grievances against the Indonesian state, it is critical that the national and local governments employ the full spectrum of hard and soft power approaches while adhering to internationally accepted human rights and the rule of law. By pursuing a strategy that includes military action where necessary, more responsive and inclusive governance, and regional and international cooperation, Indonesia will be well positioned to continue its fight against terrorism in the years to come.

Conclusion

The challenge of violent extremism and the state's response in Indonesia cannot be decoupled from the country's founding. From its beginnings as an independent nation, the secular character of the state has been challenged by religious conservatives who sought to secure a greater role for Islam in the constitution and policymaking. Darul Islam led the initial rebellion against Indonesia's secular nationalism, and was forcibly suppressed by the government in the 1960s. These events showcase the fragility of Indonesian nationalism, the early origins of Islamism, and the aggressiveness of the state's response to political challenge.

The year 1999 marked a turning point in which violent Islamic extremism in Indonesia took on a new character and ferocity. Over the last two decades, the nature of violence, the drivers of radicalism, and counterterrorism policy have evolved in subtle but important ways. From 1999 to 2009, Islamic extremism initially relied upon conventional paramilitary activity, but its defining characteristic was asymmetrical terrorist attacks against civilians. The Indonesian government, which in the 1950s and 1960s had violently repressed Islamists, devised a counterterrorism policy that used both law enforcement and de-radicalization approaches. By 2016, when the Islamic State replaced Jemaah Islamiyah as the predominant terrorist threat in Indonesia, they ushered in a more amorphous and adaptable form of extremism. Smaller, less hierarchical groups and individuals pledged loyalty to IS and its global agenda before carrying out attacks, without the level of careful orchestration that was a hallmark of Al Qaeda.

In the face of this new form of jihadism, the government has bolstered its previous counterterrorism policies while collaborating more closely with regional neighbors. Although the influence of IS has changed the methods and outward character of extremism in Indonesia, the underlying drivers are largely the same. With secularists firmly entrenched in power, the longstanding goal of Islamists for a theocratic Indonesia remains quixotic, at the same time that IS has demonstrated the possibility of building a territorial caliphate. Even as their goals remains grounded in the desire to transform the character of Indonesia's state, extremists take their inspiration and rhetorical cues from IS' construct of a global caliphate.

The risk of violent extremism is unlikely to diminish in Indonesia in the near term. While the Islamist threat is lower in Indonesia than in neighboring South and Southeast Asian countries, the government will likely be pushed to further develop its counterterrorism policy as the threat evolves. With the Islamic State in retreat in Syria and Iraq, there are already signs that IS is shifting its strategy to focus on global affiliates. This will require continuous policy innovations to meet this shifting challenge. One of the key tools to building an effective, sustainable counterterrorism policy will be through strengthening governance practices, to empower citizens and inoculate them against recruitment by extremists. Paradoxically, violent extremists exploit resentments bred by Indonesia's approach to secular democracy in order to undermine that very democracy. If Indonesia better develops a broad-based nationalism built on inclusive governance, it could win over moderate political Islamists that seek voice in the system. By combining traditional hard power approaches to countering extremism with a preventative strategy based on bolstering Indonesian democracy, the country will be well-placed to confront this evolving threat.

Notes

1 Sidel, *Riots, Pogroms, Jihad: Religious Violence in Indonesia*, 196.
2 Human Rights Watch, *Indonesia: The Violence in Ambon.*
3 Hasan, *Laskar Jihad: Islam, Militancy, and the Quest for Identity in Post-New Order Indonesia.*
4 Human Rights Watch, *Breakdown: Four Years of Communal Violence in Central Sulawesi.*
5 Sidel, *Riots, Pogroms, Jihad: Religious Violence in Indonesia*, 200.
6 Sidel, *Riots, Pogroms, Jihad: Religious Violence in Indonesia*, 203–204.
7 Perlez, "Saudis Quietly Promote Strict Islam in Indonesia."
8 The Mapping Militants Project, "Jemaah Islamiyah."
9 Sidel, *Riots, Pogroms, Jihad: Religious Violence in Indonesia*, 210.
10 Nacos, "Revisiting the Contagion Hypothesis: Terrorism, News Coverage, and Copycat Attacks."
11 Sageman, *Leaderless Jihad: Terror Networks in the Twenty-First Century*, vii.
12 Sumpter, "Countering Violent Extremism in Indonesia: Priorities, Practice and the Role of Civil Society."
13 Council on Foreign Relations, "Indonesia's Struggle Against Terrorism."
14 Butt, "Anti-Terrorism Law and Criminal Process in Indonesia."
15 Counter Extremism Project, "Indonesia: Extremism and Counter-Extremism."
16 McKay and Webb, "Comparing Counterterrorism in Indonesia and the Philippines."
17 USAID, "Indonesian and Malaysian Support for the Islamic State."
18 Council on Foreign Relations, "Indonesia's Struggle Against Terrorism."
19 Amnesty International. "Indonesia: Police Chief's Shocking Torture Admission Only Tip of Iceberg."
20 Tom Allard and Kanupriya Kapoor, "Fighting Back: How Indonesia's Elite Police Turned the Tide on Militants."
21 Hamish McDonald, "Fighting Terrorism with Smart Weaponry."
22 Sumpter, "Indonesia's De-Radicalization Blueprint."
23 Sumpter, "Reintegrating Extremist Prisoners in Indonesia: Easier Said Than Done."
24 Ibid.
25 Sukabdi, "Terrorism in Indonesia: A Review On Rehabilitation and Deradicalization."
26 Singh, *Katibah Nusantara: Islamic State's Malay Archipelago Combat Unit.*

27 US State Department, "State Department Terrorist Designation of Jamaah Anshurat Daulah."

28 GlobalSecurity.org, *Jamaah Anshurat Daulah Profile*.

29 Sing, *Katibah Nusantara: Islamic State's Malay Archipelago Combat Unit*.

30 Joe Cochrane, "More on Katibah Nusantara: Military Unit Under ISIS Linked to Jakarta Attack," *New York Times*, www.nytimes.com/live/jakarta-indonesia-explosions/background-on-katibah-nusantara-a-military-unit-under-isis-linked-to-jakarta-attacks/?mcubz=1.

31 "Jakarta Attacks: Profile of Suspect Bahrun Naim," *BBC News*, January 14, 2016. www.bbc.com/news/world-asia-35316915.

32 "Ahok Trial: The Blasphemy Case Testing Indonesian Identity," *BBC News*, February 14, 2017. www.bbc.com/news/world-asia-38902960.

33 US State Department, Counter Terrorism Bureau, *Country Reports on Terrorism 2016*.

34 Jacinta Carroll, ed. "Counterterrorism Yearbook 2017."

35 Human Rights Watch, *Indonesia: Counterterrorism Law Changes Threaten Rights*. www.hrw.org/news/2017/07/12/indonesia-counterterrorism-law-changes-threaten-rights.

36 Human Rights Watch, Revisions to the Law on the Eradication of Terrorism, 15/2003 (unofficial translation) www.hrw.org/sites/default/files/supporting_resources/indonesia_counterterrorism_bill_2017_0.pdf

37 Reuters, "Philippines eyes counter-terrorism task force with Malaysia, Indonesia," *Reuters*, September 3, 2017. www.reuters.com/article/us-militants-duterte/philippines-eyes-counter-terrorism-task-force-with-malaysia-indonesia-idUSKCN1BE0LS.

References

"Ahok Trial: The Blasphemy Case Testing Indonesian Identity," *BBC News*, (February 14, 2017) available at www.bbc.com/news/world-asia-38902960 (accessed September 14, 2017).

Allard, Tom and Kanupriya Kapoor. "Fighting Back: How Indonesia's Elite Police Turned the Tide on Militants," *Reuters*, (December 23, 2016) available at www.reuters.com/article/us-indonesia-security/fighting-back-how-indonesias-elite-police-turned-the-tide-on-militants-idUSKBN14C0X3 (accessed September 13, 2017).

Amnesty International. "Indonesia: Police Chief's Shocking Torture Admission Only Tip of Iceberg," April 21, 2016. Accessed September 1, 2017. www.amnesty.org/en/latest/news/2016/04/indonesia-police-chief-shocking-torture-admission-only-tip-of-iceberg/

BBC Monitoring. "Jakarta Attacks: Profile of Suspect Bahrun Naim," *BBC News*, (January 14, 2016) available at www.bbc.com/news/world-asia-35316915 (accessed September 11, 2017).

Butt, Simon. "Anti-Terrorism Law and Criminal Process in Indonesia." The University of Melbourne ARC Federation Fellowship. Melbourne, Asian Law Center, University of Melbourne, (2008) available at http://law.unimelb.edu.au/__data/assets/pdf_file/0010/1546327/AntiTerrorismLawandProcessInIndonesia2.pdf (accessed September 3, 2017).

Carroll, Jacinta (ed.). "Counterterrorism Yearbook 2017," Australia: Australian Strategic Policy Institute, 2017, 1–152.

Cochrane, Joe. "More on Katibah Nusantara: Military Unit Under ISIS Linked to Jakarta Attack," *New York Times*, (January 14, 2016) available at www.nytimes.com/live/jakarta-indonesia-explosions/background-on-katibah-nusantara-a-military-unit-under-isis-linked-to-jakarta-attacks/?mcubz=1 (accessed September 15, 2017).

Council on Foreign Relations, "Indonesia's Struggle Against Terrorism," (April 11, 2014) available at www.cfr.org/councilofcouncils/global_memos/p32772 (accessed September 4, 2017).

Counter Extremism Project, "Indonesia: Extremism and Counter-Extremism." New York: Counter Extremism Project, 2017, 1–9.

GlobalSecurity.org, *Jamaah Anshurat Daulah Profile*, available at www.globalsecurity. org/military/world/para/jad.htm (accessed September 2, 2017).

Hasan, Noorhaidi. *Laskar Jihad: Islam, Militancy, and the Quest for Identity in Post-New Order Indonesia.* Cornell: Cornell University Press, 2006.

Human Rights Watch, *Indonesia: The Violence in Ambon.* New York: Human Rights Watch, 1999. Accessed September 4, 2017. www.hrw.org/legacy/reports/1999/ambon/.

Human Rights Watch. *Breakdown: Four Years of Communal Violence in Central Sulawesi.* New York: Human Rights Watch, (2002) available at www.hrw.org/reports/2002/indonesia/indonesia1102.pdf (accessed September 2, 2017).

Human Rights Watch. *Indonesia: Counterterrorism Law Changes Threaten Rights.* (June 12, 2017) available at www.hrw.org/news/2017/07/12/indonesia-counterterrorism-law-changes-threaten-rights (accessed September 5, 2017).

Human Rights Watch. Revisions to the Law on the Eradication of Terrorism, 15/2003 (unofficial translation) available at www.hrw.org/sites/default/files/supporting_resources/indonesia_counterterrorism_bill_2017_0.pdf (accessed September 1, 2017).

Nacos, Brigitte L. "Revisiting the Contagion Hypothesis: Terrorism, News Coverage, and Copycat Attacks," *Perspectives on Terrorism*, 3 (3) (2009), 3–13.

McDonald, Hamish. "Fighting Terrorism with Smart Weaponry," *The Sydney Morning Herald*, (May 21, 2008) available at www.smh.com.au/world/fighting-terrorism-with-smart-weaponry-20080531-gdsfx6.html. Accessed September 11, 2017.

McKay, Scott N. and David A. Webb. "Comparing Counterterrorism in Indonesia and the Philippines," (February 27, 2015) available at https://ctc.usma.edu/posts/comparing-counterterrorism-in-indonesia-and-the-philippines (accessed September 6, 2017).

Perlez, Jane. "Saudis Quietly Promote Strict Islam in Indonesia," *New York Times*, (July 5, 2003) available at www.nytimes.com/2003/07/05/world/saudis-quietly-promote-strict-islam-in-indonesia.html (accessed September 15, 2017).

Reuters, "Philippines eyes counter-terrorism task force with Malaysia, Indonesia," *Reuters*, (September 3, 2017) available at www.reuters.com/article/us-militants-duterte/philippines-eyes-counter-terrorism-task-force-with-malaysia-indonesia-idUS KCN1BE0LS (accessed September 12, 2015).

Sageman, Marc. *Leaderless Jihad: Terror Networks in the Twenty-First Century.* Philadelphia: University of Pennsylvania Press, 2008.

Sidel, John T. *Riots, Pogroms, Jihad: Religious Violence in Indonesia.* Cornell: Cornell University Press, 2006.

Sing, Jasminder. *Katibah Nusantara: Islamic State's Malay Archipelago Combat Unit.* S. Rajaratnam School of International Studies: RSIS Publications, May 25, 2015.

Sukabdi, Zora A. "Terrorism in Indonesia: A Review On Rehabilitation and Deradicalization," *Journal of Terrorism Research* 6 (2) (2015), 36–56.

Sumpter, Cameron. "Indonesia's De-Radicalization Blueprint," *The National Interest*, (February 5, 2016) available at http://nationalinterest.org/blog/the-buzz/indonesias-de-radicalization-blueprint-15133 (accessed September 3, 2017).

Sumpter, Cameron. "Reintegrating Extremist Prisoners in Indonesia: Easier Said Than Done," *The Diplomat*, (March 16, 2017) http://thediplomat.com/2017/03/reintegrating-extremist-prisoners-in-indonesia-easier-said-than-done/ (accessed September 5, 2017).

Sumpter, Cameron. "Countering Violent Extremism in Indonesia: Priorities, Practice and the Role of Civil Society," *Journal for Deradicalization*, no. 11 (2017), 112–147.

The Mapping Militants Project. "Jemaah Islamiyah." The Mapping Militants Project, Stanford University. http://web.stanford.edu/group/mappingmilitants/cgi-bin/groups/view/251 (accessed September 6, 2017).

USAID. "Indonesian and Malaysian Support for the Islamic State." Washington, DC: United States Agency for International Development, 2016.

US State Department, "State Department Terrorist Designation of Jamaah Anshurat Daulah." Washington, DC: United States Department of State, 2017.

US State Department, Counter Terrorism Bureau, *Country Reports on Terrorism 2016.* Washington, DC: United States Department of State, Counter Terrorism Bureau, 2016.

Veitch, James. "Human tragedy in Sulawesi Indonesia: 1998–2002." In Andrew T. H. Tan (ed.). *A Handbook of Terrorism and Insurgency in Southeast Asia*. London: Edward Elgar Publishing, 2007, 122–145.

4 The Rohingya and Myanmar's counter terrorism approach

Bibhu Prasad Routray

Myanmar's Rakhine state has been in a state of boil since at least 2012, although the history of unrest goes back some decades. Approximately 700,000 of the 1.1 million Rohingya who once called Rakhine their home have been living as refugees in Bangladesh, following different phases of a brutal crackdown initiated by the Myanmar military since 2016. The crackdown ostensibly followed two series of attacks by a new terrorist group, the Arakan Rohingya Salvation Army (ARSA). This is problematic. In reality, whether it exists organizational or as a place-holder, ARSA is largely irrelevant when contextualized within long-standing discriminatory official policies and now ethnic cleansing. Ironically, a short-term, self-professed policy of "eliminating extremists" may appear as effective to the regime, but in the long-term, it will not only fuel further discontent but create conditions which could be exploited by truly radical external terrorist actors.

The attacks

Though unclear as to genesis, events appear to have been set on their present by several October 2016 attacks by ARSA cadre targeting border police personnel in Myanmar, killing nine officers. Basic weapons such as machetes, swords, and knives were used. Initial estimates suggested that a group of 90 attackers carried out raids at 1.30 am local time on a police force office in Kyiganbyin village of Maungdaw town, killing six police officers. A simultaneous attack on a border police camp in Kyeedangauk village in Rathidaung town killed one police officer and injured two others. In a third attack in Buthidaung township, two more police officers were killed.[1] Although eight terrorists were also killed, the ARSA managed to seize 62 assorted weapons and 10,130 rounds of ammunition during the attacks.

The second wave of attacks, bolder and more sustained, followed ten months later, but it is precisely at this point that the narrative becomes problematic. Best evidence supports retaliation by ARSA or others to clearing operations run amok. Regardless, on August 25, 2017, 30 police posts were targeted in a single night and were followed by ambushes on the following days. The attacks started at 1 am and continued till 4 am by "the extremist Bengali insurgents started their attack on the police post … with the man-made bombs and small weapons" the army said.[2] Quite visibly, the attacks had undergone some levels of quality upgradation and witnessed use of better quality weapons, possibly including some that were seized from the security forces during the earlier attacks. And yet, the insurgents suffered more casualties compared to their targets – at least 21 insurgents and 11 members of the security forces were killed – and the reaction by the security forces bore little connection to threat actions, regardless of genesis.

A third attack was carried out in January 2018, when a group of about 20 ARSA cadre targeted a military vehicle carrying a sick army official to the hospital in the Rakhine state. As the truck struck an improvised explosive device (IED) and came to a halt, ARSA cadres shot at it from a nearby hill. The number of injured in the attack have been differently estimated to be anywhere between three to seven. The attackers used, in the words of the Myanmar military, homemade mines and small arms. The group later claimed responsibility for the attack. As already noted, state response continued to veer from any security purpose and took on the character of what any number of sources termed "classic ethnic cleansing." The deplorable results are only too well known: some 700,000 refugees huddled (overwhelmingly) in Bangladesh at early 2018 count.

Examining ARSA

It is against this background that analysis of ARSA can proceed, if for no other reason than Myanmar – in no less person than Nobel Prize winner Aung San Suu Kyi, among others – has continued to claim that the issue is security as opposed to regime savagery. For analysis, it is pertinent to remember that most information on ARSA is deeply influenced either by regime claims or those of the group itself, both of which have a vested interest in framing events to their advantage. Analysts, similarly, tend to be divided either for their sympathy for the Rohingya and hence, for the ARSA, or are sympathetic to the Myanmar's official position.

ARSA's existence would itself to be beyond speculation. The outfit has a website, a Facebook page, and a Twitter account, hence its claims and goals are accessible. Indeed, the group has never shied away from claiming responsibility for attacks. It has justified its actions and used their very occurrence in an effort to inspire confidence that its struggle is developing. In this sense, it resembles the early actions of Liberation Tigers of Tamil Eelam (LTTE), as detailed in the final chapter of this volume. As was the early case with LTTE, so too is it now with ARSA: assessment remains divided on the extent to which ARSA poses a threat to Myanmar's security.

ARSA's predecessor, the Harakat al Yaqin (HaY), literally translating as the Faith Movement, emerged in 2013–2014, when a small band of Rohingya militants reportedly made attempts to establish training bases near the Thai border town of Mae Sot. A longer trajectory emerges if one considers the formation of the Rohingya Solidarity Organisation (RSO) in the wake of the 1978 communal violence that had sparked the first massive exodus of Rohingya from Rakhine.[3]

RSO underwent several splits. New factions continued generating funds but were not able to generate momentum against their adversaries in Myanmar. This failure was the lifeline of HaY, which originated with an aim, using the depleted yet existing RSO network, to avenge the recurrent regime atrocities committed against the Rohingya. However, HaY's efforts at setting up training bases in Thailand proved short-lived after the Thai authorities – who share close relations with the Myanmar army and are themselves faced with an Islamist separatist movement in their southern territory (see Chapter 14 in this volume), were informed by Myanmar and brought these activities to a close. It is unlikely that any form of arms training had actually occurred.

Bangladesh, on the other hand, has emerged as an important locale where ARSA apparently has been able to organize, recruit, and train. Although ARSA has made commitments that none of its violent actions will be carried out on Bangladesh soil,[4] recruitment and training of its cadre who participated in the October 2016 and August 2017 attacks could only have taken place there. Myanmar itself is an unlikely site for such preparations, since the Rohingya there were under the strict watch of the regime and Buddhist vigilantes. Clearly, one of the strategies of ARSA has been to exploit the not-so-harmonious bilateral relations between Myanmar and Bangladesh.

At least one consignment of a crate of small arms reportedly reached the militants, consisting of Chinese-manufactured Type 56 7.62 mm rifles. *Asia Times* reported, on the basis of mobile phone photographs, youths in sarongs and tee-shirts being trained in the use of the weapons by older instructors. The article speculated, "In view of the tight security lockdown imposed by the Myanmar security forces across Rakhine state … it is far more likely that the weapons reached ARSA through southern Bangladesh than from inside Myanmar."[5] It acknowledged, however, that it is impossible to confirm whether the training took place inside Rakhine state or in Bangladesh.

It is these apparent foreign connections, beyond Bangladesh, that have been leveraged in the claim that ARSA poses a threat to regional security. According again to *Asia Times*, ARSA "is understood to have a leadership council based in Saudi Arabia and local leaders with backgrounds in Pakistan. It's a support network that will almost certainly translate into donations from established businessmen in both countries."[6] The article sourced its information to an unnamed senior regional official and speculated that a group of senior Rohingya clerics based in Saudi Arabia had raised funds and facilitated money transfers through Malaysia, which provides a Muslim-friendly visa regime and hence has emerged as a transit point for militants.

The *Asia Times* article runs close to the findings of a report by the International Crisis Group (ICG), which claimed in December 2016 that ARSA was led by a "committee of Rohingya émigrés in Saudi Arabia and is commanded on the ground by Rohingya with international training and experience in modern guerrilla war tactics."[7] The ICG report identified ARSA's top leader as Ata Ullah, who it said was born in Karachi and later moved to Saudi Arabia. Intelligence analysts further identify ARSA's mentor as Abdus Qadoos Burmi, another

Karachi-based Pakistani national of Rohingya descent. Burmi has appeared in videos released on social media and has called for launching "jihad" in Myanmar. His linkages with the Pakistani terrorist group Lashkar-e-Taiba (LeT), which operates in Kashmir and Pakistan, has been discerned from the meetings he has attended with latter's leader, Hafiz Mohammad Sayyid. These conclusions, however, have been denied by ARSA. In an interview with the CNN, Ata Ullah claimed that his group is independent of influence from supporters in Saudi Arabia and has no connection to groups in Pakistan, Bangladesh, or Afghanistan.[8]

Similarly, the combatant composition of ARSA remains subject to debate and speculation. A figure of 500 seems to be favored by a majority of analysts, but whether they are volunteers or impressed villagers has been questioned. Personnel reportedly mingle with villagers and wear civilian clothes. Bertil Lintner, for example, writes, "ARSA itself may have been able to recruit angry and desperate young men among the Rohingya in Rakhine state and refugee camps in Bangladesh."[9] An article in the *New York Times* advanced the hardly startling observation that ARSA could have recruited "from the ranks of fleeing villagers"[10] after the Myanmar army's atrocities. Such enlistment could have been by persuasion or coercion. In addition to these local men, reports quoting unnamed security analysts suggest that nearly one-third of ARSA's cadre, perhaps 150 personnel, are foreigners hailing from countries including Pakistan, Malaysia, and southern Thailand, with Afghan war veterans responsible for running the outfit's training camps.

With its relative tactical success in carrying out its attacks inside Myanmar, ARSA appears to be attempting to position itself as *the* self-defense option for the persecuted Rohingya. In this, it has been spectacularly unsuccessful, as the response by the Myanmar army has been ruthless in draining the sea in which the would-be guerrilla fish have sought to swim. The repercussions of ARSA actions, then, have played themselves out in the regime's widespread human rights abuse and Buddhist mobs' atrocities.

Counter terrorism strategy

Though not completely relevant to the ARSA case, research on how "terrorism ends" is useful in emphasizing that either incorporation into the political process occurs or the group concerned is decimated by the concerted action of the security forces and intelligence agencies.[11] Rarely do military operations alone, however, result in decimation of the target. Conversely, it is almost a truism to emphasize that any group can prolong its existence by successfully eliciting support from a variety of sources. Regardless, there is historical as well as contemporary evidence to suggest that states default to the use of military force with the aim of bringing the nuisance value of a group to an early close. Even within larger supposedly multifaceted approaches, as in Afghanistan (see contributions in this volume), this privileging of what is often still termed kinetic action remains the case. This is also true in Myanmar, with nonkinetic action relegated to such a minor, tactical role as to be effectively nonexistent.

Beginning in 2012, the Rohingya in Myanmar have been subject to a campaign of persecution by radical Buddhist groups backed by the state. Although the most recent violence started that year and has continued unabated since, the history of persecution goes back by several decades. It is not our purpose here to provide details of the sustained state policy against the minority Rohingya. Suffice to say the majority society considered them to be foreigners, regardless of individual or family particulars, with overlapping factors contributing to the forced marginalization: Islamic religion, Bengali language, distinct appearance, and dark skin color. The decision of the military in 1982 to legally list national ethnicities and in the process exclude the then one million Rohingya from full citizenship was the culmination of this policy of discrimination. It made the Rohingya stateless, restricted their right to movement and employment, and severely limited their access to healthcare and education. Even with such restrictions and limited access to government support, the Rohingya remained, eking out a living while suffering from growing repression; 2012 was a culmination of sorts, as communal riots engulfed the Rakhine region, pushing thousands out of the country and almost 120,000 to squalid camps after being driven from their homes.

Myanmar's transformation into a quasi-democracy did little to alter the situation. Though civilian control was on the surface established and peace processes with a large number of ethnic insurgencies in the country's periphery initiated in 2011, with parliamentary and provincial elections in 2015 catapulting the National League of Democracy (NLD) to power, the Rohingya remained marginalized. Subject to a range of laws and arrangements that systematically stripped them of their rights to conduct trade, marry, and even hold prayers, they were pushed to the fringes. Several hundred who attempted to escape from such a regime of restrictions and state-sponsored brutality died at sea while traveling to Malaysia using the services of smugglers and traffickers, as well as in brutal jungle camps run by human smugglers. Others remained confined to squalid holding areas with little access to education, food, or clean water, awaiting a process of gradual extermination.

It is in the context of sheer hopelessness that ARSA was born. And it is amidst this backdrop of visceral hatred for the Rohingya and continuing official policies that have sought to obliterate them that the military's "counter-terrorism" policy emerged and was implemented.

The State response

Past regime campaigns against the numerous ethnic insurgencies in Myanmar have been brutal. Minorities such as such as the Kachin, Shan, Karen, the Kokang, and others, inhabiting the country's periphery, have in the past accused the *Tatmadaw* (Myanmar military) of using what amounts to scorched earth tactics to seek domination of insurgent-controlled areas.[12] Forced conscription, use of rape as a weapon, and recruitment of child soldiers have all been documented as part of the military inventory directed against the restive periphery.

While ARSA's history has been short, some of the ethnic insurgencies, such as the Kachin Independence Army (KIA) and the Shan State Army (SSA), have existed since the then-Burmese independence from the British in 1948 and boast trained manpower armed with sophisticated weapons, most of them supplied by China. ARSA thus is hardly comparable to such groups, yet the military approach to all has been of a piece.

Two phases of response were unleashed against ARSA following the October 2016 and the August 2017 attacks and sustained thereafter. The effort consisted of four distinct tactics: (1) use of overwhelming force against the group and its alleged sympathizers; (2) categorization of ARSA as a large and lethal terror group with global Islamist connections; (3) use of the civilian government, led by the NLD, to resist international pressure and seek support from neighboring countries; and (4) vigorously prosecution of journalists who challenged the state narrative. All these tactics were carried out simultaneously and were successful in insulating Myanmar from a range of pressures created by the international community, the human rights groups, and even the United Nations.

Overwhelming and disproportionate force

Immediately after the October 2016 attacks, the government declared a state of emergency in Maungdaw and in three other townships. The atrocities published gruesome photographs of alleged terrorist atrocities (and the purported actual attackers) using social media, claiming that the throats of some officers had been slit.[13] Joint counter-terror operations by the army and the police began the next day along with deployment of army helicopters to pursue the attackers. In one of the incidents, security forces loaded onto three trucks arrived in Myothugyi village and shot dead seven people. The killed included three villagers who started to run in panic after the forces arrived.[14] Organizations monitoring human rights situations in Rakhine suggested that some of the killed were unarmed and the killings are a part of the extra judicial anti-Muslim offensive in Maungdaw.[15] By early November, the attack helicopters were regularly opening fire on "attackers" and had caused more than 100 deaths. Another 500 Rohingya were arrested.[16]

In addition, imposition of a state of emergency led to severe curtailment to access to the Rohingya in Rakhine by humanitarian aid agencies, such as the World Food Progam. This initiated displacement of a large Rohingya population into Bangladesh. In February 2017, the Office of the United Nations High Commissioner for Human Rights (OHCHR) released a graphic report detailing the plight of Rohingya who had stayed in Myanmar since the communal riots broke out in 2012 and miseries they had been subjected to. Systemic persecution and blatant violation of human rights carried out against them by the Myanmar police force, *Tatmadaw*, and the Border Guard Force included summary executions, deaths due to random firing, stabbing by knives, burning, beating people to death, killing of children, rape, and other forms of physical assault.[17]

The second phase of attacks by the ARSA in August 2017 invited even harsher retaliation from the state. In the aftermath of the attacks, "area clearing operations" were initiated by the military taking help from the local police. This amounted to punishing the ARSA and all Rohingya who were seen as its collaborators, with brutal violence. In all, 471 Rohingya-inhabited villages were targeted. In the absence of intelligence, the operations did not distinguish between the civilians and insurgents, although obliteration of the dividing line appeared intentional. Reports suggest that in the months before the attacks, as many as 50 people, Muslims as well as Buddhists suspected of serving as government informants, had their throats slit or were hacked to death[18] in order to deprive the Myanmar military of intelligence in the area. If true, only ARSA could have been responsible. Nevertheless, in its operations, the military torched a number of Rohingya villages after civilians fled out of fear.

Ironically, while a large number of media accounts corroborated the involvement of security forces and Buddhist civilians in such incidents, the military blamed ARSA and the Rohingya for allegedly setting their own houses on fire. In one particular village, Tula Toli, Rohingya villagers accused the Myanmar army of shooting teenagers and adults and throwing babies and toddlers into water.[19] Summary executions were carried out by the military as they herded adults into open spaces and subsequently shot or decapitated them using machetes. Such graphic accounts, firmly denied by the military, were corroborated by independent media investigations and also by eyewitnesses which included security force personnel who were part of the raiding team.[20] In another instance, in the village of Gu Dar Pyin, the Associated Press published a report, using photographs and testimony from refugees in Bangladesh, on the use of at least five mass graves to conceal systematic executions of Rohingya prisoners.[21] Mortar and machine-gun fire was reported to have been used on Rohingya attempting to cross into Bangladesh.[22] Doctors treating some of the refugees said they had seen dozens of women with injuries consistent with violent sexual attacks.

On September 1, 2017, the Myanmar military said 370 ARSA militants had been killed in its retaliatory operations. On September 12, the government spokesperson announced that of the 471 villages targeted for clearing operations, 176 were empty and 34 partially abandoned.[23] Even after the displacement of 700,000 Rohingya into Bangladesh, the military continued to project ARSA as the source of permanent threat to the country's security. In March 2018, a United Nations official said that Myanmar forces had changed their tactics from "the frenzied blood-letting and mass rape of last year to a lower intensity campaign of terror and forced starvation."[24] The official anti-Rohingya narrative facilitated continued militarization of the border with Bangladesh. Refugees were continuously intimidated, even as both countries negotiated and declared they had reached an agreement to repatriate the refugees who had crossed over into Bangladesh.

In one instance, in early March 2018, Myanmar increased its troop deployment in the Tambru area and used it as a base to threaten refugees settled in the no-man's land between the two countries. Some troops were reported to have crossed over the border fence into the no-man's land using ladders and threatened refugees with eviction from the camps. Even as Dhaka protested and asked for immediate pullout of troops and military assets from the area, Myanmar government spokesperson defended the action, saying that it was based on fresh inputs on ARSA activities and was not directed against Dhaka.[25] Yet the tactic was clearly directed toward ensuring, first, that the 700,000 Rohingya who had left the country could never return; second, those who remained in Myanmar were either compelled to leave or submit to becoming victims of a process of "slow genocide."[26] Such assessment was more than confirmed by the statement of September credited to Myanmar's Army Commander, Sr. Gen. Min Aung Hlaing, who noted that the campaign against ARSA was "unfinished business,"[27] dating back to the Second World War.

ARSA presented as a large jihadist organization with global links

ARSA maintained that the raison d'être of its existence was Rohingya self-defense and dignity. On September 14, 2017, it announced that it had no "links to Al Qaeda, the Islamic State in Iraq and Syria, Lashkar-e-Taiba or any transnational terrorist group."[28] However, regime statements repeatedly labeled ARSA as a jihadist threat posing a religious challenge to Buddhist-dominated Myanmar. The government claimed ARSA had fighters who had trained with Pakistani Taliban and that it sought to impose Islamist rule over a portion of Rakhine State. While independent media reports did appear to establish at least some links between ARSA leadership and the likes of LeT and Taliban, there was little tangible aside from a September statement by al-Qaeda urging Muslims to support the Rohingya cause and "make the necessary preparations – training and the like – to resist this oppression."[29]

International Crisis Group, in the aftermath of the October 2016 attacks by ARSA, concluded that ARSA was guided by Rohingya emigres in Saudi Arabia and commanded in the field by overseas-trained guerrilla fighters.[30] Yet assessments quoting an ARSA cadre to effect that the group had 5,000 fighters[31] were obviously exaggerated and served only to strengthen the military's threat narrative. To the contrary, experts opined that the number of hardcore, armed ARSA members did not exceed 500.[32] Myanmar experts such as Bertil Lintner even doubted if all the 400 persons killed during area clearing operations were actually ARSA, which was certainly correct. Lintner assessed that ARSA was at best a local outfit which used conscripted villagers to launch a new type of insurgency unknown to the Myanmar soldiers similar in techniques to those used by the Maoists in Nepal and the left-wing extremists in India.[33] Even this seemed in many ways to go beyond what was actually occurring on the ground.

ARSA, on September 9, declared a month-long unilateral ceasefire, purportedly to enable aid groups to reach Rohingya refugees and to avert a full-blown humanitarian crisis. Although the move could have been a mere public relations exercise and a ploy to bring the military campaign to a halt, it brought the operational weakness of the ARSA overground. Nevertheless, it was rejected by the military, which vowed to continue its onslaught. Even as analysts argued that ARSA's actual military capability was limited and disproportionate to its larger-than-life presence on social media, the regime appeared to be in an unforgiving mood.

As part of its tactic to vilify ARSA, Myanmar – particularly the army but joined by civilians, especially elements of the Buddhist clergy – unleashed a publicity campaign highlighting alleged ARSA attacks on Hindu and Buddhist villages in Rakhine. On September 27, 2018, it announced the discovery of a purported mass grave in Rakhine containing the bodies of 28 Hindus, including women and children. The killings were blamed on ARSA.[34] A select group of local and foreign media persons was flown to the spot, which resulted in a number of favorable reports describing atrocities committed by ARSA. Such claims, though, were not verifiable, given that access throughout the affected areas was severely restricted by the military. On the other hand, media reports quoted a number of Hindu female refugees arriving in Bangladesh to the effect that the Myanmar army had killed their fathers and husbands "for their reluctance to partake in Muslim killing in Rakhine."[35]

Using the civilian government as a shield

Though questions remain as to the extent the Myanmar military has ceded decision-making power to the elected civilian government, there is little doubt that the former still holds ultimate power in all matters of more than minor importance. As per the military-drafted 2008 constitution, it controls 25 seats in both houses of the national parliament, as well as in provincial assemblies, thus allowing it to veto any constitutional change. Moreover, the military constitutionally is in charge of three key ministries, Home, Defence, and Border affairs, which not only gives it primacy in any counter-terrorism effort but also perpetuates historically lopsided civil-military relations in its favor. One of NLD's top goals, as stated in the 2015 election, was to amend the constitution. This has not been pushed in Parliament. To the contrary, Aung San Suu Kyi, Myanmar's state counselor and the NLD's supreme leader, apparently aspires to break free from the limitations imposed by the constitution not through challenging the military but by becoming their mouthpiece.

Most grievously, she has become an apologist for the regime. In April 2017, she alleged that the conflict was about "Muslims killing Muslims as well."[36] In early September, days after ARSA carried out its attacks and the military launched its brutal crackdown, she said sympathy for the Rohingya was being generated by "a huge iceberg of misinformation calculated to create a lot of problems between different communities and with the aim of promoting the interest of the terrorists."[37] In imperceptible statements on the crisis, she appeared to repeat the charge of the military that ARSA's killing of Rohingya who collaborated with authorities had precipitated the refugee movement into Bangladesh. As pointed out earlier, there could have been some minor elements of reality in this accusation – that is, ARSA has in fact killed at least some informers – but this was at most a ripple in a storm-tossed sea. The astonishing lack of sensitivity, and even awareness of the enormous crisis at hand, could only reflect poorly on both her situational awareness and moral engagement.

Even as the United Nations high commissioner for human rights accused Myanmar of carrying out "a textbook example of ethnic cleansing"[38] against Rohingya Muslims, NLD officials conceded that pushing back on the Rohingya issue had served to unite the party, the army, and Buddhist civilians.[39] In the midst of international criticism, the NLD indeed feared that speaking against the *Tatmadaw* would seriously disrupt the democratic transition. As a result, not only did NLD see its own position decimated, but the military remained insulated from scathing attacks by human rights activists, who instead directed all their frustrations against the civilian government. This hapless group thus served both as a shock absorber for the military and as shield for continuing ethnic cleansing. The ostensible counter-terrorism campaign went on.

Persecute the messenger

The fourth component of the CT strategy was to intimidate journalists who strayed from the official line. As Myanmar struggled to battle intense negative reporting in the Western media, the regime severely restricted the movement and activities of journalists and NGOs in the Rakhine. In January 2018, two journalists working for Reuters news agency were formally charged with violating the Official Secrets Act[40] for possessing confidential government documents. Both journalists had been arrested on December 12, 2017, after which Reuters published a graphic report they had contributed detailing the massacre of ten Rohingya men by the Myanmar security forces and Buddhist villagers at Tula Toli. Both face up to 14 years in prison if convicted. Citing security reasons, the military barred media persons from reporting from Rakhine. Only occasional conducted tours of the region were allowed and obviously intended to endorse official position.

It is debatable whether the tactic of muzzling the press to conceal information is useful in the age of social media, satellites, and citizen journalism. Regardless, the Myanmar government has systematically ignored the concerns of the international community and attempted to cultivate and coerce the domestic press to conform to the official line. The effort has been partially successful. Some media sources, earlier critical of the government, have published reports that vilify the Rohingya as illegal migrants and ARSA as a "jihadi" terrorist organization. The government has further attempted to muzzle dissent by proposing amendments to the Peaceful Assembly and Procession Law. The amendments, which have been passed in the upper house of the Parliament,

> stipulate jail terms for those convicted of provoking or exhorting others to organize or participate in demonstrations by bribing or paying them or doing anything else with the intention of harming the stability, rule of law, peace and tranquility of the community.[41]

Activists term the provision as being so broadly written as to likely stifle political dissent. Certainly, that is the point.

The future

The overall impact of such application of brute force, in military terms, has indeed been a success. The sea has been drained, and ARSA's already limited capabilities have been neutralized. The qualitative jump apparent between the October 2016 and August 2017 attacks has been relegated to the past. From August 2017 until the present, only sporadic attacks, few in number, have been carried out by the group. The January 2018 attack on the military vehicle, which injured five security force personnel and was promptly claimed by ARSA, involved the use of "home-made mines and small arms."[42] The Myanmar military has further accused ARSA of carrying out 30 attacks between August 2017 and February 2018, but this has been contested by analysts who accuse Myanmar of either trying to amplify a minor threat or to falsely blame ARSA for all acts of violence in order to justify the regime's brutal campaign. On December 8, 2017, for instance, four Myanmar army soldiers were injured when their convoy hit three IEDs in Myebon township.[43] The attack was blamed on ARSA, although attribution neither could be established nor was claimed by ARSA.

Yet, as has been noted was the case with LTTE in Sri Lanka or the Islamists of Barisan Revolusi Nasional (BRN)-Coordinate in southern Thailand, ARSA will remain a potential factor precisely because the state has now created the conditions from which can emerge the threat its misjudgment and mis-statements has created. The astonishing brutality of the regime's "counter-terrorism" effort has been directed at the very social base ARSA claims to have tapped but with which it hitherto had minimal links. Even though a majority of the Rohingya are undoubtedly not subscribed to support for the group or certainly to violence, the deep sense of alienation and anger at what has happened has opened a pathway for radical access.

The prospect of Myanmar revisiting its counter-terrorism approach and moderating its behavior to mediate the concerns of the Rohingya (and other ethnic minorities) is minimal. The ruthless approach of the post-independence years will continue as long as the military dominates the national security decision-making process, though the civilians have not conducted themselves in such manner as to give grounds for hope that astute, ethical decision making would follow an unlikely military exit. Any change in approach hence would seem possible only through significant external pressure, but this seems as unlikely as the previous possibility of civilians gaining the lead. Whatever pressure traditional external sources of moral and economic investment might at one time have been able to apply is now quite negated by the role of China. For China, it hardly needs reiteration, the Myanmar approach is wrong-headed only in its lack of subtlety. A more sophisticated "CT plan" would, for Beijing, proceed along the lines of societal engineering through synthetic repression as is unfolding in Xinjiang. Technically and psychologically, this may simply be a bridge too far for regime. Thus we are likely to see more of the same.

Notes

1 "Myanmar insurgents kill at least 17 people in targeted attacks on border in Rakhine State," *ABC News*, October 10, 2016. Accessed December 9, 2017. www.abc.net.au/news/2016-10-10/police-killed-in-myanmar-attacks-near-bangladesh-border/7917382
2 "At least 32 killed in Myanmar as Rohingya insurgents stage major attack," *CNBC*, August 25, 2017. Accessed December 8, 2017. www.cnbc.com/2017/08/25/at-least-32-killed-in-myanmar-as-rohingya-insurgents-stage-major-attack.html
3 Swami, "The vicious circle."
4 Mahmud, "ARSA: We will not carry out attacks inside Bangladesh."
5 Davis, "Foreign Support Gives Rohingya Militants a Lethal Edge."
6 Ibid.
7 International Crisis Group, *Myanmar: A New Muslim Insurgency in Rakhine State*.
8 Wright and Watson, "Inside the Rohingya Resistance: The Rebels Who Provoked Myanmar's Crackdown."
9 Murdoch, "Rohingya Pay Steep Price for Emergence of ARSA."
10 Beech, "Desperate Rohingya Flee Myanmar on Trail of Suffering: 'It Is All Gone'."
11 Jones and Libicki, *How Terrorist Groups End: Lessons for Countering al Qa'ida*.
12 Useful background, not just for the case of the title but for the ethnic insurgent dynamic in general, may be found in Thomas A. Marks, "The Karen Revolt in Burma," *Issues and Studies* 14, no. 12 (December 1978), 48–84.
13 Moe, "Dozens Believed Killed as Violence Erupts in Myanmar."
14 Ibid.
15 Ives, "Violence Mounts in Restive Myanmar State, Leaving a Dozen Dead."
16 Wright and Watson, "Inside the Rohingya Resistance: The Rebels Who Provoked Myanmar's Crackdown."
17 OHCHR, "Interviews with Rohingyas Fleeing from Myanmar Since 9 October 2016."
18 Lintner, "The Truth Behind Myanmar's Rohingya Insurgency."
19 Holmes, "Massacre at Tula Toli: Rohingya Recall Horror of Myanmar Army Attack."
20 McPherson, "Witness to a Massacre: The Former Myanmar Soldier Who Saw His Village Burn."
21 Klug, "AP Finds Evidence for Graves, Rohingya Massacre in Myanmar."
22 "Myanmar troops open fire on civilians fleeing attacks," *Al Jazeera*, August 27, 2017. Accessed March 3, 2018. www.aljazeera.com/news/2017/08/myanmar-violence-traps-rohingya-bangladesh-border-170826101215439.html
23 Holmes et al., "Myanmar Says 40% of Rohingya Villages Targeted by Army Are Now Empty."
24 "UN: Myanmar's Ethnic Cleansing of Rohingyas Continues," *Voice of America News*, March 6, 2018. Accessed March 7, 2018. www.voanews.com/a/un-rights-official-myanmar-ethnic-cleansing-of-rohingyas-continues/4282702.html
25 "Myanmar: Troop Build-up on Bangladesh Border Over ARSA Movement," *Dhaka Tribune*, March 2, 2018. Accessed March 3, 2018. www.dhakatribune.com/world/south-asia/2018/03/02/myanmar-troop-bangladesh-border-arsa/
26 New York Times reporter Nicholas Kristof said, after he, on a tourist visa, discreetly toured five Rohingya villages in February 2018, that although the phase of counter terrorism is over, Myanmar is slowly killing the remaining Rohingya in the country by making them live in "concentration camps" where health and education facility are absent. Nicholas Kristof, "I Saw a Genocide in Slow Motion."
27 On September 1, 2018, speaking in the capital Naypyitaw, he said the army was pursuing its patriotic duty to preserve Myanmar's borders and prevent Rohingya insurgents carving out their own territory in northern Rakhine State. He referred to communal violence in the area in 1942, when ethnic Rohingya who sided with the retreating British forces clashed with local ethnic-Rakhine Buddhists, who aligned themselves with the Japanese. Tens of thousands of people died in a failed attempt to

create a Rohingya state. James Hookway, "Myanmar Says Clearing of Rohingya Is Unfinished Business From WWII."

28 "Rohingya Militants Deny Links with Global Terror," *The Star*, September 15, 2017. Accessed February 12, 2018. www.thestar.com.my/news/regional/2017/09/15/rohingya-militants-deny-links-with-global-terror/

29 "Al Qaeda warns Myanmar of 'punishment' over Rohingya," *Reuters*, September 13, 2017. Accessed March 10, 2018. www.reuters.com/article/us-myanmar-rohingya-alqaeda/al-qaeda-warns-myanmar-of-punishment-over-rohingya-idUSKCN1BO0NI

30 International Crisis Group, *Myanmar: A New Muslim Insurgency in Rakhine State.*

31 "ARSA Well Organized to Wage Insurgency, Says Self-Claimed Cadre in Bangladesh," *Radio Free Asia*, November 14, 2017. Accessed March 10, 2018. www.rfa.org/english/news/myanmar/arsa-myanmar-11142017194030.html

32 Subir Bhaumik, "Myanmar Has a New Insurgency to Worry About."

33 Bertil Lintner, "The Truth Behind Myanmar's Rohingya Insurgency."

34 "Graves of 28 Hindus found in Rakhine: Myanmar Army," *The Hindu*, September 25, 2018. Accessed March 2, 2018. www.thehindu.com/news/international/graves-of-28-hindus-found-in-rakhine-myanmar-army/article19750909.ece

35 Hasnat, "Who Really Attacked the Rohingya Hindus in Myanmar's Rakhine State?."

36 "Aung San Suu Kyi Denies Ethnic Cleansing of Rohingya Muslims in Myanmar," *Guardian*, April 5, 2017. Accessed March 3, 2018. www.theguardian.com/world/2017/apr/05/myanmar-aung-san-suu-kyi-ethnic-cleansing

37 "Myanmar's Suu Kyi Slams 'Iceberg of Misinformation' Over Rohingya," *The National*, September 6, 2017. Accessed March 2, 2018. www.thenational.ae/world/asia/myanmar-s-suu-kyi-slams-iceberg-of-misinformation-over-rohingya-1.626063

38 Cumming-Bruce, "Rohingya Crisis in Myanmar Is 'Ethnic Cleansing.'"

39 Shoon Naing and Yimou Lee, "In a First, Myanmar's 'Ethnic Cleansing' Unites Suu Kyi's Party, Army and Public."

40 "Reuters Journalists Covering Rohingya Crisis Charged Under Official Secrets Act," *Guardian*, January 10, 2018. Accessed March 8, 2018. www.theguardian.com/world/2018/jan/10/two-reuters-journalists-rohingya-court-myanmar

41 Zaw, "NLD Lawmakers Debate Pros, Cons of Proposed Protest Law Amendments."

42 "Rohingya Insurgents Ambush Myanmar Military Truck, Five Wounded," *Reuters*, January 5, 2018. Accessed February 11, 2018. https://in.reuters.com/article/myanmar-rohingya/rohingya-insurgents-ambush-myanmar-military-truck-five-wounded-idINKBN1EU1FR

43 "Seven Wounded in Landmine Blast on Military Vehicle in Myanmar's Rakhine State," *Radio Free Asia*, January 5, 2018. Accessed February 11, 2018. www.rfa.org/english/news/myanmar/seven-wounded-in-landmine-blast-on-military-vehicle-in-myanmars-rakhine-state-01052018154708.html

References

Beech, Hannah. "Desperate Rohingya Flee Myanmar on Trail of Suffering: 'It Is All Gone'," *New York Times*, (September 2, 2017) available at www.nytimes.com/2017/09/02/world/asia/rohingya-myanmar-bangladesh-refugees-massacre.html?mcubz=3 (accessed February 12, 2018).

Bhaumik, Subir. "Myanmar Has a New Insurgency to Worry About," *South China Morning Post*, (September 1, 2017) available at www.scmp.com/week-asia/geopolitics/article/2109386/myanmar-has-new-insurgency-worry-about (accessed February 12, 2018).

Cumming-Bruce, Nick. "Rohingya Crisis in Myanmar is 'Ethnic Cleansing,' U.N. Rights Chief Says," *New York Times*, (September 11, 2017) available at www.nytimes.

com/2017/09/11/world/asia/myanmar-rohingya-ethnic-cleansing.html (accessed February 12, 2018).

Davis, Anthony. "Foreign Support Gives Rohingya Militants a Lethal Edge," *Asia Times*, (August 15, 2017) available at www.atimes.com/article/foreign-support-gives-rohingya-militants-lethal-edge/ (accessed February 15, 2018).

Hasnat, Mahadi Al. "Who Really Attacked the Rohingya Hindus in Myanmar's Rakhine State?," *Scroll*, (October 2, 2017) available at https://scroll.in/article/852527/who-really-attacked-the-rohingya-hindus-in-myanmars-rakhine-state (accessed February 16, 2018).

Holmes, Oliver. "Massacre at Tula Toli: Rohingya Recall Horror of Myanmar Army Attack," *Guardian*, (September 7, 2017) available at www.theguardian.com/world/2017/sep/07/massacre-at-tula-toli-rohingya-villagers-recall-horror-of-myanmar-army-attack (accessed February 16, 2018).

Holmes, Oliver, Murphy, Katharine and Gayle, Damien. "Myanmar Says 40% of Rohingya Villages Targeted by Army Are Now Empty," *Guardian*, (September 13, 2017) available at www.theguardian.com/world/2017/sep/13/julie-bishop-says-myanmar-mines-in-rohingya-path-would-breach-international-law (accessed February 17, 2018).

Hookway, James. "Myanmar Says Clearing of Rohingya Is Unfinished Business From WWII," *Wall Street Journal*, (September 2, 2017) available at www.wsj.com/articles/myanmar-army-chief-defends-clearing-rohingya-villages-1504410530 (accessed September 12, 2017).

International Crisis Group. *Myanmar: A New Muslim Insurgency in Rakhine State*, Report No. 283, (December 15, 2016) available at www.crisisgroup.org/asia/south-east-asia/myanmar/283-myanmar-new-muslim-insurgency-rakhine-state (accessed February 18, 2018).

Ives, Mike. "Violence Mounts in Restive Myanmar State, Leaving a Dozen Dead," *New York Times*, (October 12, 2016) available at www.nytimes.com/2016/10/13/world/asia/sectarian-violence-myanmar-rakhine.html (accessed February 12, 2018).

Jones, Seth G. and Martin C. Libicki. *How Terrorist Groups End: Lessons for Countering al Qa'ida*, (2008) RAND. www.rand.org/content/dam/rand/pubs/monographs/2008/RAND_MG741-1.pdf (accessed February 12, 2018).

Klug, Foster. "AP Finds Evidence for Graves, Rohingya Massacre in Myanmar," *Associated Press*, (February 1, 2018) available at www.apnews.com/ef46719c5d1d4bf98cfe-fcc4031a5434/AP-Exclusive:-AP-confirms-5-unreported-Myanmar-mass-graves (accessed February 21, 2018).

Kristof, Nicholas. "I Saw a Genocide in Slow Motion," *New York Times*, (March 2, 2018) available at www.nytimes.com/2018/03/02/opinion/i-saw-a-genocide-in-slow-motion.html (accessed March 12, 2018).

Lintner, Bertil. "The Truth Behind Myanmar's Rohingya Insurgency," *Asia Times*, (September 20, 2017) available at www.atimes.com/article/truth-behind-myanmars-rohingya-insurgency/ (accessed February 12, 2018).

Mahmud, Tarek. "ARSA: We Will Not Carry Out Attacks Inside Bangladesh," *Dhaka Tribune*, (February 2, 2018) available at www.dhakatribune.com/bangladesh/2018/02/02/arsa-will-not-carry-attacks-inside-bangladesh/ (accessed February 21, 2018).

McPherson, Poppy. "Witness To a Massacre: The Former Myanmar Soldier Who Saw His Village Burn," *Guardian*, (February 5, 2018) available at www.theguardian.com/world/2018/feb/05/witness-massacre-tula-toli-rohingya-myanmar-soldier-village (accessed February 21, 2018).

Moe, Wai. "Dozens Believed Killed as Violence Erupts in Myanmar," *New York Times*, (October 10, 2017) available at www.nytimes.com/2016/10/11/world/asia/myanmar-attack-rakhine.html (accessed February 12, 2018).

Murdoch, Lindsay. "Rohingya Pay Steep Price for Emergence of ARSA," *The Courier*, (September 22, 2017) available at www.thecourier.com.au/story/4943514/rohingya-pay-steep-price-for-emergence-of-arsa/?cs=5 (accessed March 5, 2018).

Naing, Shoon and Yimou Lee. "In a First, Myanmar's 'Ethnic Cleansing' Unites Suu Kyi's Party, Army and Public," *Reuters*, (September 14, 2017) available at www.reuters.com/article/us-myanmar-rohingya-suukyi/in-a-first-myanmars-ethnic-cleansing-unites-suu-kyis-party-army-and-public-idUSKCN1BP205 (accessed February 23, 2018).

OHCHR, "Interviews with Rohingyas Fleeing from Myanmar Since 9 October 2016," (February 3, 2017) available at www.ohchr.org/Documents/Countries/MM/Flash Report3Feb2017.pdf (accessed March 5, 2018).

Swami, Praveen. "The Vicious Circle," *Indian Express*, (September 24, 2017) available at http://indianexpress.com/article/india/the-vicious-circle-myanmar-bangladesh-rohingya-muslims-refugees-4858209/ (accessed March 5, 2018).

Thomas, Elise. "Australia to Train Myanmar Military Despite Ethnic Cleansing Accusations," *Guardian*, (March 5, 2018) available at www.theguardian.com/australia-news/2018/mar/06/australia-to-train-myanmar-military-despite-ethnic-cleansing-accusations (accessed March 5, 2018).

Wright, Rebecca and Watson, Ivan. "Inside the Rohingya Resistance: The Rebels who Provoked Myanmar's Crackdown," *CNN*, (February 3, 2017) available at http://edition.cnn.com/2017/02/03/asia/myanmar-rohingya-resistance/index.html (accessed February 12, 2018).

Zaw, Htet Naing. "NLD Lawmakers Debate Pros, Cons of Proposed Protest Law Amendments," *Irrawaddy*, (March 9, 2018) available at www.irrawaddy.com/news/burma/sri-lanka-arrests-10-anti-muslim-violence-towns-smolder-2.html (accessed March 12, 2018).

5 Emerging violent radical Islamism in the Maldives

Kirklin J. Bateman

Introduction

The islands of the Maldives Archipelago are some of the most beautiful in the world. Not surprisingly, this tropical paradise, lush with green vegetation and crystal-clear waters of the Indian Ocean, relies heavily on the tourism industry (along with fisheries) to drive its economy.[1] Yet there is a danger lurking just below the surface of the jeweled waters, hiding amongst the shadows of the towering palm trees. Violent extremism threatens to upset the delicate balance wherein a Muslim nation allows a Western lifestyle in the approximately 100 tourist resorts scattered across the archipelago even as it seeks to remain aloof from the activities of those enclaves. This has proved increasingly problematic. In 2007, terrorists detonated an improvised explosive device at Sultan Park in the capital city, Malé. No one was killed, but 12 individuals were wounded. Over the ensuing decade, extremism has continued to grow and has become a central concern, a challenge for which the country is ill-prepared.

In a June 2017 article in the *New York Times* titled "Maldives, Tourist Haven, Casts Wary Eye on Growing Islamic Radicalism," Kai Schultz writes:

> The killing in April [2017] of Yameen Rasheed, a strong voice against growing Islamic radicalization, has amplified safety concerns – particularly for foreign tourists, a highly vulnerable group and one that the islands' economy depends on. It is no idle threat, in a country that by some accounts supplies the world's highest per-capita number of foreign fighters to extremist outfits in Syria and Iraq.[2]

This attack was the latest in an escalation of violence and threat of violence that the government of the Maldives has struggled to address. Particularly noteworthy was the 2015 assassination attempt of President Abdulla Yameen, during which his wife, Fathimath Ibrahim, was seriously injured, together with several aides and a body guard.[3] Disputed accounts as to severity of the injuries notwithstanding, the brazenness of the attack is noteworthy in the small world that is the Maldives.

Astonishingly, despite a growing number of such episodes and a visible rise of fundamentalism and accompanying violent extremism, the Maldivian government has hesitated to acknowledge the severity of the threat. Indeed, this is a common problem among many nations around the world faced with the increasing threat of violent extremism – especially that which is rooted in Salafi Jihadist ideology. While an Islamic state like the Maldives may believe it needs to tread lightly around issues it identifies with religion, avoiding the problem or minimizing it cedes the narrative to threat groups and radicalized individuals. This reactive stance means it is the extremists who control the message and public perception of what is happening. In the digital battlespace where much of this fight is occurring, the Maldivian government is constantly on the defensive.[4]

It should come as no surprise, therefore, that the country is also faced with a growing number of individuals who, having become radicalized, have left to join the Islamic State in Syria or Al Qaeda-associated groups such as Jabhat al Nusra. There are reports of Maldivians making *hijra* through Sri Lanka, India, and Pakistan, ultimately to Syria. Indeed, the Maldives has more foreign fighters per capita in Syria than any other country.[5]

This chapter addresses this emerging threat, the danger it represents to the Maldives, and how the country is currently crafting and implementing its response. Also assessed is the efficacy of counter strategies the Maldives is employing and whether the country can expect to continue to measure up to the motto of its tourism industry, "Maldives: The Sunny Side of Life."

Defining the problem

Obviously, any extremist non-state effort that uses violence to further its political project is committing terrorism. Yet the radicalized individuals and groups in the Maldives are using terrorism as a method with the desired goal of creating an Islamic state in the Maldives governed under *Sharia* (Islamic law), thus are committed to insurgency.[6] Salafi-Jihadism as an ideology for expressing societal alienation and demanding redress has grown steadily, but it was the 2007 bomb attack in Sultan Park that focused attention upon the violent dimension. Immediately thereafter, violence seemed to subside but has steadily increased since 2013.[7] Most attacks have occurred in the capital, Malé, but the tourist resort islands are increasingly at risk. Any such attack would likely cripple the tourist industry that drives the Maldivian economy.

The purpose of violent activity then is to facilitate organizational growth of a challenge to the existing democratic polity and its market economy. There is also increasing evidence of radicalization occurring on remote islands far from the capital, away from government influence. The archipelago is so geographically dispersed that it is easy for radicalized individuals to exploit opportunities for proselytization. This occurs both in the virtual space through social media and in the physical space of mosques.[8]

Ultimately, this is a political problem that requires a political solution. While the Maldives has taken steps to protect the nation from terrorism and violent extremism, much of its focus has been on application of the law enforcement and military instruments of national power. Of the eight elements in the June

2016 presidential document, "State Policy: Terrorism and Violent Extremism," only one seeks to address social issues that are the key drivers of alienation among the population.[9] Thus, the government is struggling to address the root causes of the problem, focusing instead on symptoms. Without a more comprehensive effort to address the drivers of alienation, to include shaping the narrative among the people, the Maldives will remain challenged in its efforts to counter growing radicalization and extremism.

Roots of conflict

The Maldives has fallen under Dutch, Portuguese, and British sway since the sixteenth century, but mostly was left to its own rule by a series of sultans and sultanas. It gained independence from the United Kingdom in 1965 and became a republic in 1968. It is an Islamic country, and its constitution does not recognize any other religion; all Maldivians are expected to be Sunni Muslims. The Maldives' proximity to the Indian sub-continent reflects its South Asian ethnicity – most directly related to the Sinhalese in Sri Lanka. Arab traders in the late twelfth century brought Islam to a Maldives that was Buddhist, which results in the syncretic nature of Maldivian Islam.[10] This reality is central to the current struggle between traditional national practice of the faith and tenets set forth by those who have embraced Salafi Jihadist ideology.

In the 1970s, the Maldives began developing its tourist industry as the government encouraged the development of resorts on uninhibited islands. To this day, the resorts are kept separate from Maldivian-inhabited islands and are staffed overwhelmingly with third-country nationals. Only recently have Maldivians been allowed to join their ranks, but most resorts prohibit Maldivian employees from visiting while off-duty. This separation of economy and society enabled tourism, even as it supported the continued practice of conservative Islam, but, as one source notes: "This system worked well as long as the longest serving dictator of the Maldives, Maumoon Abdul Gayoom, was in power."[11]

The 1980s were a decade of rapid growth and change as Gayoom consolidated his power and the new tourism industry thrived. Nevertheless, the period was not without conflict. In 1988, the Maldives foiled a coup attempt involving Sri Lankan Tamils from the insurgent group Liberation Tigers of Tamil Elam (LTTE). There are varying reports as to why the Tamil Tigers would support such a coup, but most likely they were acting as mercenaries and hoped to reap financial reward for their participation. One report indicated LTTE received as much as US$2 million for its work. India, already deploying peacekeepers in the Tamil-controlled areas of Sri Lanka, sent naval assets and paratroopers to the Maldives to end the effort.[12]

The exceptional nature of a coup attempt only serves to highlight the political turmoil that has characterized the period since independence. The country has had six different constitutions. The most recent, in 2008, enacted a multiparty democracy that continues to struggle for legitimacy. In fact, "the first democratic government headed by former president Mohamed Nasheed could not complete its full term and ended when Nasheed resigned under pressure on February 7, 2012. Since then, the Maldives has been continuously in political turmoil."[13] It is this troubled past that indelibly colors the present situation: a democratic Islamic country, struggling with fulfilling the promises of democracy while also promoting Islam as the official state religion, all funded by a bifurcated economy, most of which is off-limits to its citizens. Salafi-Jihadism has acted as an intervening variable, which, in challenging the manner Islam is practiced, has also called into question other facets of existing economic, social, and political practice.

Grievances concerning the Maldives' economy are behind much of the alienation among the populace. Tourism and fisheries comprise 40 percent of the Maldives' annual economic output[14] but do not provide commensurate employment. Though the services industry as a whole accounted for earnings of nearly 4,800 MVR million in 2016,[15] outsiders appear to have been the principal beneficiaries.

Historic terms of reference have resulted in most revenues associated with tourism flowing offshore. Those that enter state coffers, which is also the case with fisheries, benefit national finances but have, until relatively recently, done little for the average Maldivian. Recent steps have dramatically increased the number of Maldivians directly employed, mainly in tourism. The state has also allowed for an expansion of lodging to populated areas, with Maldivians themselves as business owners. Nevertheless, these have not yet impacted employment in a substantial manner. As noted earlier, where employment is allowed, the actual strictures are such as to severely limit social adaptation and growth.

A consequence is that even as un- and under-employment remain chronic, third-country nationals, who already make up the preponderance of the work force, continue to flow into the country[16] (most recently for work in the construction industry, which in turn is financed through tourism receipts). Immigrants, both legal and illegal, account for nearly a third of the national population.[17] Many of these are construction workers, mostly South Asians, especially from Nepal, Sri Lanka, and Bangladesh, who do not receive promised wages from the companies that hire them and then find themselves stranded in the Maldives, unable to return home. Numerous workers are caught up in human trafficking networks, further trapping them in a cycle of abuse that brings their problems to the Maldivian doorstep.[18]

A critical shortage of affordable housing coupled with the staggering population density in the capital, Malé, has compounded the situation, which the government is unable or unwilling to address. Indeed, Malé itself is only two square kilometers with a population of approximately 130,000. The resulting population density is oppressive.[19]

As a consequence, faced with an 8.35 percent unemployment rate among youth ages 18–25 (2016 as reported in May 2017), the Maldives has a growing population of idle young people with little incentive or opportunity to engage in education or employment.[20] Adrift in a South Asian culture that encourages them to remain at home, these young people are easy targets for the radicalization efforts of extremist imams in mosques and equally extremist messages disseminated through social media. The government has documented a number of Maldivians who have gone to Pakistan and Saudi Arabia for madrassa education who, upon return, spread dogmatic and extremist ideology. Coupled with the sprouting of numerous foreign-funded Salafi Jihadist mosques across the archipelago, these factors have created a situation where the Maldives' population, especially young people, is under increasing threat of radicalization and operationalization for violent extremism.[21]

To speak out in such an environment has become dangerous, as the murder of human rights activist and blogger Yameen Rasheed illustrates. As previously detailed, Rasheed was a persistent and outspoken critic of the growing violent extremism in the Maldives and the simultaneous government repression of dissent. He was particularly vocal in trying to "find out the truth of the disappearance of his journalist friend Ahmed Rilwan, who worked for the *Maldives Independent*."[22] The Prosecutor General's Office has charged seven suspects in the murder (with six on trial in September 2017), with the police acknowledging the violent radical Islamist ideology the alleged murderers believed. They considered Yameen to be sacrilegious in his views and blog posts.[23]

Rasheed's family reportedly has secured representation from Perseus Strategies, LLC, a private consulting firm in the United States that provides consultation on legal services, international human rights, and government affairs. Jared Genser, managing director of Perseus Strategies, wrote a May 11, 2017 letter to the UN High Commissioner for Human Rights requesting an independent and impartial investigation into Rasheed's murder.[24] On September 17, 2017, the Maldives Civil Court denied a lawsuit Rasheed's family had filed in May that accused "the Maldives Police service of negligence in protecting Yameen prior to his brutal killing … stating that Police had not responded appropriately to reports of the numerous death threats Yameen had been receiving for years."[25]

In these events there is considerable irony – and more than enough cause for concern. On the one hand, the threat from those radicalized by the intrusion of Salafi-Jihadism continues to grow. On the other hand, there is an increasingly hostile media environment that the state has created. Freedom House, a nonprofit organization that defends "human rights and promotes democratic change, with a focus on political right and civil liberties,"[26] assessed the situation of the press in the Maldives as "not free" in its 2017 report. This is a decline from 2016 where Freedom House assessed the Maldives as "partly free." It based its assessment on several factors. "The government further tightened its control of the media, including the passage of new legislation that criminalizes defamation." This new legislation, "combined with ongoing police harassment and arbitrary arrests … contributed to increased self-censorship among journalists."[27] This marks a continuation of an already hostile environment to the press. In April 2016, the Maldives Police arrested a group of journalists from six different media outlets during a press freedom rally. Using pepper spray, police "roughed up reporters who were staging a sit-in protest outside President Abdulla Yameen's office in the capital Malé, according to several outlets including the *Maldives Independent* website."[28]

Driving the government's aggressive response to issues of transparency in the judiciary and freedom of the press is that a primary topic of criticism are concerns over alleged government corruption. Indeed, corruption emerges as a central issue, as the recent Transparency International ranking of the Maldives as 95 out of 176 countries in its 2016 index illustrates. The Maldives score of 36 was 7 points higher than the global average of 43 and is indicative of the endemic corruption in the country.[29] This was revealed in stunning detail in *Al Jazeera's* controversial 2016 investigative documentary, *Stealing Paradise*, which detailed the top government officials and others involved in selling licenses and real estate for personal gain. *Al Jazeera* reported that a corrupt political class under President Yameen "sold off paradise islands for its own personal enrichment."[30]

Though the Maldives Government denied the report and prompted a fierce backlash, there is increasing evidence that political corruption is a much deeper problem than many analysts previously thought. There was much optimism for democracy and civil liberties in the Maldives in 2008 when, in a free and fair election, Mohamed Nasheed became the president and ended the 30-year rule of Maumoon Abdul Gayoom; but this proved short-lived. The opposition conducted a bloodless coup in February 2012. Under the guise of reform, Yameen – Gayoom's half-brother – and his administration have focused on the interests of political players and ultimately created a flawed democracy that is so entangled in corruption that focus upon the pressing problems of livelihood and emergence of radical alternatives has proved difficult.[31] While the current government is struggling with these challenges, there is a deeper legacy of corruption and patronage dating to Gayoom's regime that Nasheed's shortened presidency was unable to address.

These developments have had a particularly negative impact upon the half of the population that is female. Johanna Higgs, an anthropologist and founder of Project Monma, has grappled with this process and detailed the decline of democracy in the Maldives, especially as it pertains to women. She recounts her visit there in July 2016:

> I first went to meet with a small NGO called Hope for Women. We met in their small office in the centre of Malé. "The move to democracy was difficult for the people of the Maldives," they told me. "People didn't know what to do with freedom of speech." It has also had a negative impact on women. Before there were limits on who could preach Islam, now anyone can.[32]

The irony of a democratic system resulting in alienation of women is not lost on Higgs. "A system that is supposed to promote freedom has instead allowed individuals with an agenda against women to deny women freedom."[33] While there are many in the Maldives who dispute Higgs' assessment of women's civil liberties, the perception of grievances has nearly the same political fallout as if the grievances were real.

There is other evidence of enduring gender inequality.

> Women face little discrimination in basic aspects of life such as primary
> education, health, and survival – unlike in much of South Asia. This eco-
> nomic and social progress has yet to be fully inclusive, however, and gender
> inequality endures, despite constitutional guarantees to the contrary.[34]

Challenges in gender issues are not confined solely to women. Young men are
experiencing a high rate of unemployment and are increasingly alienated from
society and family structures. A 2016 World Bank report titled "Understanding
Gender in Maldives: Towards Inclusive Development" found that:

> this alienation, combined with a lack of strong alternative social structures
> to replace the traditional family structures whose breakdown has
> accompanied Maldives' development trajectory, appear to be propelling
> young men towards greater social conservatism, participation in gangs, drug
> use, and violence.[35]

It is thus the contextual factors discussed above that are continually impacting
individual Maldivians to produce grievances, both real and perceived, which the
state is challenged to address. In this contested space, people who are increasingly facing limited or closed access to opportunity to seek outlets. Ideally, the
national fabric of the Maldives would provide these from within. Corruption and
its accompanying authoritarianism close off this pathway. As a consequence, the
alienated and marginalized look elsewhere. A powerful social movement advocating change is much in evidence, but its repression clears the playing field of
secular alternatives and leaves only religious extremism as a viable outlet. A
growing element within this extremism embraces violence.

Losing the battle of the narrative

Radicalized individuals and groups in the Maldives use the same diagnostic,
prognostic, and motivational frames[36] associated with both the Islamic State and
Al Qaeda. The diagnostic frame seeks to define the problem: Western governments and apostate regimes have defiled Islam and prevented Muslims from
practicing their faith in accordance with the way the Prophet Muhammad
intended. In the Maldives, this is manifested in beliefs that the people are engaging in un-Islamic activities, especially those associated with the tourism industry. Consumption of drugs and alcohol, together with men and women failing to
adhere to conservative styles of dress, to include the hijab for women, are illustrative elements within this frame.[37] For radicalized individuals and groups, these
and other issues associated with democratic governance have flourished in the
absence of Sharia, evidence of a world increasingly at odds with its Muslim
faith. It is this intersection of Salafi-Jihadist ideology and the Maldives' desire
for a moderate, inclusive society that is one of the key drivers of the problem.

Consequently, the prognostic frame that radicalized individuals and groups employ advances the incompatibility of democracy and Islam and hence requires destruction of the former and elevation of the latter. Anything associated with democracy – freedom of speech, press, or assembly – is unacceptable and requires elimination. Islam, already the constitutionally mandated religion for all Maldivian citizens, must be elevated through requiring Sharia, especially restrictions on women, such as mandatory wearing of the hijab.[38] There is also growing intolerance for what is regarded as "un-Islamic" behavior, particularly as it pertains to women but extending to non-conservative dress and consumption of alcohol.

This leads increasingly to violent methods to bring about desired change, and thus creation of a new societal structure. Though the improvised explosive device (IED) or knife attacks that have already occurred are the most serious illustrations, a further manifestation of this violence is the veering into supposedly authentic forms of personal regulation and punishment, such as "Taliban-style public flogging as a punishment for adultery."[39] In January 2017, Islamic scholars in the Maldives, at an annual gathering entitled "Do Not Overstep the Limits Set by Allah," called for written rules at the various island resorts to educate tourists on how not to violate Islamic culture. "Tourism companies must teach visitors how to conduct themselves while holidaying in the Maldives 'in order to address actions by tourists that violate Islamic culture and social norms'."[40]

As traditional Maldivian lifestyles have hitherto failed to embrace rigid adherence to societal codes based solely upon ostensible religious rules, the motivational frame that radicalized individuals and groups use is also reminiscent of the universe of violent radical Islamism. Motivational frames typically involve images and are designed to incite individuals to participate. Images are powerful tools for motivating individuals and spurring them to engage in illegal and violent activities beyond their normal behavior. The black flag of the Islamic State, for instance, has emerged more frequently in the Maldives as a symbol of increasing support for radical Salafi Jihadist ideology. This is coupled with an emphasis on men growing beards and women wearing the hijab. This emergence and increasing use of motivational frame has only occurred over the last few years.

> One of the earliest incidents demonstrating organised support for ISIS in Maldives was a protest conducted by about 200 people on 5 September 2014. Some of the protesters were carrying ISIS flags, calling for the full implementation of the Sharia and to put an end to secular rule in the Maldives.[41]

Despite this framing of the problem from radicalized individuals and groups, a 2015 report from the International Institute for Counter-Terrorism assessed a direct attack against one of the resort islands as still unlikely. Quoting former President Nasheed, the report stated:

They [extremists] don't want to hit the tourism industry because they are getting such good "milk" out of it. They are able to launder their money through it. They are able to recruit people. The government wants the money out of tourism. Everybody wants the money out of that. How the tourists behave on their uninhabited islands is nothing to do with us apparently. They are not worried about the hypocrisy of it. Not all worried – they think it's very clever, and it is. They have two tracks going. You have your money on one track and then you have religion on another track. They think that they have found an excellent model.[42]

Though one would like to see the information upon which this assessment is based, it does highlight the security challenge violent radical Islamism poses, a challenge which is both tangible and intangible. In dealing with the former, the unusual approach to tourism by the Maldives, where single islands historically have contained but a single resort, "has also meant that entire islands without robust security teams are vulnerable to being seized." Indeed, "security experts say many resorts are ill equipped to fend off an attack on par with those that have occurred in places like Tunisia and Bali, Indonesia." This is the official government position as well. "Abeer Ismail, the information officer at the Ministry of Tourism, said that as far as he knew, no safety concerns had been raised officially by any resorts."[43]

Simultaneously, there is the growing intangible challenge of message as embodied in the Salafi-Jihadist framing and narrative construction. This speaks to alienation consequent to psychic and physical marginalization and thus the government cannot simply use counter-messaging to counter. Absent structural reform, such as addressing corruption, rival state narratives will remain ineffective.

Threat strategy

Thus, the Maldives must make strategic choices, balancing tangible and intangible approaches. The country struggles with this challenge, because it does not fully understand the political nature of it. While the problem of radicalized individuals and groups is manifesting itself in the Maldives as terrorist violence, it is not simply violence for the sake of violence. Rather, it is violence in service of building a new world to supplant the existing world. Ultimately, these actors are seeking to establish an Islamic Caliphate in the Maldives governed by Sharia. Thus, the threat is an insurgency that is advancing along predictable, albeit still developing, lines of efforts with attendant campaign architecture.[44]

Preliminary evidence identifies the present violent radical Islamist challenge as yet comprised of individuals and small groups that are operating on a political line of effort to mobilize manpower and resources. This recruitment is both for organizational construction in the Maldives and participating in the Islamic State state-building project in Iraq and Syria. As noted previously, the Maldives has the highest number of individuals per capita going to fight for the Islamic State, with estimates reaching 200 from the nearly 400,000 Maldivians in the national population. Those who have made this move are generally viewed at home as heroes. While the government denies it, "nearly every Maldivian has a brother, a cousin or a friend in Syria. While the world watched the Olympics in August 2016, in the Maldives they watched the battle for Aleppo. Cheering on al-Qaeda."[45]

The violent radical Islamists are tapping into deep-seated contradictions with roots in the economic, social, and political deprivation in the country. Indeed, there are two Maldives: the one seen on tropical destination websites and frequented by tourists, and the other, the island capital, Malé. A recent report from the Norwegian Refugee Council provides a stark description of what life is like for most Maldivians in Malé:

> On the island, there are just a few shops. A school. A soccer field. Sometimes there is no electricity. But for whatever you need, you come to Malé. It looks like countless other cities; but it barely stretches two square miles to house its 130,000 registered residents. Its actual population is probably more than double that. In Malé, every inch is inhabited. [In one small house like so many others] there are 16 people here altogether. They wear rags and worn out shoes. The walls are patched up with jute and metal sheets. The kitchen is a camp stove. There are no tables, chairs, windows. Body odour pervades throughout. But then a strange sight: on the wall hangs a plasma TV, received in exchange for votes during the last election. An average salary here is 8,000 rufiyah, approximately US$510. For a shack like this, rent costs around 20,000 rufiyah per month, and electricity can run up to 7,000 rufiyah.[46]

Not surprisingly, gang activity and drug use are rampant in Malé. "About 30 gangs rule Malé, each of them with up to 500 affiliates. If we take the highest estimate, that's one-tenth of the residents and one-fifth of youth."[47] Many of these youth, both out of school and unemployed, feel caught between two worlds – one foot in each – belonging to neither one. This divergence between how tourism portrays the Maldives and what it is in reality provides ample stimulus for mobilization for a quest to seek an authentic identity suffused with meaning. Salafi-Jihadism provides the road-map. The struggle in the Middle East and that in the Maldives are thus linked.

Saudi Arabia, in particular, is funneling money into the archipelago not only for infrastructure development, but also for mosques and madrassas.

> Saudi Arabia has for decades spread its conservative strand of Islam in the Maldives by sending religious leaders, building mosques and giving scholarships to students to attend its universities. The Saudis are building a new airport terminal, and have pledged tens of millions of dollars in loans and grants for infrastructure and housing on an artificial island near the capital, Malé.[48]

This serves to both strengthen alternative state-building efforts and to exacerbate the situation.

The role of violence is integral to the violent radical Islamist project. It is an individual line of effort that serves to shape the human terrain, to eliminate obstacles to creation of the growing Salafi-Jihadist world, a world that exists both tangibly, in terms of followers, and intangibly, in terms of fundamentalist ideas and strictures. The specifics of the primary battle space, Malé, necessitate that terrorism is the actual method of violence. Attacks to date have used mainly knives augmented by threats of other violence, such as attacks against tourist resorts. The April 2017 attack against the activist and blogger Yameen Rasheed is the latest and most high-profile incident. The victim had received numerous death threats over the years from violent radical Islamists seeking to silence him and instill fear in other activists.[49]

Maldivians traveling to Syria to fight for the Islamic State are often utilized as suicide bombers, fighters, and support personnel. While there are no documented instances of Maldivian foreign fighters returning from Syria, there is growing concern that such individuals will inevitably appear and target the numerous tourist resort islands.[50]

Physical "return" is not necessary for the non-violent line of effort to operate. Violent radical Islamists, as discussed earlier, have worked extensively to shape the narrative in both real and virtual spaces. As highlighted in the discussion of women, with the 2008 Constitution and the Maldives' transition to democracy, "space opened up for greater religious expression, and conservative ideologies like Salafism cropped up."[51] With this new freedom, public preaching of Islam outside of mosques became more prevalent as did the ease with which radicalized individuals and groups could spread extremist ideology while a hamstrung government could only look on. By presenting efforts at legal imposition of Sharia requirements as but normal democratic process (even while using violence against opponents), radicals have been able to turn the law itself into a weapon to facilitate their effort.

An allies line of effort which has sought to woo the unwary and weak-minded even while facilitating the avarice of the corrupt all through the cultivation of allies within the polity has assisted the radicals in this effort. Tactical compromises are deemed acceptable as long as they advance the cause strategically, a methodology championed by not just Lenin but bin Laden.

Finally, radicalized individuals and groups have undertaken efforts to internationalize the struggle. This has proved a particularly potent line of effort, because it exploits the intense division that exists within the domestic political scene as exacerbated by abuse of international processes and organizations, such as those concerned with law enforcement. Criminalizing opposition both at home and abroad (for instance, by seeking international warrants based upon alleged transgressions in the Maldives) has served to divert attention from the more serious problems of grievances and accompanying radicalization. While former President Nasheed has been sounding the alarms concerning increasing radicalization and advocating action by the West, the administration of President Yasheem has claimed that the former is simply trying to discredit the latter's administration ahead of the 2018 presidential election.

Present Maldivian response

Globally, there is ample historic precedent for domestic infighting enmeshing political processes to such an extent that violent challenges are pushed aside as the least pressing matter in a world of complex challenges. This is perhaps the most neutral manner in which to describe the situations in the Maldives. The country has struggled to fully understand the threat and hence to conceptualize and implement a strategic counter. Currently, the response has been tactical in treating the challenge as one of "terrorism" without comprehending the political challenge a violent political Islam advancing in an integrated manner along multiple lines of effort poses. Even the simplest of schematics would make obvious the requirement for a whole of government response, with instruments of power serving as a means to facilitate the operational ways of the strategic concept.

To the contrary, the government has framed the threat as but a cat's paw concocted by former President Nasheed (residing in the United Kingdom after going there in early 2016 for medical treatment and then receiving asylum) to discredit the Yameen administration, damage the economy, and adversely affect tourism. This is a conceptualization of the challenge so skewed as to relegate to the side lines the true nature of the problem, the emergence of a challenge to the very legitimacy of the democratic order. That order seems as much threatened by counterterrorism response as protected by it. The 2015 Prevention of Terrorism Act, for example, was the first comprehensive act, since the previous 1990 effort, to revision of legal instruments ostensibly to enable address of the threat. Yet in its actual implementation, its casualties have been more civil liberties and less violent radical Islamists.

Similarly, in its most recent update to response, the Office of the President in 2016 issued the document "State Policy: Terrorism and Violent Extremism." Therein, only one of the seven "pillars of response" addresses economic, social, and political drivers of alienation: "Taking measures to understand and address social issues stemming from terrorism and violent extremism."[52] If the Maldives continues to focus solely on law enforcement and security responses, it will only address the symptoms rather than the underlying causes of the problem it faces from violent radicalized individuals and groups.

Conclusion

Much work remains, and the Maldives has requested help from the West on numerous occasions. Even so, the current U.S. administration is still trying to gain its footing in foreign policy and has yet to respond. Tactical "transactionalism" is currently driving American foreign policy and is most likely to continue throughout the current administration, even structurally, the Maldives is second priority in an embassy structure which sees U.S. representation in Colombo, Sri Lanka also charged with serving the Maldives.

Recent events in February 2018 have further complicated democratic and political reform in the Maldives. The Supreme Court vacated the 2015 Prevention of Terrorism Act conviction of former President Mohamed Nasheed at the beginning of February, setting off a fresh wave of political instability. The Supreme Court decision pointed to violations of the "Constitution of the Republic of the Maldives and human rights treaties that the Maldives is party to" as the underlying factors in politically motivated investigations and prosecutions.[53] In the aftermath of this decision, political turmoil has ensued as demonstrations, detentions and arrests, and violations of the rule of law now threaten to further destabilize the country. The Maldives is currently (March 2018) under a state of emergency as President Yameen struggles to maintain control. Some regional and international powers such as India and China are seeking to leverage the instability for hegemonic advantage. Others, such as the United States and Great Britain, are cautiously standing by, prepared to assist, while seeking to remain disentangled from another South Asian problem.

Nevertheless, there is hope. Numerous individuals across the Maldivian government and civil society are increasingly engaged in various ways to prevent the spread of extremist views and violence within Islam. In order to build upon this foundation, the Maldivian government must enhance these efforts through structural reform while simultaneously protecting and encouraging civil liberties. Only within such an approach will security efforts achieve their objective of safeguarding the population and the economic, social, and political infrastructure.

Notes

1 "The World Factbook – Central Intelligence Agency."
2 Schultz, "Maldives, Tourist Haven, Casts Wary Eye on Growing Islamic Radicalism."
3 "Maldives Vice-President Held over Bomb."
4 Ningthoujam, "Maldives Is No Longer a 'Paradise'."
5 Bearup, "ISIS Threat to Maldives Paradise."
6 Wieviorka "Terrorism in the Context of Academic Research."
7 Dharmawardhane, "Maldives," 130.
8 Ningthoujam, "Maldives Is No Longer a 'Paradise'."
9 The President's Office, "State Policy: Terrorism and Violent Extremism."
10 Metz, *Maldives: A Country Study.*
11 Kumar, *Multi-Party Democracy in the Maldives and the Emerging Security Environment in the Indian Ocean Region.*
12 Auerbach, "Coup D'etat Attempted in Maldvies."
13 Kumar, *Multi-Party Democracy in the Maldives and the Emerging Security Environment in the Indian Ocean Region,* 2.
14 "The Economy of the Maldives."
15 "Maldives GDP | 1980–2017 | Data | Chart | Calendar | Forecast | News."
16 World Population Review estimates there are 70,000 foreign nationals and 33,000 illegal immigrants in the Maldives. "Maldives Population 2017." World Population Review.
17 "Maldives Population 2017 (Demographics, Maps, Graphs)."
18 "Maldives: Office to Monitor and Combat Trafficking in Persons."

19 "Maldives Population 2017 (Demographics, Maps, Graphs)."
20 World Bank, "Youth Unemployment Rate for Maldives."
21 Dharmawardhane, "Maldives," 64.
22 Transparency International Secretariat, "Transparency International Calls for a Full Investigation into Murder of Activist Yameen Rasheed."
23 "Yameen Rasheed Murder Trial Begins with Secret Hearing | Maldives Independent."
24 Genset and Santiago, "Request for Independent and Impartial Investigation of Murder of Yameen Rasheed, Citizen of the Maldives."
25 Shaahunaaz, "Court Quashes Police Negligence Case in Yameen Rasheed Murder."
26 "Freedom House."
27 "Freedom of the Press 2017: Maldives."
28 "Maldives Police Arrest Reporters at Press Freedom Rally."
29 "Corruption Perceptions Index 2016."
30 *Al Jazeera Investigates: Stealing Paradise*, Investigative Broadcast, 2016, www.aljazeera.com/investigations/stealing-paradise/. A useful "report on the report" is *Asia Wired*, "After Stealing Paradise," available at: www.youtube.com/watch?v=yjANdnTAb8g.
31 Jordan, "Maldives: 'The System Has Failed'."
32 Higgs, "Unveiling Violence in the Maldives: How the Rise of Islam and the Decline of Democracy Are Affecting Women."
33 Ibid.
34 Pande and El-Horr, "Understanding Gender in Maldives: Towards Inclusive Development."
35 Ibid.
36 Della Porta and Diani, *Social Movements: An Introduction.*
37 Dharmawardhane, "Maldives," 65.
38 The Maldives has a tropical climate with annual temperatures moderate but generally above 80°F. See "Weather and Temperature Averages for Maldives, Maldives."
39 Dharmawardhane, "Maldives," 64.
40 Saeed, "Maldives Clerics Roll Out Wide-Ranging Religious Agenda."
41 Dharmawardhane, "Maldvies," 64.
42 Ningthoujam, "Maldives Is No Longer a 'Paradise'."
43 Schultz, "Maldives, Tourist Haven, Casts Wary Eye on Growing Islamic Radicalism."
44 Marks, *Maoist People's War in Post-Vietnam Asia.*
45 Borri, "Where Jihadists Are Heroes."
46 Ibid.
47 Ibid.
48 Moosa and Anand, "Inhabitants of Maldives Atoll Fear a Flood of Saudi Money."
49 Moosa and Schultz, "Outspoken Maldives Blogger Who Challenged Radical Islamists Is Killed."
50 Bearup, "ISIS Threat to Maldives Paradise."
51 Schultz, "Maldives, Tourist Haven, Casts Wary Eye on Growing Islamic Radicalism."
52 The President's Office, "State Policy: Terrorism and Violent Extremism."
53 "Supreme Court Orders Nasheed's Release in Landmark Ruling | Maldives Independent."

Bibliography

Al Jazeera Investigates: Stealing Paradise. Investigative Broadcast, (2016) available at www.aljazeera.com/investigations/stealing-paradise/.

Auerbach, Stuart. "Coup D'etat Attempted in Maldvies." *Washington Post*, (November 4, 1988) available at www.washingtonpost.com/archive/politics/1988/11/04/coup-detat-attempted-in-maldives/ec8ab147-2317-4213-b942-40585e95fc0e/.

Bearup, Greg. "ISIS Threat to Maldives Paradise." *The Australian*, (August 29, 2016) available at http://at.theaustralian.com.au/link/7a80bff385fb3e27ade26a72ce43b3e6?d omain=theaustralian.com.au.

Borri, Francesca. "Where Jihadists Are Heroes." Norwegian Refugee Council, (March 14, 2017) available at www.nrc.no/perspectives/2017/where-jihadists-are-heroes/.

"Corruption Perceptions Index 2016." Transparency International, (January 25, 2017) available at www.transparency.org/news/feature/corruption_perceptions_index_2016.

Della Porta, Donatella, and Mario Diani. *Social Movements: An Introduction.* 2nd ed. Malden, MA: Blackwell Publishing, 2006.

Dharmawardhane, Iromi. "Maldives." *Counter Terrorist Trends and Analysis*, South Asia Annual Threat Assessment, 7, (11) (December 2015), 130.

"Freedom House," available at https://freedomhouse.org/our-work (accessed October 3, 2017).

"Freedom of the Press 2017: Maldives." Freedom House, (April 27, 2017) available at https://freedomhouse.org/report/freedom-press/2017/maldives.

Genset, Jared and Nicole Santiago. "Request for Independent and Impartial Investigation of Murder of Yameen Rasheed, Citizen of the Maldives," (May 11, 2017) available at http://perseus-strategies.com/wp-content/uploads/2017/05/Complaint-and-Letter-to-UN-and-SRs-on-the-Case-of-Yameen-Rasheed-5.11.17.pdf.

Higgs, Johanna. "Unveiling Violence in the Maldives: How the Rise of Islam and the Decline of Democracy Are Affecting Women | HuffPost UK," available at www.huffingtonpost.co.uk/johanna-higgs/maldives-democracy_b_10913556.html (accessed August 15, 2017).

Jordan, Will. "Maldives: 'The System Has Failed'." *Al Jazeera*, (September 7, 2016) available at www.aljazeera.com/news/2016/09/maldives-system-failed-160904133451601.html.

Kumar, Anand. *Multi-Party Democracy in the Maldives and the Emerging Security Environment in the Indian Ocean Region.* New Delhi: Pentagon Press Institute for Defence Studies and Analyses, 2016. www.idsa.in/system/files/book/book_democracy-maldives-ior.pdf.

"Maldives GDP 1980–2017 Data Chart Calendar Forecast News." https://tradingeconomics.com/maldives/gdp (accessed September 24, 2017).

"Maldives: Office to Monitor and Combat Trafficking in Persons." U.S. Department of State. www.state.gov/j/tip/rls/tiprpt/countries/2016/258815.htm (accessed March 14, 2018).

"Maldives Police Arrest Reporters at Press Freedom Rally." *Al Jazeera*, (April 4, 2016) available at www.aljazeera.com/news/2016/04/maldives-police-arrest-reporters-press-freedom-rally-160403133751586.html.

"Maldives Population 2017 (Demographics, Maps, Graphs)." http://worldpopulation review.com/countries/maldives-population/ (accessed October 2, 2017).

"Maldives Vice-President Held over Bomb." *BBC News*, (October 24, 2015) sec. Asia, available at www.bbc.com/news/world-asia-34625558.

Marks, Thomas A. *Maoist People's War in Post-Vietnam Asia.* Bangkok, Thailand: White Lotus Press, 2007.

Metz, Helen Chapin, ed. *Maldives: A Country Study.* Washington, D.C.: GPO for the Library of Congress, (1994) available at http://countrystudies.us/maldives/.

Moosa, Hassan, and Geeta Anand. "Inhabitants of Maldives Atoll Fear a Flood of Saudi Money," *New York Times*, (March 26, 2017), sec. Asia Pacific, available at www.nytimes.com/2017/03/26/world/asia/maldives-atoll-saudi-money.html.

Moosa, Hassan and Kai Schultz. "Outspoken Maldives Blogger Who Challenged Radical Islamists Is Killed," *New York Times*, (April 23, 2017) available at www.nytimes.com/2017/04/23/world/asia/yameen-rasheed-dead-maldives-blogger-dead.html?_r=0.

Ningthoujam, Alvite Singh. "Maldives Is No Longer a 'Paradise'." International Institute for Counter-Terrorism, (February 4, 2015) available at www.ict.org.il/Article/1372/Maldives-is-No-Longer-a-Paradise.

Pande, Rohini Prabha and Janna El-Horr. "Understanding Gender in Maldives: Towards Inclusive Development." The World Bank, (May 8, 2016) available at http://documents.worldbank.org/curated/en/448231467991952542/Understanding-gender-in-Maldives-towards-inclusive-development.

Saeed, Xiena. "Maldives Clerics Roll Out Wide-Ranging Religious Agenda." *Maldives Independent*, (January 25, 2017) available at http://maldivesindependent.com/politics/maldives-clerics-roll-out-wide-ranging-religious-agenda-128530.

Schultz, Kai. "Maldives, Tourist Haven, Casts Wary Eye on Growing Islamic Radicalism," *New York Times*, (June 18, 2017) sec. Asia Pacific, available at www.nytimes.com/2017/06/18/world/asia/maldives-islamic-radicalism.html.

Shaahunaaz, Fathimath. "Court Quashes Police Negligence Case in Yameen Rasheed Murder." *Mihaaru.Com*, September 18, 2017. http://en.mihaaru.com/court-quashes-police-negligence-case-in-yameen-rasheed-murder/.

"Supreme Court Orders Nasheed's Release in Landmark Ruling | Maldives Independent." http://maldivesindependent.com/politics/supreme-court-orders-nasheeds-release-landmark-ruling-135502 (accessed March 14, 2018).

"The Economy of the Maldives," available at www.maldivesholidays.org/economy-maldives (accessed September 24, 2017).

The President's Office. "State Policy: Terrorism and Violent Extremism." State Policy. Male, Republic of Maldives: The President's Office, (June 8, 2016) available at www.presidencymaldives.gov.mv/Documents/4560_ee6e0576-8_.pdf.

"The World Factbook – Central Intelligence Agency," available at www.cia.gov/library/publications/the-world-factbook/geos/mv.html (accessed September 4, 2017).

Transparency International Secretariat. "Transparency International Calls for a Full Investigation into Murder of Activist Yameen Rasheed." Transparency International, April 23, 2017. www.transparency.org/news/pressrelease/transparency_international_calls_for_a_thorough_investigation_into_murder_o.

"Weather and Temperature Averages for Maldives, Maldives," available at www.holiday-weather.com/maldives/averages/ (accessed October 16, 2017).

Wieviorka, Michael. "Terrorism in the Context of Academic Research." In Martha Crenshaw (ed.). *Terrorism in Context*, 597–606. University Park: The Pennsylvania State University Press, 1995.

World Bank. "Youth Unemployment Rate for Maldives." FRED, Federal Reserve Bank of St. Louis, (January 1, 1991) available at https://fred.stlouisfed.org/series/SLUEM1524ZSMDV.

"Yameen Rasheed Murder Trial Begins with Secret Hearing | Maldives Independent." http://maldivesindependent.com/crime-2/yameen-rasheed-murder-trial-begins-with-secret-hearing-132477 (accessed March 14, 2018).

Part II

Cautious optimism – or false dawn?

6 Back to the future

Nepali people's war as "new war"

Thomas A. Marks

Considerable irony attends the case of Nepal. It was regularly examined during the overt 1996–2006 conflict for the reason that the strategic and operational approach of the insurgents, the Communist Party of Nepal (Maoist) or CPN(M) [*sic*], was something of a throwback in time, a self-proclaimed, classic people's war insurgency that continued into the twenty-first century. Yet the manner in which *overt* struggle ended with a November 2006 peace agreement obscured the reality that there followed a second, equally interesting period (2006–present) driven by the same strategic doctrine but with emphasis upon different operational lines of effort. In this period, covert struggle continued and arguably resulted in insurgent victory. For reasons to be discussed below, the political dust has settled in a rather more complicated fashion than any concerned might have anticipated, but this does not change the salience of the case as an illustration of "new war" as presently waged by rebels who seek power.

"New war" is best conceptualized rather than defined. In the post-Cold War era, the term engaged security professionals as they sought to address conflict wherein the role of force was directed not at traditional targets, such as, most saliently, enemy forces, rather at the facilitation of societal change or even capture. The debate went in any number of directions, with state militaries embracing terms such as hybrid warfare or "the gray zone" to describe as much as analyze emergence of war fought in ways that obscured attribution and confused response. When applied to irregular conflicts, though, "new war" seemed slow to recognize that it was neither particularly new nor a closely-held approach. To the contrary, it was but the unrecognized second coming of people's war. As such, a political challenger proceeded on multiple lines of effort, only one of which was violence, with each line comprised of constituent campaigns unfolding both tangibly and intangibly.

As noted, this can hardly be considered "new." Sophisticated nonstate actors seeking to seize state power have at least since the American Revolution advanced in such manner. Best known in recent history have been the theory and practice of the Chinese and Vietnamese, but they are unique only in the extent to which their approaches were committed to writing and thus study and transmission. What the post-Cold War decades have produced in "new war" is the leap inherent to what the Chinese have more recently termed *unrestricted warfare*. Violence moves from being the action supported by shaping to a position as the shaping mechanism itself; that is, violence is not decisive but preparatory for use of other tools. Ironically, no context is better suited for such an approach than "peace." To this end, nonstate actors globally have increasingly opted to continue their struggles by other means within targeted systems themselves. Events in Nepal offer one such illustration.

Maoist methodology: the period of overt war (1996–2006)

Political violence in Nepal emerged from the imperfections of the polity. Having evolved in the post-Second World War years from a kingdom closed to the outside world and dominated by a hereditary prime minister, to a constitutional monarchy reigning in uneasy partnership with a parliament, Nepal was by the February 13, 1996 outbreak of insurgency a troubled formal democracy.[1] A rapidly growing population and geographic realities (the country is both land-locked and dominated by mountainous terrain) resulted in a rigid, stratified struc-ture of political opportunity defined by community (a mix of linguistic, ethnic, and Hindu caste divisions), class (economically, one of the poorest countries on earth and as close to a zero-sum game as can be imagined), dysfunctional pol-itics (near-fatally wounded by corruption and the universal practice of demo-cratic centralism by all major political parties), and patriarchy.[2] Into this mix was thrown the determination of a small Communist Party of Nepal (Maoist) or CPN(M) [*sic*] to make a revolution.[3] Led throughout by Pushpa Kamal Dahal *aka* Prachanda ("Renowned," though "Fierce One" is most common in Western media), its objective was to seize power in order to institute a version of Maoism that overtly held the pinnacle of political progress was the Chinese Great Prole-tarian Cultural Revolution, generally regarded as one of history's most signi-ficant crimes against humanity, not only by outsiders but by the Chinese themselves.[4] Perhaps predictably, there has been an increasing tendency in the past decade to read back into the *overt* conflict (1996–2006) an ostensible Maoist ideological moderation which simply was not there at the time. In late 2008, for example, two years after the formal end of hostilities but with terrorism continu-ing, Maoist military leader, Nanda Kishor Pun *aka* Pasang, asserted:

> For the extermination of the old order, the People's War (PW) was initiated on 13 February, 1996 under the efficient leadership of Chairman Com. Prachanda by CPN (Maoist). The people's war is advancing continuously in the wave-like forms horrifying the domestic and foreign enemies and creat-ing a hurricane to shake the world. Awaking the world in the 21st century, the PW is advancing forward ideologically and politically for a revolu-tionary social, economic and cultural transformation following the Chinese Great Proletarian Cultural Revolution (GPCR), and is oriented to develop Marxism-Leninism-Maoism.[5]

Maoist ideology provided not only the objective but the strategy, people's war: the violent creation within the state of a rival state – a counter-state – in order ultimately to overthrow and replace the old-order with the new revolutionary-order. Terrorism eliminated human and organizational obstacles, focusing upon individuals who hindered its achievement of local political domination. Guerrilla warfare destroyed the armed local presence of the state, the police; and a People's Liberation Army (PLA), armed and equipped as much like state forces as possible, engaged in battle with the Royal Nepal Army (RNA; Nepal had but a single service in its military). In this strategy, terrorism was instrumental, a tactic, and served to open up political space. Whether the obstacle to be removed was a citizen supporting a rival political party or an actual politician belonging to a rival political party, or an NGO worker or a teacher, or a citizen who would not pay his Maoist taxes – or, ultimately, a policeman or a member of the military – the glue that held the polity together was stripped away, and the pieces were reassembled by the insurgents as they achieved first local, then regional, and finally national power. In normal politics, votes decide; in insurgency, violence is the key intervening variable that delivers prospective "voters" for indoctrination.

Democratic empowerment as realized through political institutions and process (in particular, elections) was the major foe; and Maoist violence caused all attempts to hold elections during the 1996–2006 period to collapse. The resulting crisis ultimately led to the declaration of direct royal rule, which itself led to further crisis. Strategically, then, an attack on a democracy such as Nepal required that the insurgents co-opt or destroy the individuals and organizations who made democratic governance a reality, as well as eliminating their protectors, the security forces. This point was repeated by all high-ranking Maoist personalities, regardless of group, in my recent fieldwork. As put directly on August 26, 2016 by the only woman who became a Maoist brigade commander and who remains active in party activities, when asked why local democratic governance was targeted by the Maoists during the 1996–2006 period:

> Even though they were elected at local level, they were controlled and directed by the central government. It was like they had given power to vote at local level, but they were being controlled by the center. So this was more of a contradiction. It seemed like all of institutions were part of the center [and thus had to be destroyed].[6]

Targets were considered contextually by the Maoists, both as to selection and mode/level of violence: A teacher might resist having his students subjected to propaganda and mobilization, and thus would be kidnapped, beaten, and warned; but another might speak directly against the ideology of the insurgent movement and continue even after being warned, thus requiring assassination. Public display of the victim was used when messaging was required, simply dumping the body when the killing alone was sufficient to make the point. An effective development worker would be targeted, because he enhanced the legitimacy of the state by making people's lives better; a corrupt, ineffective development worker served the same purpose if left in place – he destroyed the legitimacy of the state by being what he was. A policeman, because he was charged with law and order, became a target – and in the process, his seized weapons served to enhance the insurgent arsenal. Members of rival political parties or simply those who exercised their rights as citizens through voting were, simply by their actions, key mobilizing agents the polity, hence targets. To the extent individuals had overlapping attributes – e.g., they were economic, social, *and* political personalities – they increased the necessity for their neutralization. To the extent they were all these things and *effective* in what they did, they vaulted still higher in the need to be removed from the locality targeted.[7] Indeed, the most iconic photo of the entire war is that of an effective and popular teacher, Muktinath Adhikari, headmaster and 10th Grade instructor at Padmini Sanskrit Higher Secondary School in Lamjung district, who was murdered on January 16, 2002 for refusing to provide funding for the Maoists and continuing to teach Sanskrit, which the Maoists had deemed a language of Indian imperialism.[8] So fearful was the community that the body-as-message remained hanging for more than a day as no one would venture to cut it down. As the conflict unfolded, more than 300 teachers were to suffer similar, heinous attacks.

Localities, in turn, were targeted for a variety of reasons, both instrumental and ideological. A police station that was repeatedly attacked in Rukum during 1996–2006 stood at a key trail junction and hence instrumentally needed to be eliminated to ensure insurgent mobility. Another police station, though, of no tactical value, was attacked during the same period for the ideological reason that its garrison had been labeled "abusive of the population." Eliminating it redounded to the moral favor of the Maoists, in the same manner the Maoist banning alcohol in villages was highly popular with women, who were the principal victims of widespread abuse by inebriated husbands (a major problem in Nepal). Timing and severity of actions responded to such a host of factors that quantitative attempts at prediction were unconvincing at even the macro-level. Not surprisingly qualitative assessment determined that leadership decisions responding to strategic-operational-tactical imperatives offered the best explanations for who and what was targeted. At the local level, though, decisions to attack a human target often stemmed from impulse or individual motives such as revenge or even hatred.

The crucial point is that in the Nepali Maoist people's war, the need to neutralize opposition responded to general policy guidance, with the specifics of implementation delegated to local authority. This led to a staggering level of destruction and brutality as the use of terrorism increasingly spread. For example, as related in an article discussing the victim of a typical attack upon a Nepali Congress cadre, which occurred in Rukum (see Figure 6.1 for district locations[9]):

> Gopal Prasad Sharma, a Mahasamiti [steering committee] member and district-level leader of the Nepali Congress in Rukkum [*sic*], lost his eyesight after [Maoist] cadres sprinkled acid in his eyes during the insurgency. They also tortured him by sprinkling kerosene on his body and then setting

him on fire. "After the incident, they chased me away from my village in Kholagaun-9. They have not returned my seized land yet," said Sharma, who was also a VDC chairman then. He had become a target of the Maoists for being actively involved in politics as an NC leader.[10]

Similarly, as related in an article discussing the victim of a typical attack upon a teacher in Dang district, to the far south of Rukum:

He is paralyzed on the left side of the body. He cannot walk without crutches. A scar on his head reminds him of the grim days of the Maoist insurgency. Kiran Yogi of Dang was harassed and thrashed by the Maoists for being a "bourgeois" teacher.

He was eventually abducted by a group of Maoist rebels on 21 August 2003 while taking classes at Nepal Rastriya Secondary School, Bardiya. A well-wisher of Nepali Congress, he was targeted by the Maoists for refusing to become a Maoist supporter. He was taken to a field near the school. "I was shot in the head by the Maoists," he said. He was in a coma for 23 days and did recovered [*sic*] later. But doctors could not remove the bullet lodged in his head and this has left him paralyzed.[11]

Such actions were routine. As early as mid-2003, the number of those *brutalized* by the Maoists reached approximately 15,000. My published figures, based upon fieldwork and examination of data available at the time, state that as of August 20, 2003, the Maoists had killed 870 civilians, injured 4,669, and attacked a

Figure 6.1 Nepal districts.

further 10,943. "Killed" included beheadings and torture resulting in death; "injured" included everything from breaking both legs and feet of a "big land-lord" with hammers,[12] to mutilation with *kukris*, the curved knives ubiquitous in Nepal and associated with the legendary Gurkha infantry; and "attacked" was all manner of assault.[13]

The relatively low number of those killed compared to those tortured and maimed highlights the effectiveness of terrorism in achieving the Maoist objective: to remove from local areas those who opposed the Party in any manner and could serve as rallying points. Only several dozen Village Development Committee or (VDC) (county) chairmen, for example, had been killed by September 2002, but the manner in which they had been mutilated and displayed caused a majority of the 3,913 VDC chairmen to flee their posts.[14] District offices were also attacked. Combined with this decimation of personnel was the devastation of the structure of governance. One-third of the VDC offices had been physically destroyed within the year following the November 2001 attack on the army in Ghorahi (Dang district), which the Maoists saw as initiating their push for strategic stalemate. Inside such structures and hence normally lost was the documentation necessary for the correct functioning of society and the state. Whatever its flaws, Nepal was a democracy that was not predatory. Most observers felt that the Maoist violence was unleashed at the very moment when real progress was visible, above all in local democracy at the county (i.e., VDC) level.

Definitive statistics were hard to come by, a product of both the irregular nature of the conflict and the imperfections of the organizations that sought to produce tallies. Often, researchers could not even stay alive. Groups that proved particularly effective were rendered inoperative through the assassination of their key individuals. For instance, the reality of the situation was for a time well documented by the Nepal Maoists Victims Association (MVA; also translated by the organization itself as "Association of the Suffers from the Maoist Nepal," or ASMIN). An NGO, MVA kept excellent files (which I and other scholars used in our work) and was vocal in condemning the Maoists. Its work was a thorn in the side of the Maoist effort. The Maoist response that played itself out over a period of time signally highlights the ruthless use of terrorism that lurks behind a revolutionary project's efforts to remove all individuals and institutions that stand in its way. On Sunday, February 15, 2004, the Maoists assassinated Ganesh Chilwal [also rendered as Chiluwal], the MVA president. The killing of Chilwal was carried out at 5:30 in the evening in a busy area of the capital, Kathmandu. On the Friday just two days before, Chilwal had directed the public burning of effigies of the Maoist chairman, Dahal, and his deputy, Baburam Bhattarai *aka* Laaldhwoj ("Red Flag," also transliterated as Lal Dhoj). In August, a second key member of the MVA was murdered in his Kathmandu home; and on January 4, 2005, the MVA vice-president, Jay Khadka Rawal, was gunned down in a Kathmandu market. The organization thereafter struggled to remain viable in the effort against Maoist terrorism. It resurrected itself following the November 2006 peace agreement (to be discussed below) but was again attacked (again, in Kathmandu). Its elected head, Amrit Mishra, was compelled to flee abroad. Needless to say, the organization ceased to be a source of statistics on Maoist depredations.

Maoist methodology was the same, regardless of place and the period of time under consideration; the only distinction lay in the *level* of violence involved. This methodology has been discussed conceptually; Maoist terrorism operationalized it. Each district was comprised of its VDCs and municipalities. Each of these was characterized by its unique distribution of power as made up of key local players and infrastructure. As in any political campaign, a mapping of what has come to be called "human terrain" served to guide targeting, with the objective being population and resource domination.[15] To advance their control of areas and population, as previously discussed, the Maoists sought to neutralize all opposing political, administrative, and civil society activity at the local level, through (1) physical elimination; or (2) forcing the victim to flee; or, the preferred option (3) pressuring individual opponents (real or imagined) into their ranks so as to benefit from the psychological impact of converts (actual or coerced) crossing over to their side – and then sending them forth to co-opt their own network of associates. Violent action was facilitated by the use of numerous front organizations to mobilize individuals into cause-oriented groups that engaged in actions parallel to the formal work of the Party. These were either explicitly allied with the Party or ostensibly independent. Regardless, in reality, all operated according to Party orders and thus further served to neutralize opposing political impulse. Either directly (by joining in assault) or indirectly (by identifying targets, both human and financial).[16]

Any hope of protection by the armed local representatives of the state was small, because police presence was (and remains) sparse, as well as often corrupt and ineffective. Police, in any case, regardless of individual or group efficacy, were a prime target of the insurgents, because once they were eliminated, the population could be absorbed into the counter-state and mobilized. Rukumkot in Rukum district illustrated this well. It had the largest police garrison outside the Rukum capital of Musikot. Following extensive Maoist preparation, the police station was attacked on April 1, 2001. Though reinforced to some 80 officers, it was quickly overrun with 32 dead, 18 wounded, and most of the survivors captured. All weapons and equipment were seized by the Maoist force of about 1,000 (300 combatants, 120 support personnel, and 550 civilian auxiliaries). These last two numbers in the tripartite force come from the Maoist commander of the operation and are significant, because such individuals would have come from the local area. In all likelihood, then, they would have been involved in front activity and in the intelligence work that made the attack possible – and in the local elimination of resistance and generation of taxation. At the end of overt hostilities, they remained active as Maoist enforcers; at the time of the attack, the police post was already effectively under siege and was not conducting even local patrols. The next day, the smaller police post in Mahat VDC, to the south, was abandoned since it was clear it also could not be held.[17]

This was the pattern that occurred throughout Rukum and other districts in the country. "Riding to the rescue," even when a possibility for the security forces, faced guerrilla ambush in the difficult terrain (Nepal had minimal air capability, and RNA at the height of the insurgency possessed just four troop-transport helicopters). In this manner, entire areas were stripped of government presence. Ultimately, even regular army units found themselves attacked, and isolated units were always in danger of being overwhelmed. This "strategic offensive" (i.e., military assault) phase of the war was initiated by the Maoists in November 2001 with an assault upon the Royal Nepal Army (RNA, now Nepal Army or NA) cantonment in Ghorahi, Dang district, a city of 43,000 inhabitants.[18] Government military forces were caught in a conundrum. Only dispersed deployment of small units could generate the presence required to secure and protect the population; yet such a posture increased the dangers of annihilation as the Maoists could mass similar large units of multiple battalion strength (approximately 600 individuals per battalion). Even district capitals were at risk.[19] In such circumstances, the level of brutality and bloodshed increased dramatically, because no protection could be afforded to any of the population save *partially* in district capitals and major urban centers. Throughout this period, though, they – and even the nation's capital – had been penetrated by Maoist operatives.

The extent to which this was true was on vivid display in Kathmandu when, on January 26, 2003, a Maoist urban partisan group in Kathmandu ambushed and killed the commanding officer of Nepal's paramilitary Armed Police Force (APF), Inspector General Krishna Mohan Shrestha. Shot at close range while he took his morning walk, Shrestha was accompanied by his wife and bodyguard, who also died. Information obtained later revealed that the hit-team had made *16* aborted prior assassination attempts upon Shrestha, commencing in July 2002, all in the most heavily garrisoned and guarded area in the country. At the time, APF was playing an increasing role in the effort against the insurgents. Shrestha was the highest ranking officer to die in the entire conflict, and the Maoists used the shock of his death as their stepping stone to an offer of a ceasefire and negotiations.

The negotiations, though, were but cover for regroupment, resupply, and repositioning of combatants. On August 27, 2003, the Maoists ended the ceasefire. Simultaneously with their proclamation, they attacked a number of government positions using forces they had moved duplicitously during the talks. Further, their urban partisans began a campaign of targeted killing against important counterinsurgent figures stationed in Kathmandu.[20] This effort accelerated the building of a Maoist counter-state within the very boundaries of the legal Nepali state. Ultimately, some 70–80 percent of the population, located primarily in the rural areas outside the 75 district capitals and the several major urban centers, was controlled by the Maoists. Regardless of methodology or area, the Maoist posture was built upon terrorism, which was used in a ruthless campaign of intimidation and assassination. The longer the Maoists dominated an area, though, the less need there was for the more brutal mechanisms of control. Numerous individuals fled, while others who did not desire to embrace the new order lived in fear. Killing of alleged spies was a regular occurrence throughout the country.[21]

It may be noted that to the absence of police in much of the country must be added the frequent ineffectiveness of a force simply overwhelmed by the demands placed upon it by a Maoist challenge that rapidly proved capable of taking on even the army. For nearly all victims throughout the conflict, therefore, this was reality. Lodging a complaint was not even an option for a majority, because after 2001 *most* areas had little to no police presence; and where they were present, the police, whatever their professionalism, were increasingly hard-pressed simply to survive. It hardly needs telling, incapable of protecting even key figure counterinsurgency figures, the police, whatever their motivation, were incapable of providing security to the citizenry. Flight of targeted individuals could lead to but tenuous security. Numerous cases testify to what to outsiders seemed an almost pathological insistence upon pursuing victims who sought either to continue their activism or only to flee, which proved difficult in the small country that is Nepal.[22] It is North Carolina in area – if the Himalayas are excluded – with 25–28 million people living in an aggregate of small worlds, the 3,913 VDCs, most of which are hill/mountain/river valleys tied to agriculture and animal husbandry. Even towns are of a dispersed configuration, aggregates of smaller settlements, and thus not true urban areas. In this context, there was very little that remains hidden or unknown for long.

What ensured that the conflict in Nepal would become savage stemmed from the complex human geography of the country, wherein some 126 recognized ethnic and caste groups sought to survive in a society which economically was one of the poorest on earth and socially was dominated by the realities of the Hindu caste system (i.e., stratification is the norm). Given the resulting ineffi-ciency and corruption of Nepali politics, the Maoists were able to systematically exploit the issues of community, caste, gender, and class; and thus sought to enflame every group against the others. Children were a particular target, with widespread recruitment and impressment occurring myriad times. Schools, it has been noted previously, were a major site for such effort, with entire student bodies regularly kidnapped and held for several days while subjected to propa-ganda. Outside such locations, in the villages themselves, a demand for one member of each family to enter Maoist service, often a young son or daughter, was normal in all areas where the Maoists were able to establish dominance. Age was never a disqualifying factor.[23] The Maoist movement overall thus was char-acterized by older, radicalized adults leading younger, malleable manpower, both mobilized by a variety of local concerns (or kidnapped) but inspired by a warped construction of "the other" and a promise of material gain as the ultimate reward for their efforts. The Party surfed the resulting wave of violence and destruction. The Maoists have essentially been given a pass as concerns this aspect of the conflict. The topic of their use of child soldiers has been little dis-cussed and certainly not subject to the sort of activist and legal interventions that one associates with the various conflicts in Africa.[24]

It was this ugly underside, driven by the zero-sum game which is often evident in Nepal (a consequence of scarcity as a way of life), that caused the U.S. Embassy[25] to warn regularly of a potential Khmer Rouge mindset upon the part of the insurgents – referring to the tragedy which engulfed Cambodia when radical Maoists took over there in 1975 and ultimately killed an estimated one-fourth to one-third of the entire population.[26] It is significant that Maoist pronouncements of all factions, as noted earlier, cite the concept of "continuing revolution" as embodied in the Cultural Revolution in China (1966–1976) as necessary to ensure a just order; that is, the overthrowing of hierarchy must be institutionalized. In China, the resulting horror and brutalization of this strategy led to the present turn away from communism (though not an end to Party dictatorship). In Cambodia, it was an effort to be more Maoist than the (Chinese) Maoists that led to the Khmer Rouge "auto-genocide" (as it was astutely termed by one analyst). In all interviews I have conducted with Maoist leadership figures, this brutal and disastrous legacy is denied and a counter-claim advanced that all evidence of past Maoist crimes has been manufactured by "hostile elements" (led by the U.S.).[27] Those individuals I interviewed during the *overt* phase of the conflict are the same individuals in leadership positions today during the *covert* phase, the time of "peace."[28]

The reality of the Nepali Maoists' purported new society, in other words, was anything but utopia. Making the situation worse was that the dynamic of the insurgency unleashed countless freelance actors who engaged in criminal activity even as they pursued their tasks as Maoists. This introduced a hopeless situation for victims. On the one hand, the Maoists were conducting an orchestrated campaign of terrorism; on the other hand, much of the execution was decentralized to local organs over whom the central insurgent authorities did not even try to exercise authority. This may be conceptualized thus: If brutalizing targets such as those discussed above was Maoist policy, the precise mode was left to local insurgent authority or even to the individual perpetrators. Not only was terrorism routinely used to force compliance to Party demands in local areas, but, from the earliest days of the insurgency (to the present), there was in place a ruthless, highly efficient effort to track those who had resisted the movement and thwarted its designs.

Such ferocious violence against the innocent was in a sense predictable. The strategic vision implemented initially by the Nepali Maoists (for perhaps five to eight years after 1996) was modeled upon that of the brutal Shining Path (*Sendero Luminoso*) of Peru specifically[29] – and other Maoist groups generally.[30] These included especially the Maoist insurgents in the Philippines,[31] where widespread use was made of urban partisans or "sparrows"[32] not unlike those used by the Maoists in their Kathmandu assassination campaign, and India. Thereafter, the Maoists acted on their own initiative. It is significant that the models that shaped the Nepali approach were among the more ruthless and savage to appear in the Cold War era and immediate period thereafter, especially Shining Path (*Sendero Luminoso*), which convulsed Peru from 1980 to the mid-1990s, creating an internal war that cost more than 69,000 deaths (more than half killed directly by Shining Path) and property damage equivalent to one-third to one-half of the gross domestic product (GDP).[33] What all of these groups shared was the approach previously noted: the use of terrorism to eliminate, first, group or individual resistance, whether of those who were directly active in local politics and civil society or those who were involved in the processes of administration and governance; second, government presence so that the population in Maoist-controlled areas could be organized as a rival to the state in a contest for political power. To fund their effort, the Nepali Maoists relied upon "revolutionary taxation" (a profile of illicit action dominated by looting, extortion, and kidnap-for-ransom).[34]

Throughout the events discussed above, Nepal's response was clumsy and often chaotic. At the time of the April 2006 ceasefire that launched the negotiations culminating in the November 2006 peace accord, statistics were shocking. In addition to more than 17,000 deaths in the conflict,[35] the numbers maimed, kidnapped, and displaced ran to an order of magnitude higher.[36] Various efforts at negotiation had come to naught, and even signed agreements had little effect upon Maoist behavior. Indeed, even prior to the declaration of the 2006 ceasefire, a "12-Point Agreement" between the Maoists and its seven legal (but alienated from the government) party partners all committed "to carry out the political activities without any hindrance."[37] A more comprehensive and explicit "Code of Conduct" followed the next month and committed the signatories

> not to make any hindrance and give any mental and physical pressure from either side to the workers of political parties and members of social organizations or individuals to disseminate their opinion, to conduct meetings and assembles, to conduct the act of extending organizations through movement around any part of the country.[38]

None of these provisions were observed.

Maoist methodology: the period of *covert* war

For the leopard did not change its spots: The Maoists continued such actions as detailed above despite public commitment to peaceful politics. In reality, there was not a decision for peace. Instead, at a September 2005 strategy meeting held at Chunwang (also rendered Chungwang), Rukum the Maoist leadership outlined to its senior figures a plan to agree to peace publicly while *covertly* continuing toward revolution using front activity and terrorism (rather than guerrilla or military action) as the main weapons. As to motivation, documents and interviews with former insiders indicate that the Maoists entered into the April 2006 ceasefire talks because a shift to subversion under the cover of peace was the least costly, most logical way to complete the neutralization of government power and to push the revolutionary struggle through to completion. Available materials state that the purpose of peace was to enter the areas of government control (called "white areas" in standard Maoist terminology) in order to mobilize a general insurrection. Dahal's presentation to the Chunwang meeting states directly as its first military goal of five (translation from the Nepali): "To extensively militarize the party, authority, party members, and people and seek

to configure, specialize, and train the People's Liberation Army to take necessary action in the cities, center, region, districts, and capital" (i.e., to prepare
forces for urban insurrection).[39] The state in effect had sued for peace, and the
Maoists adroitly moved to exploit the opening.

When a formal end of hostilities commenced in November 2006, the Nepalese military was confined to its barracks, supposedly along with the Maoist
PLA, which was sent to regroupment camps. In reality, the bulk of the Maoist
forces in the camps were poorly trained local militia and recent recruits.[40] Key
combatants moved into an alleged youth organization – the Young Communist
League (YCL) – which expanded dramatically and functioned as the storm
troopers for the Maoist political arm, using terrorism to eliminate political
opponents in local areas as had been done during the conflict. Such activity
extended to fundraising through illegal activity and to forced organization of
workers in various sectors of society. There was no mutually exclusive membership between Maoist organizations, and operations were often joint operations of
various groups, with individuals often holding multiple memberships; e.g., in
YCL and ANNISU-R (see above), particularly since any number of the latter
were not actually students. Neither, it needs mentioning, did one need to be
"young" to be in the YCL.

In a typical action,[41] the main offices of Nepal Dairy, the leading private dairy
and food processing industry in Nepal, with four dairy-based food outlets and
more than 300 affiliated retail outlets, was invaded on August 27, 2006 – in the
period of ceasefire prior to the formal signing of the November peace accords –
by 25 YCL members in T-shirts but camouflage trousers and weapons concealed
in bags. They abused management and staff as they attempted to force workers
to form a Maoist union, despite the fact that existing procedures were in place,
which satisfied all. Though the workers were held against their will, police
refused to intervene. Only when a hostile crowd gathered did the Maoists depart.
Though threatened, the owner/operator, Heramba Rajbhandary, filed the appropriate reports with both police and government authorities. Consequent escalating threats culminated in a September 16, 2006 assault upon the 74-year-old
Rajbhandary and his son in their home in Lalitpur, a major location in Kathmandu. Their injuries required treatment at a hospital. Threats were again levied
not to report the episodes, but workers' friends and the workers themselves went
to their own contacts and provided details. Such YCL incidents were widespread
throughout Nepal.

YCL existed as a Maoist front organization during the conflict, began to function operationally after the September 2005 Chunwang strategy meeting, and
was formally launched as an action arm of the Maoists after November 2006.
Initially, it was manned predominantly by PLA combatants transferred to the
new organization, was organized in military fashion, and maintained its own
camps or barracks. It continued to conduct training in violent activities. As it
recruited new members, the parameters of organization, basing, dress, and equipment became more informal. The chain of command always functioned,
however, and individuals at all times considered themselves to be "Maoists,"

even as they would often wear items that identified them as YCL. Of the result, there was no question: YCL unleashed a veritable reign of terror that included murder even as it engaged in substantial and ubiquitous extortion and intimidation.

Anecdotal evidence of Maoist depredations and attempts at construction of lists of their violent activities do not do justice to the full nature of their menace and terrorism. Most reports, regardless of source, understated the level of violence, because it was *covert* and/or took place in rural areas which were difficult for the media, to access and observe, limited as it is in personnel and resources. An action such as following an individual and terrorizing his family and friends in an effort to locate him would rarely if ever appear in any formal tabulation. This was doubly the case, because often reports which were disseminated appeared via radio or in the vernacular press, thus were simply missed.[42] There were limited and generally ineffective operations by the police to shut down such efforts. Even as individuals were terrorized, there was effectively nowhere to turn. Instances abound of complainants actually being assaulted by the police for the effrontery of lodging a complaint. The Maoists had thoroughly infiltrated the police and had access to all such records. For their part, the police had little motivation or ability to deal with such activities. They were, in fact – as evidenced by numerous complaints and laments from professional officers – routinely instructed not to become involved in "political" matters, which in practice meant any activities carried out by political parties, in particular by Maoist members. Only episodes that were so blatant as to provoke a popular uproar and a demand that "something" be done stood a chance of being acted upon. The result – the very essence of "new war" – was that by August 2008, the Maoists controlled the government.

Irony was on display. On the one hand, surface developments could be seen as a steady process, whereby a "better tomorrow" appeared to loom compared to the ugly realities of the day. On the other hand, each step toward normalcy, whether a ceasefire or a signed agreement to move toward conflict resolution, invariably resulted in dashed hopes.[43] Nothing illustrated this more than the years which immediately followed the formal declaration of peace in November 2006. To gain power, the Maoists had entered into an arrangement with anti-government but legal political parties, which ultimately placed Maoist representatives in the legislature (still carrying their weapons) and Maoists in control of various government agencies. Part of the power-sharing agreement stated that elections would be held for a combined Constitutional Assembly (CA) and Parliament. Twice (in June 2007 and November 2007), they were postponed, both times due to Maoist subversion – and a Maoist need (as they perceived the situation) to continue eliminating local resistance prior to participating in any open vote. A final attempt to hold a vote, April 2008, was successful. Not surprisingly, given their destruction of their rivals in most areas of the country, the Maoists emerged as the largest parliamentary party, though they had less than 40 percent of the seats (with a lower percentage of the actual popular vote). They were hence unable to form a government until August 2008, when they brought Marxist-but-non-Maoist politicians into a coalition.

During the post-war period under discussion, terrorism went on throughout Nepal. Incidents numbered in the thousands and included not only armed assaults and kidnapping but torture-rapes and murders. Since the objective, as stated previously, was to systematically and illegally prevent rival parties from reestablishing themselves in the 70–80 percent of the population already controlled by the Maoists, while simultaneously making inroads into government-held areas, the main targets were rival political party cadre. Especially targeted where those who attempted to crack the physical cordon the Maoists put in place around their captive vote-bank, as well as various civil society actors who challenged (simply by their social welfare actions) the Maoist monopoly of power and domination of a narrative that held only Maoists legitimately served the interests of the people. Since Nepali Congress remained the single largest party, it was the main target, especially in traditional areas of strength which the Maoists saw themselves as "owning." One of the most pernicious facets of the aggressive Maoist effort was a renewed campaign to seek out those who had earlier been targeted but had fled to government-held areas, such as Kathmandu, areas that was now be more thoroughly penetrated by Maoist informants and operatives than had been the case during 1996–2006. With their assumption of actual power in August 2008, and with the army still confined to barracks, the Maoists could use the state itself as the cover for their continued terrorism.

Far from moderating their behavior when faced with the task of actually ruling, the Maoists continued in much the same vein as already discussed. Shakti Khor camp, for example, south of Kathmandu and the major Maoist regroupment center, became a nerve center for criminality and death squad activity. Best evidence suggests, however, that it was but a first among equals in such a profile. Concurrently with neutralization of targeted individuals, the Maoists dramatically increased generation, through illegal activity, of the resources necessary for maintenance of their political party.[44] Such actions reached enormous proportions and were always accompanied by actual or threat of violence. On the surface, democratic norms were observed – at least to a point – while beneath the surface, *covertly*, intimidation and attacks were routine. Numerous terrorist incidents that occurred during Dahal's tenure, August 2008 to May 2009, bear witness to the extent of Maoist depredations. Given Maoist control of the levers of power, these often involved police complicity, to include the release of apprehended suspects in torture and murder cases after the police had received orders from Maoist officials.[45]

Only when the Maoists completely overplayed their hand did they stumble. In May 2009, a sensational videotape surfaced of Maoist leader and then-Prime Minister Dahal *aka* Prachanda openly boasting to a PLA gathering at Shakti Khor camp of his use of the peace process to subvert parliamentary democracy and of his future plans for eliminating all rivals to the Maoist people's republic.[46] The section dealing with both PLA numbers and the role of the YCL is worth quoting in full for the insight it gives into "new war." Dahal states:[47]

> You say our numbers have decreased. That's not true. Our army has grown significantly. Where is the shrinking? You must understand strategy and

tactics. Tell me, how many of us were there earlier? Speaking honestly, we were few before the compromise. We were at 7,000 to 8,000. If we had reported that, we would have had 4,000 left after verification. Instead, we claimed 35,000, and now we have 20,000. THIS IS THE TRUTH. [emphasis in original transcription] We cannot tell others, but you all and I know the truth.

How can anyone say our numbers have decreased? Look how wisely our leadership took a 7,000-person army and made it a 21,000-person regular army. That is what you are now. We have not shrunk; we have grown. And on the outside, we have created the YCL infrastructure, and we have thousands in the YCL. So we have built a lot, and are still building. It is true that there some complexities, but they are still a strength. About our friends who did not make it through the verification process, there is a fear that they are done. But arrangements will be made.

The outrage was such that the highly fragmented opposition, supported by number of external players, mustered a parliamentary majority and became the government. The overall result was that the Maoists retreated tactically to the role of opposition party, which allowed greater opportunities for continued *covert* violence and extortion, free from the scrutiny and irksome oversight that came when they were forced to operate in the open as the government. Indeed, they promptly retaliated by using their substantial numbers to block all normal business and filled the streets with YCL and rural thugs, bused in to man demonstrations designed to bring normal governance to a halt. A nationwide general strike was begun May 10, 2010, and anyone attempting to cross the Maoist lines was met with violence.

Those of the press who energetically pursued stories on continued Maoist crimes were menaced and at times murdered. Among the most prominent cases was that of Uma Singh, a correspondent for vernacular print and radio media in Janakpur, Dhanusa district (in the central *tarai*; essentially the southernmost strip of districts in Figure 6.1). Ms. Singh was hacked to death on January 12, 2009 after her detailed stories exposed the continued criminal and terroristic activity of a Maoist Politburo (the inner ruling group of the Party) member, Matrika Yadav, who at the time was a minister in the Dahal government and was continuing to lead a campaign of violent land seizures.[48] Though a great deal is known about her killers, the case history remains checkered, with a lack of resolution.[49] Yadav first continued to operate openly as the leader of an extreme Maoist splinter, then, after the split in the "mainstream" Maoist organization, functioned publicly as a leader of a breakaway radical group committed to violence.[50]

Nepali media, after pointed warnings from the Maoists, largely retreated to self-censorship, which has made gathering comprehensive data difficult.[51] Yet some Maoist acts were so astonishingly brutal that they have been impossible to keep under wraps. The case of Uma Singh is mentioned above. Shortly thereafter, in December 2009, Tika Bista, journalist for *Rajdhani* daily, was savagely beaten in Dang and left for dead after a column she wrote for a local weekly, *Jantidhara*, criticized the YCL. Among the weapons used to assault her were razor blades; the fatal assault on Uma Singh was carried out using *kukris*. The legendary curved knife of the Nepali hills, it is a weapon which has figured prominently in Maoist terrorism and training that was conducted openly by the dissatisfied radicals.

Even as such assaults went on, in a dramatic turn, a longstanding post-Dahal political gridlock, in which no party could form a majority government (in Nepal's semi-restored parliamentary system), was broken on August 28, 2011 via a back-door deal, which brought the mainstream Maoists again to power and Dr. Bhattarai (the second ranking Maoist) to the prime ministership. The practice of Leninist democratic centralism dictated that Dr. Bhattarai answer to the Maoist Politburo, which continued to hold closed door meetings to plot its next steps.[52] Once in power, one of Dr. Bhattarai's key initiatives, on February 27, 2012, was to table legislation pardoning all Maoists for past crimes. Though the action was temporarily sidelined by various Supreme Court interim orders, the government persistently reintroduced the legislation, thus making it quite clear that it would not stand in the way of terroristic activity. Likewise, in one particularly heinous case, the five Maoists accused of murdering *Radio Nepal* journalist Dekendra Thapa by burying him alive, who had been apprehended when one of the five, in a fit of conscience, went to the authorities, were released after orders were received by the police "from above" – which additional scrutiny determined was an order issued by Bhattarai. Though the five were finally convicted of "second degree" offenses and given sentences ranging from six months to two years each, it was only in late 2015 that the sentences were upheld, by which time they effectively were moot.[53] Such was the norm, and Maoists of all factions vociferously threatened further violence if even the feeble efforts to bring past instances of terrorism to trial were continued.

Making the situation still more fraught with peril, throughout these events, the Maoists were locked in an escalating intra-Party strategic debate that stemmed directly from the agreements made in the 2005 Chunwang strategy meeting previously discussed: what *form* should the use of violence take to further their revolutionary political project – and how intense should be its application? Party radicals, led by a dogmatic veteran of the movement, Mohan Baidya *aka* Kiran ("Ray of Light"), in the company of a recognized party firebrand, Netra Bikram Chand *aka* Biplab (also rendered as Biplav; "Big Thunder" is often advanced as the most useful translation), had grown increasingly displeased with what in their assessment was a lack of urgency in sweeping out the old-order – despite the Maoists twice having seized the top position in the polity. They advocated immediate, *overt* use of terrorism and urban combat to the extent necessary to destroy all remaining vestiges of Nepal's *ancien régime* in order to implement *revolution*.[54] The opposing "moderates" in such debate (i.e., the mainstream Maoists, who were the government at that time) claimed in party leadership sessions that the ongoing course of action – i.e., *covertly* using terrorism to neutralize all opposition while openly participating in politics – was the preferred and least costly approach. This mainstream faction could point to dramatic progress in advancing to a position of absolute power.[55] To resolve the increasingly nasty factionalism, therefore, the mainstream faction placed party unity above party discipline. The radicals were allowed to use *systematic terrorism* – and to prepare forces for "revolt" – even as the mainstream advocated *opportunistic terrorism* against its foes.

Such distinction was lost on the local-level perpetrators (and their victims), with the result that political violence against the innocent escalated both as a consequence of party policy and as per individual direction of dissident leadership figures (i.e., those who did not agree with Party strategy). In effect, use of terrorism was accepted as the price of Party unity, even as outbidding went on. Each faction was concerned that appearing soft would lead to its hard men defecting to the other faction. Hence there was minimal effort to limit local terrorism for the sake of strategic consistency. The result was a robust element committed to provocative, violent action – terrorism.[56] Most revealing was the December 27, 2011 document, "On Problems of the Party and Their Resolution," denouncing the mainstream faction, which was presented in the Central Committee by Mr. Baidya. Noting that party leader Mr. Dahal claimed there was no intra-party disagreement upon "going for mass insurrection with especial [*sic*] focus on national liberation and federal republic," only that "contradictions and disputes have often taken place on issues concerning the tactical steps to achieve the above-mentioned goals, which have attracted and had impact on party's overall ideology on certain conditions," Mr. Baidya scathingly replied, "...Chairman Comrade has stated that mass insurrection could not immediately materialize despite repeated and conscious attempts. But this is not true. This is only his claim. In fact, he had never been effortful and oriented towards that direction."[57]

It was therefore, to his mind, the duty of the radicals to forge ahead – which they did. The objects of their terrorism necessarily were those who still stood in the way of revolutionary project. Numerous incidents during this period attest to the reality that the radical Maoists were seeking to neutralize not only their immediate foes but those from the past who had gone into hiding. Still, such violence was relative. In a stunning development, even by the standards of the conflict, the February 2012 issue of *Lalrakshak*, the mainstream Maoist house organ – the editorial policy of which is controlled by Mr. Dahal – ran as its cover story, "These Are the People's Enemies," a vitriolic attack on three of the country's most distinguished commentators (with their photos), Kanak Mani Dixit, Kul Chandra Gautam, and Subodh Raj Pyakurel. As the three quite correctly noted in an op-ed:

> The attack on us three citizens seems to be part of a plan by the senior-most Maoist bosses to instill fear in the broader civil society, an attempt to force submission among independent citizens including journalists, rights activists, intellectuals and local leaders in the districts.[58]

That it certainly was.

Dixit in particular had aroused Maoist ire due to his authoring a comprehensive indictment of Maoist duplicity and violence, *Peace Politics of Nepal*.[59] The book was withering in its treatment not only of Maoist duplicity but that of various international actors involved in providing a veneer of legitimacy for Maoist terrorism by minimizing publicly its nature and extent, all the while actually knowing what was occurring. Significantly, the tenuous Maoist intra-Party arrangement just described, based on an agreement to disagree concerning the *form* the use of terrorism was to take, came apart when political events forced the differences to a crisis point. First, the turmoil surrounding the Constitutional Assembly (CA), within which the Maoists controlled the largest block as a consequence of the violent vote manipulation discussed above, was discharged by the president on May 27, 2012 after failing to meet its mandate to produce a new constitution. Bhattarai announced that he would continue to lead an interim government and that elections for a new CA would be held. Initially, only the first of these came to pass, so Bhattarai ruled by mandate. The Baidya faction saw this chain of events as final validation of the bankruptcy of the mainstream approach and formally split in late June 2012 to form its own party, the Communist Party of Nepal-Maoist or CPN-M.

The new Maoist party, which included the ultra-radical faction led by Chand, was openly committed to strategic use of violence – i.e., committed to a position which argued that *overt* versus *covert* terrorism constituted the only viable approach to seizing all power and pushing through revolutionary reordering of the state. Both the mainstream and the Baidya group maneuvered vigorously to bring Party organizations and manpower to their respective sides, with the radicals in particular seeking to incorporate alienated combatants of the PLA and YCL. Though violent clashes between the two factions increased throughout the latter half of 2012 and throughout 2013, both were constrained by their need to move carefully lest they give opponents in the legal political spectrum, who themselves were quite divided, grounds to unite. Such a posture was particularly difficult for the breakaway party, because it had to present a public face which did not excite alarm and provoke response even as it led its membership in appropriate violent action, such as land seizures and violent labor actions. Especially favored were the ubiquitous *bandas* – i.e., general strikes suddenly declared by an organization and which, in the Nepali context, *always* involved enforcement through menace and terrorism.[60] Ultimately, the Baidya faction sought to provoke repression which would facilitate a renewed people's war.

It was this situation that led to a U.S. decision in September 2012 to drop the Maoist TEL designation (referenced previously) in order to buttress the "moderates" of the mainstream faction.[61] The mainstream, to be clear, continued to agree to the use of *opportunistic terrorism* as long as it was kept from breaking into the open. The U.S. action was taken in much the manner that the Provisional Irish Republican Army (PIRA) was suspended from the FTO list when it appeared that it was willing to work toward an end to conflict. In Nepal, the U.S. (and the West in general) had a considerably weaker hand to play than in Northern Ireland. In any case, dropping the mainstream faction from the TEL designation in no way changed the situation of targeted individuals, because the most violent Maoist elements had moved into the breakaway, radical Baidya party.

In the larger picture, continued resistance to the Maoist-controlled status quo from civil society and non-Maoist political parties, aided by Indian intelligence,[62] resulted in an agreement on March 14, 2013 which placed the country's Supreme Court Chief Justice, Khil Raj Regmi, as interim prime minister and charged him with holding November 19, 2013 elections for a new Second Constitutional Assembly (CAII), elections the radical Baidya group declared it not only would boycott but would use violence to thwart.[63] There thus commenced an astonishing period which ultimately saw the entire security forces (to include the long-sequestered army) deployed, in conjunction with tens of thousands of local volunteers accompanied by INGO members, to resist the effort of the Baidya Maoists to block the holding of elections. They focused their attention, however, upon the polling stations themselves, not upon providing security for the general public.

Nepal was in many ways again at war. The massive security force deployment (more than 224,000 personnel were deployed) and considerable foreign intervention kept casualties and damage in the period described above relatively low as per official tally – three dead and 26 wounded, with 109 vehicles torched/ damaged – despite the fact that 54 Bombs/IEDs exploded and another 430 were defused or exploded by the authorities[64] with 34 instances of small arms firing.[65] In reality, these figures considerably understated the level of local terrorism directed at those whom the Baidya Maoists identified as their opponents. Episodes in "out of sight, out of mind" locales were ubiquitous and often quite violent. Detailed descriptions and photos fill the better part of the nearly 260-page monitoring report compiled by the Citizen's [*sic*] Campaign for Clean Election, working with the major cause-oriented statistical body operating in Nepal today, INSEC (Informal Sector Service Centre).[66] It could hardly be otherwise, since Maoist cadre were actively rallied to the attack. Still, the results of the election were heartening, as the mixed first-past-the-post and proportional (from a national party list) seats, selected through 78 percent turnout achieved due to the unprecedented security and foreign involvement, gave the traditional political party powers, Nepali Congress (NC) and Communist Party of Nepal (Unified Marxist-Leninist) or UML, 61.7 percent of the total CAII 601 seats, earning 196 and 175 seats, respectively. The mainstream Maoists, having chosen to participate, saw their continued odious behavior rewarded by finishing a distant third with but 80 seats (only 26 of which were obtained in direct first-past-the-post competition).[67] When allies of the dominant "big two" parties were factored in, the ruling coalition had more than the two-thirds necessary to pass any decisions.

As had happened previously, Dahal again was caught on tape (in the final days before voting), urging the members of his inner circle to use whatever means were necessary, legal or illegal, violent or no, to ensure victory.[68] Many of his followers took him at his word and committed acts of terrorism. When early returns indicated the scale of the security-enabled debacle in the making, the mainstream pulled all election observers, demanded that the vote-counting be stopped, said they would boycott CAII, and claimed that a national and international plot had been engineered. For his part, rival Baidya, leader of the breakaway radicals, sought to take credit for the mainstream faction's precipitous defeat and vowed to continue with the radical strategy of terrorism and engineering "urban revolt." His attempt to implement a nationwide ten-day *banda* before the voting had failed but nonetheless had packed a punch.

Baidya's radicals, in fact, had demonstrated, in Thawang VDC of northern Rolpa, their still considerable ability to use terrorism to thwart democratic process. For decades, Thawang had been a Maoist stronghold and remained under radical control. Though the Nepali Congress had previously been a force in the area, that ended when its local leader, Ramji Gharti, 54, was kidnapped in late January 1997 by the Maoists, taken south to Mawang VDC, and there tortured and murdered. His body was found on February 1, 1997 with his head thrown some 40 meters from the torso.[69] In the election, not a single local ballot was cast in the first-past-the-post system (of 1,876 registered voters) once the radicals announced they would attack any individuals who sought to vote and used personnel to do just that (with even mainstream Maoists being attacked).[70]

In this confluence of circumstances is on display the heart of the matter as it pertains to the Maoist conception of peace in a "new war" context. A large portion of the Maoist manpower has over the past several decades taken its key leaders at their words and adopted violence as an integral form of political discourse – i.e., adopted terrorism as a routine option in dealing with others with whom they disagreed. So ingrained was this methodology that when the mainstream sought to introduce a more subtle use of terrorism, the most radical element among the Maoists, as highlighted above, split so as not to be bound by what it saw as the betrayal of revolutionary ideals. Yet, mainstream or radicals, the issue was *how* to use terrorism, not *whether* to use it. Violent manpower that had become too well known merely moved laterally into new or affiliated groups, such as the violent front organizations (especially the YCL and various "student" groups). Menace, extortion, and terrorism continued; the Maoists still attacked rivals, be they individuals or party members or even religious groups.[71]

It was such a context – with Maoist leadership consistently fanning the flames of resentment even while urging discretion – that frustration boiled over amongst the Maoist local operatives (the foot soldiers), who had invested their lives in terrorism but saw their hopes of a people's dictatorship slipping away. In late November 2014, the most extreme elements of the Maoist spectrum denounced the Baidya group as too cautious and as producing no results in returning the country to a state of war as planned when the group broke away from the mainstream. On December 1, 2014, therefore, the Communist Party of Nepal (Maoist) or CPN(M) – the original acronym of the Maoists – declared its existence under the leadership of Chand, who had prepared the way with a series of clandestine meetings in Rolpa. Chand promptly observed that the form of violence this group would adopt would depend upon circumstances, but he called for a "people's retaliation" against those who had betrayed the cause of people's war.[72]

Part and parcel of this effort was seeking out enemies of the Party who had taken refuge elsewhere in the country or abroad. The continued vitality of the Maoist intelligence network, which, though divided between contending groups, nonetheless continued (and continues) to function well, the different groups sharing information (and even weapons) and coordinating as necessary against perceived threats.

Maoist methodology: *covert* war becomes "new war"

When a destructive April 2015 earthquake hit Nepal, the country was locked in the contentious (and at times violent) debates of the CAII process and thus was still, nearly a decade after the formal end of hostilities, without a constitution. Terrorism continued. The Chand group was deep into a campaign of aggressively seizing land and dwellings from ostensible enemies for redistribution to its followers and was establishing parallel governance structures in areas it had selected as best suited for rebuilding the counter-state.[73] Reports demonstrated that it was also collecting weapons, even as it located and moved against those who had previously resisted the Maoists, whether these individuals had resisted the unified movement or its splinters. State response was characterized by distraction and denial. Within this context, the mainstream grappled with restless manpower and renewed factionalism. Thus the mainstream Party was under intense pressure *to do something*. That something was a decision to move deeper into the system, maintaining an ambiguous position on terrorism while agreeing to let the constitution become a reality. This, though, served to split the mainstream yet again.

Throughout the history of the movement, there had been a marriage of convenience and much jostling between the two senior figures, Dahal (a former school teacher with an MA; born 1954) and Bhattarai (also a former teacher but with a PhD; also apparently born in 1954). Both were active in CAII, because – though relegated to a distant minority position in popular voting – their mainstream Maoist faction remained important, particularly because it was invariably ready to turn to terrorism even as it maneuvered constantly to reunite its estranged, radical factions. Anxious to keep this reunification from happening, CAII leaders (from the traditional majority parties) kept both Dahal and Bhattarai in gateway positions in the drafting process (i.e., committees through which constitutional drafts flowed) disproportionate to the actual Maoist mainstream Party count of delegates. This facilitated the completion of the constitution-writing process and the promulgation of the new constitution on September 18, 2015.

Yet faced with the task of selecting – from within the already seated CAII membership – the first officials to govern under the new constitution pending its gradual implementation, the longtime non-violent wing of Nepali Marxism, the Communist Party of Nepal (Unified Marxist-Leninist) or UML – who were a slight second in CAII votes to the more centrist governing party, Nepali Congress – cut a deal on October 11, 2015 with the mainstream Maoists (and several smaller parties) to provide their crucial swing votes to create a UML majority government. In return for their support, they gave the Maoists a number of key Cabinet positions, but these went to supporters of Dahal's faction within the mainstream party to the exclusion of Dr. Bhattarai and his followers. As a result, Dr. Bhattarai resigned from the Party to form a "New Force."[74] This left the Dahal group as the official Maoists.

The Cabinet positions given to the Maoists included the all-important Home Affairs Ministry, which controls the police. The individual named to the position, Shakti Basnet, was the long-time personal secretary to Dahal to include during the latter's term as prime minister. That term, as discussed above, had been characterized by extensive use of terrorism. Further, as already noted, once in office, Dahal used his position to neutralize the already minimal police response to Maoist depredations. Best evidence is that Basnet was an important figure in these crimes. The position of Home Minister, to be clear, was reserved for the Maoists (as were the others) as the price for their support. Basnet was named by Dahal to be the Party's nominee, thus was appointed. Not surprisingly, also rewarded in the new Cabinet distribution of power was Ganesh Man Pun, the head of the YCL, which had played such a prominent role in the post-November 2006 terroristic violence. He was named by the Maoists to head the Commerce and Supplies Ministry, an apparently ideal position for siphoning funds to the Party. Altogether, eight ministries were "gifted" (the Nepali terminology) to the Maoists. The apex post, though, was the reward bestowed on the previously mentioned former head of the Maoist armed forces (the PLA), Nanda Kishor Pun *aka* Pasang (also known Nanda Bahadur Pun), who was elected Vice President (by the now-dominant majority coalition). His first recorded act was to call on Dahal to thank him for naming him as the Party's candidate in the prearranged vote.

If this was the foreground, the background was violence that threatened to engulf the country in communal strife. For the constitution was a series of tortuous compromises between traditional and radical positions, with the clear loser being minority populations that had been mobilized by the Maoists with the promise that benefits would flow from seizure of power. The ethnic fronts discussed previously, for instance, had each been promised autonomous zones. The most salient minority population that had received similar promises was clustered in Nepal's traditional breadbasket, the *tarai* (also rendered as *terai*), the narrow flatlands bordering India. When the promised benefits did not materialize and grievances were further exacerbated by a decade of "more of the same," an explosion of discontent erupted that has only recently died down.[75] Matters were made substantially worse due to an unofficial blockade of Nepal instituted by India pursuant to its own geopolitical concerns.[76] By the time the blockade ended and the protests diminished in early 2016, the UML coalition was on such shaky ground that the mainstream Maoists saw an unparalleled opportunity to regain power through advocating a unity government – which they would head as the distant third party holding the balance between the Nepali Congress and UML. The result was that on August 3, 2016, partnering with Nepali Congress, Dahal began his second term as prime minister – the third by a mainstream Maoist. Needless to say, he received no support from the breakaway Maoist parties, who continued to demand that the revolution go violently forward, and thus continued to use terrorism.

Beyond opportunism, the willingness of past mortal enemies Nepali Congress and the Maoists to make common cause stemmed in large part from the shared imperative to neutralize the increasingly pressing issue of transitional justice.[77] It was especially important for the Maoists to avoid a true airing of the crimes of the 1996–2006 period, because this would open the door to what had been going on during the 2006–present period. Demands for a settling of legal accounts had quite naturally looked in the first instance at indiscipline by the security forces;[78] which for much of the in 1996–2006 period had answered to a Nepali Congress government. This made Nepali Congress ultimately responsible for the crimes committed, which had rarely if ever been investigated during the conflict years.[79] It did not take long, though, for further inspection to be directed toward the insurgents. For the Maoists, the possibility that the process would go further was a potential disaster, because their criminal acts were not violations of prevailing norms and laws, as was the case with the government, but pursuant to Party policy. Even during the bitter fragmentation of the Party, therefore, Maoist splinters periodically came together at public sessions to denounce attempts by victims to pursue justice through the criminal justice system. Such pursuit appeared increasingly likely as a series of Supreme Court decisions provided mechanisms to do so.[80] Gaining control of the government would place the Maoists in the position to deal definitively with the situation through perverted application of the law. This they did.

Once the mainstream Maoist faction again was in power, with the attendant ability to distribute positions and resources, some of the leading estranged radical leaders returned to the fold. Still, many did not; and a burning question amongst Nepalis during my recent fieldwork was whether the violent splinters actually intended to move forward with their plans for renewed assault upon the system. Interviews with individuals at the leadership level who were semi-underground, show that it was clear they did. It was the Chand faction which has assumed the lead in this radical effort. It had retained many of its weapons from the 1996–2006 period[81] and has also obtained supplies of explosives. It continued to drill personnel and to assault individuals as it deems necessary. It regularly carried out acts of terrorism (though they are not labeled as such in the Nepali press). A small sampling serves to illustrate. In June 2016, the Chand group destroyed ten cell-phone towers; in September 2016, a similar set of attacks targeted private schools within Kathmandu with bombs, alleging crimes against the people.[82] Shortly thereafter, on December 6, 2016, four members of the group were apprehended moving a substantial lot of semiautomatic rifle ammunition and magazines.[83] In February 2017, the Chand group held a national congress at which its semi-underground status and active formation of paramilitary units were discussed, as well as the necessity for raising funds for the radical party.[84] On March 28, 2017 in the Nalsing Gad area of Jajarkot district, the Nepali Congress district head, who had interfered with the Chand fundraising effort, was assassinated.[85] On May 9, 2017, in Chitwan district, two leadership cadre of the Chand group were apprehended with explosives they were distributing for use against candidates in the effort to conduct the first local elections in two decades.[86] Only a month later, on June 7, 2017 in Rautahat district, a further two Chand cadre were arrested with an even larger batch of explosives.[87] Even more recently, on September 25, 2017, the prominent Shangri-la Hotel in Kathmandu was targeted in a thwarted bombing attempt.[88] These 2017 incidents were but the most prominent of a string of explosives-related attacks and attempted attacks that culminated in well over 100 bombings (without assaults being counted) in late November–early December 2017. These have been followed by regular bombings, to include attacks on February 10, 2018 on the country's major hydroelectric project in Diding, Sankhuwasabha district, and February 12, 2018 on communications towers in Rolpa, Nuwakot, and Chitwan districts.[89]

The purpose of this and other acts of terrorism is to painstakingly re-establish local political domination, to regain the overwhelming position which the Maoists held in November 2006. The strategy, however, is now to "build people's revolt on the foundation of people's war." This position has been widely disseminated within Chand group circles and even in interviews with Nepali media. As a leadership figure stated flatly (in translation): "At this point in time, we need to focus our attention on the reality that we [the Party] did not say [in opting for 'peace'] that the people's war was unnecessary or wrong. Rather, we were opting [through the united front] for people's revolt. If that failed, then we would continue with the people's war."[90] Her point was that the mainstream had betrayed the strategy that had been agreed upon beginning with the September 2005 Chunwang Plenum and subsequently several more times in hard-fought meetings of the Party's central leadership.

What this strategy meant on the ground is illustrated by the acts of actual and planned terrorism discussed above, which in practical terms meant mainly targeting individuals. Violence in the form of terrorism was to continue as the driver for the Maoist effort. As was the case during the overt 1996–2006 phase of hostilities, there is little the state can do (even assuming it is willing) to counter it and provide security for ordinary citizens. This was driven home in a typical but noteworthy incident that occurred on April 27, 2017, when – unable to lay hands on a rival party activist many times attacked and long sought by the Maoists – they kidnapped the victim's father in Dang district and assaulted his mother. Significantly, in the attack, at least some of the Maoists were armed, a reality which has surfaced in similar incidents. Though the subject of major coverage in the vernacular press, nothing has come of the case. The father, who refused to give his son's location, remains missing. The perpetrators were originally with the Baidya group, though it is unclear to whom they presently give allegiance.[91]

There was, in fact, an ominous Chand group ideological silence after their February 2017 congress in Rolpa noted above. It was a silence that should have prompted renewed government focus upon the group, for all documents discussed at the meeting remained close-hold, and previous sources willing to discuss party positions as late as August 2017, had gone silent. Party members already in 2016 considered themselves to be "semi-underground" (the term they used), but at least some appeared to have taken the significant step of entering a clandestine state. It now seems clear the Chand group was preparing for a significant assault inspired by continued use by the mainstream Maoists of taking over the system from within rather than assaulting it from without.

On June 7, 2017, the mainstream Maoists and Nepali Congress, responding to an earlier "gentleman's agreement," swapped the lead positions in their coalition. War-era leader Sher Bahadur Deuba returned for his fourth stint as prime minister. Nothing changed policy-wise. It appeared that this arrangement would continue to dominate Nepali politics, but with local elections completed, Dahal spotted an extraordinary opportunity *en passant*. By offering a classic united front union with the UML, he reconstituted the "alliance of the left" or "Left Alliance," with the tremendous difference that the two parties, hitherto bitter rivals despite both being communist, would not merely cooperate but would field a unified slate of candidates, divided 60–40 percent in favor of UML. After their expected victory, they would unify and rule jointly under a new party name.

Victory in fact came to pass. In a two-phased election held November 26, and December 7, (2017) – marked by a level of violence that rivaled that of the 2013 CAII election – the Left Alliance emerged with nearly a two-thirds margin. Indeed, with the true contests being overwhelmingly UML versus Nepali Congress battles, the Maoists were able to emerge as the second strongest party in first-past-the-post seats.[92] Party list seats (i.e., those selected by proportional representation) altered the final toll to make NC the second largest party but did not in any way affect the lopsided margin. In the final tally, UML had 121 seats and the mainstream Maoists 53, while NC had just 63 seats; the remaining seats were filled by two Madhesi parties (RJPN, 17 and FSFN, 16) and independents (five seats for five parties that did not make the 3 percent threshold for party list representation). To make matters worse, the much smaller upper house, selected from the seven provinces, was completely Left Alliance (six provinces) and Madhesi (one province). Considerable irony attends these stark figures. Contrary to the picture painted by the seat totals was the popular vote, which saw the UML and NC run neck-and-neck, 3,173,494 to 3,128,389, respectively (of a registered total of 15,427,731, with 68.63 percent turnout). The Maoists were a distant third at 1,303,721 votes. Regardless, NC has been effectively wiped out electorally, thus validating Mr. Dahal's strategy.

Alongside the stunning political developments was significant terrorism, with the Chand group identified in most instances. More than 800 Chand cadre were arrested (and most just as quickly released). Though no figures have been released, the three weeks leading up to December 7 alone saw more than 100 bombs exploded or neutralized prior to detonation as the Chand faction sought to stymie the elections. Explosives themselves were overwhelmingly from mining stocks in India; delivery mechanisms were the same IEDs that were ubiquitous during the 1996–2006 period (and since); e.g., wired rice cookers packed with explosives and shrapnel.[93] The substantial damage yet relatively low loss of life and wounded was a result of Chand instructions that terrorizing through property damage was to take precedence over creating human casualties in order to avoid a popular backlash.

This did not keep key NC figures from being targeted, to include Prime Minister Deuba and local candidates.[94] In fact, attacks were focused upon NC personalities and facilities (with UML and others a very distant second). Non-lethal attacks upon individuals were so numerous as to be relegated virtually to background noise. No source reported on them in any systematic fashion.[95]

In the doling out of positions, it was clear that the Maoists were in a position to dominate the polity. Among other key positions, they were again given the Interior Ministry, which controls the police, thus law and order. Ultimately, Dahal expects to again become prime minister, from which position he will be a major gatekeeper in whatever parliamentary alignment emerges. Most crucially, he will be able to shape the process of transitional justice to ensure that it effectively is meaningless.

Conclusion

At this point, it hardly needs reiteration that this is "new war" in spades. The intense debate within the Maoists that led to the various party splits was occasioned by just what was on display: how aggressively and in what form to use violence in the post-2006 (i.e., post-war) period. The original Maoist agreement upon strategy for the era of peace had been understood by radicals such as Chand (and Baidya before him) as attack from within an infiltrated old-order. Dahal would agree with this assessment, but he would emphasize just what has played out: use "other means" – alternative lines of effort in strategic parlance – as the cloak to provide concealment for violence as necessary. He has claimed all along that the use of opportunistic, *covert* terrorism was producing results, while the radicals favored systematic, open assault such as marked this election cycle. Ultimately, in their plan, such terrorism as they unleashed during the election should have been accompanied by the use of urban mobilization and guerrilla warfare linked to action in the countryside along the lines pursued in 1996–2006. From Dahal's perspective, not only has this approach been proved wrong, he has been proved correct in spades. It is he who stands in the corridors of power.

Especially revealing as to the Chand group's position on violence is a February 2017 video displaying the normal agitprop used at all Chand rallies to energize Party followers. The featured dances include the use of weapons, especially *kukris*, as well as prominent movements slitting enemy throats and slashing limbs.[96] The accompanying lyrics are unabashed in their violence. This reality may be further illustrated by the text of a typical Chand front group poster displayed in Kathmandu:[97]

> *Let the laborers of the world unite.*
> *Marxism-Leninism-Maoism – Hail.*
> *Unified People's Revolution* – Hail.*
>
> *Let's not fall into the trap of delusion set by the reactionary, corrupt parliamentary system.*

Let's march forward to achieve scientific socialism through unified people's revolution. *

Nepal Communist Party welcomes you.
Courtesy: All Nepal National Independent Students' Union (Revolutionary)
Kathmandu
University Area Coordination Committee

To clarify: "Unified people's revolution" (at two points where I have inserted *) is the use of urban terrorism combined with rural action (as already achieved) to seize power. Such posters are numerous, not only in the capital but throughout the country. In recent fieldwork, I found them displayed in even the most obscure spots. It is the disseminators of such sentiments – the work force, if you will – who were the eyes and ears of the Maoist intelligence and assassination network during the *overt* conflict and remain as such during this, the *covert* conflict. The evidence is extensive that they have continued to function now in the same manner they did during outright warfare. Now, however, they support terrorism as opposed to terrorism-reinforced-with-guerrilla-and-main-force (i.e., military) action.

In his latest actions, Chand may finally have crossed a line. A warrant has been issued for his arrest, but the execution has been indifferently pursued. This is hardly surprising. A permeable membrane separates the various Maoist factions. Whatever their positions on the use of terrorism, the most Chand can be held accountable for is an excess of zeal. He remains, in Dahal's eyes, on the right side of history. In fact, Dahal himself has consistently refused to condemn political violence, even when holding the position of prime minister. At an event in December 2016, bringing together communist parties in the country on the birthday of Karl Marx, the once-again prime minister was reported as opining that "the fundamentals of Marxism cannot be ignored including the armed conflict as a tool to capture the state power."[98] Violence, in other words, has a place in the normal functioning of politics.

This allows us to conclude by returning to the savage attack upon journalist Tika Bista, considered above. Not discussed was the manner in which she came to be attacked, and it is this that is germane here, because the assault upon her is typical of thousands that have occurred in the decade of "peace." Originally from Rukum, Ms. Bista moved to Kathmandu to attend school; thus she came to work for *Rajdhani* daily, a Nepali language publication. Her experiences in Rukum led to coverage of Maoist terrorism in the post-war period that was prolific and scathing in its assessments of Maoist perfidy. She was a consistent defender of Ms. Tirtha Gautam, the widow of a Maoist-murdered Rukum politician in whose name the wife had won a seat in the first Constituent Assembly (April 2008). Eviscerating the Maoist effort to impugn Ms. Gautam by questioning her sexual propriety, Ms. Bista found herself targeted. A Maoist surveillance network tracked her movements nationwide until the moment, on December 8, 2009, when she was vulnerable while working on a story in Dang district, at which time a hit team attacked and left her for dead. As related by one source, Ms. Bista ...

> ... was mercilessly slashed with razor blades by masked individuals before being thrown into the forest in a far-flung district of Nepal [Dang], she said. She was airlifted to Kathmandu for treatment after journalists and police found her unconscious. Her condition was critical. She had blood clots in her skull and injuries in her spine, her left shoulder, the fingers of her right hand, and in her foot. She has recovered well, but still needs a couple of months for a complete recovery. The doctors have suggested she undergo physiotherapy sessions and psychological counseling.[99]

The police were "unsuccessful" in tracking down the assault team, and the Maoists themselves, who claimed to have conducted their own investigation, asserted that she had attempted suicide to impugn them.[100]

Thus is the lot of the citizen in the "new war" end-state of Nepal.

Notes

1 Converting Nepali to Gregorian dates requires the use of complicated tables, a task that has been made easier by computers. The country uses a lunisolar rather than Gregorian calendar, with the former about 56.7 years ahead of the latter. A straight conversion cannot be made, however, because in the Nepali calendar, days begin at dawn; thus they do not directly correspond to Gregorian calendar-dates. Each Nepali-calendar day encompasses two Gregorian calendar-days, resulting in two possible dates for any conversion, a task made still more difficult by the simultaneous use of at least three different systems for rendering dates. It is noteworthy that one of the most important episodes of the war, the attack upon the army in 2001 that caused the conflict to dramatically increase in intensity, is rendered in the Maoists' own documents at three different dates, with as much as a month separating the versions.

2 Being locked in a shed during monthly menstruation (*Chhaupadi*), for example, regardless of time of year, was widespread and persists in certain areas of the country (though it was made illegal during my most recent fieldwork, on August 9, 2017); so, regrettably, does rape.

3 Thorough discussion of the genesis of this voluntarism and the individuals involved may be found in Benoît Cailmail, "A History of Nepalese Maoism Since its Foundation by Mohan Bikram Singh," *European Bulletin of Himalayan Research* 33–34 (2008–2009), 11–38; available at: http://himalaya.socanth.cam.ac.uk/collections/journals/ebhr/pdf/EBHR_33%2634_02.pdf (accessed December 31, 2017). The party changed its name to the Unified Communist Party of Nepal-Maoist (UCPN-M) after the end of *overt* hostilities (1996–2006). Later, when various splinters returned to the fold, the mainstream altered its name yet again, becoming Communist Party of Nepal (Maoist Centre) or, alternatively, CPN Maoist Centre, CPN-Maoist Centre, or CPN (Maoist Centre). It has recently announced that its alliance with the communists of the Unified Marxist-Leninists or UML will lead to yet another renaming, with the new entity to be termed simply the Communist Party of Nepal.

4 See e.g., Tan Hecheng, *The Killing Wind: A Chinese County's Descent Into Madness During the Cultural Revolution.*

5 Pun *aka* Pasang, *Red Strides of the History: Significant Military Raids of the People's War*, iii. For similar sentiments, see "Nepal: Interview With Comrade Baburam

Bhattarai," December 12, 2009, available at: http://marxistleninist.wordpress.com/2009/12/12/baburam-bhattarai-on-nepals-social-revolution/#more-4155 (accessed January 20, 2018). Therein, Bhattarai, then the mainstream Maoist second figure – while repeatedly lauding the Cultural Revolution – responds to a radical questioner thus:

> Yes we think the Great Proletarian Cultural Revolution was the pinnacle of revolution not only in the 20th century but in the whole history of the liberation of mankind. It is the pinnacle of the development of revolutionary ideas. So all the revolutionaries must make the Cultural Revolution their point of departure and develop the revolutionary idea and plan further.

6 Interview conducted during fieldwork, in Nepali (as simultaneously translated); particulars omitted at subject's request.

7 A voluminous body of academic work on this dynamic has accumulated since the Second World War. Certainly the premier treatment remains Pike, *Viet Cong: The Organization and Techniques of the National Liberation Front of South Vietnam*, which provides in considerable detail – citing insurgent documents and persons – the necessity that a new order accumulate strength by controlling the organizations and ideas through which the population goes about normal life. This allows all activity – from schooling to medical treatment to economic activity to administration of justice to festivals to clothing – to be channeled as necessary into support of revolutionary activity. My own early-academic career contribution to this discussion, which remains in print, is *Maoist Insurgency Since Vietnam*, which compared four conflicts (Thailand, the Philippines, Sri Lanka, and Peru); Nepal was added to the cases under discussion in my later *Maoist People's War in Post-Vietnam Asia*. Peru was again examined for the same reasons it is discussed in this statement: it was an inspiration and model for Maoist movements in Nepal and elsewhere.

8 Photo may be found in Aryal, "Open Wounds." Best treatment of the killing itself is Luitel, "How Maoists Killed Muktinath Adhikari."

9 Available at: https://en.wikipedia.org/wiki/Rukum_District#/media/File:NepalRukumDistrictmap.png (accessed December 31, 2017).

10 See Krishna Prasad Gautam, "Endless Pain for War Victim," *Kathmandu Post*, March 18, 2012; available at: http://kathmandupost.ekantipur.com/printedition/news/2012-03-17/endless-pain-for-war-victim.html (accessed December 31, 2017).

11 Devendra Basnet and Ganesh B.K., "A Conflict-era 'Bourgeois' Teacher Victim," *Republica*, August 15, 2016, 4; available at: www.myrepublica.com/news/3842 (accessed December 31, 2017).

12 Fieldwork photo during interview, 2002.

13 Fieldwork photos obtained from confidential source, 2003.

14 At August 2006, 68 percent remained displaced. For figure, see Donini and Sharma, *Humanitarian Agenda 2015: Nepal Country Case Study* (Medford, MA: Feinstein Intl Center, August 2008), 17; available at: www.research.ed.ac.uk/portal/files/12595908/Humanitarian_Agenda_2015_Nepal_Country_Case_Study.pdf (accessed January 20, 2018). In the new local governance structure, VDCs, which never recovered from their destruction, 1996–2006, are folded into townships.

15 This is a methodology used on both sides of irregular conflicts. For U.S. forces in Iraq and Afghanistan engaged in counterinsurgency operations, the imperative for information concerning the population was met by teams of individuals (normally about half a dozen) generated by the Human Terrain System (HTS) instructional program located at Ft. Leavenworth, Kansas and deployed to each deployed combat brigade. Its personnel were primarily young academics, often anthropologists, hired on contract to construct societal mapping not unlike that conducted in-house by intelligence personnel in the Vietnam War era. For this very reason, it was highly

controversial, the argument being that the U.S. government was compromising academic integrity by involving academic personnel in generating detailed understanding of relevant localities (often down to the smallest level). The Maoists would have found such a debate, at best, irrelevant. If the U.S. purpose was to protect the population, the Maoist design was targeted violence to enable seizing and controlling the population.

16 Among the more important Nepali Maoist national-level fronts were (all still exist): All-Nepal National Independent Students Union (Revolutionary) or ANNISU-R, Nepal Trade Union Federation (Revolutionary), the All- Nepal Women's Association (Revolutionary), the All-Nepal Janajati Federation, the All Nepal Peasants Association (Revolutionary), the All-Nepal Teacher's Organization (Revolutionary), the Nepal National Intellectuals Organization, and the All-Nepal People's Cultural Union. Additionally, at least a dozen important ethnic and regional fronts were formed, to include the Magarat National Liberation Front, the Tamuwan National Liberation Front, the Tharuwan National Liberation Front, the Tamang National Liberation Front, the Thami National Liberation Front, the Majhi National Liberation Front, the Madhesi National Liberation Front, the Newa Khala, and the Nepal Dalit Liberation Front. Such front activity is integral to all communist efforts but assumed special salience in the Maoist people's war approach to seizing power. The subject is extensively treated in my *Counterrevolution in China: Wang Sheng and the Kuomintang, passim.*

17 See again the discussion in Pasang, *op.cit*, 90–99.

18 For the Maoist assessment of the 2001 attack, see ibid., 112–123. In addition to establishing the vulnerability of the military and bringing it into the conflict, substantial quantities of advanced weaponry and munitions were captured (requiring half a dozen trucks to remove), together with NPR 50 million in cash (US$660,000 at the time).

19 The first attack upon a district capital was September 24, 2000 against Dunai, Dolpa district (see Figure 6.1); for attack details, see ibid., 76–89. A force of 150 regular police was based there, reinforced by some 60 field force police – the strength of a reinforced company. It was attacked by a battalion-size PLA force of 416 combatants and 150 support personnel. The government apparatus was destroyed and the police mauled, with some 15 dead and 48 wounded; a large number of prisoners were eventually released. Considerable weapons and equipment were captured, as well as at least NPR 50 million (US$660,000). A nearby army company remained out of the fight, the military having yet to be legally committed in support to civil authority. Though the town (with a population of 2,136 in a district of only 29,545) was held, the action established just how vulnerable even towns were to massed Maoist combat power.

20 The Maoists claimed that provocation caused the collapse of the negotiations. Yet their attacks in Kathmandu against government and security forces personalities were launched within hours of the calling off of the ceasefire, indicating the break in hostilities and the relatively free movement afforded the Maoists had been utilized to prepare for assassinations and bombings. Among the targets were two RNA colonels holding important operational responsibilities (one was killed, the other crippled with bullets to the spine). The Maoists subsequently announced that they had 217 names on a "hit list." See Campbell Spencer, "Rebels Target Kathmandu," *New Zealand Herald*, September 5, 2003; available at: www.nzherald.co.nz/world/news/article.cfm?c_id=2&objectid=3522024 (accessed January 20, 2018).

21 In my fieldwork in Ghartigaun, Rolpa in early 2003, the local postal official had been killed on such a charge. His family spoke to me (through interpretation) only after it had posted lookouts to warn of Maoist approach. It was because of such activity that my colleagues (both Nepali journalists) and I were eventually threatened with execution if we did not depart, which we did quickly. At the time, I estimated that perhaps

20 percent of the major homes in the area where we billeted were padlocked, since their owners had fled.

22 The "pathological" characterization is my own as I can think of no other way to describe the pursuit of targeted individuals across time and space –that is a hallmark of this conflict (whether during the 1996–2006 period or 2006–present). An extensive effort to find *any* commentary that would allow relating this behavior to observed characteristics of Nepali society has proved unsuccessful. A much-experienced, linguistically fluent (in Nepali) foreign observer has in the past publicly observed that in his experience Nepalis "never" forgive a perceived slight or insult which has passed a certain threshold. This opinion provoked considerable negative reaction from Nepalis, but in the absence of any documentation suggesting a tradition of revenge, one is forced to give it consideration.

23 In my fieldwork during July–September 2016, I found that of my 45 in-depth interviews with former female Maoist combatants of officer rank, 17 had been recruited during this period when 15 or under, 24 had been recruited when 16–18. This computes to 38 percent as unquestionably "child soldiers" when first entering Maoist ranks, a full 91 percent of the sample being "very young people."

24 The subject is treated to an extent in Human Rights Watch, *Children in the Ranks: The Maoists' Use of Child Soldiers in Nepal*. First mention of the "one family, one child" policy is on p. 5. During the conflict, it was impossible to conduct fieldwork in the hills without encountering heart-rending cases of young people, forcibly recruited, who, in unobserved moments, would ask for assistance through my guides/translators.

25 U.S. presence in Nepal historically has been focused upon development. During the conflict, however, the U.S. was involved in efforts, carried out with (in particular) Britain, to improve the quality of Nepal's security forces. U.S. training included specific courses, both in Nepal and abroad. Neither the U.S. nor Britain was involved in direct operational or intelligence operations with Nepali security forces but considered them friendly; sent training personnel to Nepal to work with individuals and units; provided weapons and equipment; and assumed a strong role in educating Nepali personnel to better perform locally in facing the challenges posed by terrorism and insurgency. A counterargument that British intelligence to some extent participated in the conflict may be found in Thomas Bell, *Kathmandu* (London: Random House, 2014), 356–383.

26 See especially Ben Kiernan, *The Pol Pot Regime: Race, Power, and Genocide in Cambodia Under the Khmer Rouge, 1975–1979*, 3rd ed. (New Haven, CT: Yale University Press, 2008).

27 While the Nepali state – in its parliamentary, market economy form – was identified throughout the 1996–2006 *overt* struggle as the immediate enemy, the Maoists regularly claimed that Indian imperialism was the most pressing threat to the Nepali nation and that the United States and capitalism were the ultimate enemies of mankind. The Maoists were placed on the State Department's Terrorist Exclusion List (TEL), which enumerates terrorist organizations for immigration purposes, and were one of the groups designated in Executive Order 13224, "Blocking and Prohibiting Transactions With Persons Who Commit, Threaten to Commit, or Support Terrorism." Among the victims of the Maoists were Nepalis employed in U.S. Embassy security.

28 I was able to conduct such interviews and to do fieldwork, because one of the peculiarities of the *overt* conflict was that at no time did the Maoists – except by happenstance – attack foreigners. Now, in the *covert* phase, with radical Maoist splinter groups active in certain areas, the precise situation for outsiders remains unclear. In my recent fieldwork, though, it was not I who was threatened (when the Maoists felt a need to move in that direction) but my Nepali researchers/translators (in Nepali, as later related by them to me in English).

29 Comparison of the Peruvian and Nepal cases, as assessed in the midst of the *overt* violence period, may be found in Marks and Palmer, "Radical Maoist Insurgents and Terrorist Tactics: Comparing Peru and Nepal." I have conducted fieldwork in Peru and published and spoken extensively on the Peruvian Maoists.

30 For additional information on the approach, see my "Mao Tse-tung and the Search for 21st Century Counterinsurgency," *CTC Sentinel* 2, no. 10 (October 2009), 17–20; available at: www.ctc.usma.edu/posts/mao-tse-tung-and-the-search-for-21st-century-counterinsurgency (accessed January 20, 2018).

31 Though nearly destroyed by government action in the 1980s and early 1990s, the group remains active and supports the Nepali Maoists (and vice versa). I have conducted extensive fieldwork in the Philippines and published and spoken often on the Philippine Maoists.

32 Discussed in depth in my *Maoist Insurgency Since Vietnam*, Chapter 3, "The Role of Terror: The Case of the Philippines," 151–173.

33 Shining Path's example was central in the debate on the timing for launching the people's war and the strategy to be followed, as related in an August 22, 2016 interview (with author) by Shobhakar Acharaya, a lifelong Party member and editor of the Maoist weekly newspaper, *Jama Uphar*, in 1999–2000. The Nepali Maoists continue to proclaim solidarity with Shining Path and would, under proper circumstances, not hesitate to reprise their December 2010 sponsoring of the son of imprisoned Shining Path leader, Abimael Guzman, as the guest of honor at the Party's youth organization's 18th National Convention (held in Kathmandu).

34 Each insurgent group has its own unique funding profile driven by the context at hand. The Revolutionary Armed Forces of Colombia (FARC), for instance, while initially emphasizing the same funding activities as the Nepali Maoists, was increasingly able to exploit the lucrative opportunities that came with participation in the drug trade. Nepal being the impoverished country it is, there were no such windfall sources available. The Nepali Maoists thus ultimately relied overwhelmingly upon extortion. In the mechanics of such extortion, of course, there is little to differentiate groups.

35 In many older sources, a figure of 13,000 is used, since this was the accepted tally at the time overt conflict ended. Continuing efforts to arrive at a more accurate assessment have caused this to be raised to the 17,000+ total used here, with some counts approaching 18,000.

36 To this day, the Maoists deny that such attacks occurred and claim the statistics, which were documented by victims' groups, have been concocted. Efforts to bring assailants to justice have been met with resistance, to include declared "general strikes" enforced with violence. To my knowledge, no successful prosecutions have occurred to date except in several cases on collateral charges. The two bodies specifically formed to push forward individual justice – the Truth and Reconciliation Commission (TRC) and the Commission for Investigation of Enforced Disappeared Persons (CIEDP) – have been effectively thwarted from tabling a single case for resolution.

37 See Paragraph 5 of "12-Point Understanding Reached Between the Seven Political Parties and Nepal Communist Party (Maoists)," available at: www.satp.org/satporgtp/countries/nepal/document/papers/12_Point_Understanding.pdf (accessed January 20, 2018). It may be noted that the agreement was signed in India, where the top Maoist leadership had resided throughout most of the conflict. Kathmandu, which was monitoring their interaction with personalities and units in Nepal, requested their apprehension (to include providing locations) but was ignored.

38 See Paragraph 7 of "The Code of Conduct for Ceasefire Agreed Between the Government of Nepal and the CPN (Maoist) on 25 May 2006 at Gokarna," available at: http://peacemaker.un.org/sites/peacemaker.un.org/files/NP_060525_25%20Point%20Ceasefire%20Code%20of%20Conduct.pdf (accessed January 20, 2018).

39 Document examined during March 2017 fieldwork in Nepal. Particularly useful explanation for the logic behind the change may be found in a detailed interview given previously (mid-2009) to World People's Resistance Movement (Britain) by the Maoists' then-chief ideologue and "Number Two," Baburam Bhattarai, wherein he stated directly that all accommodation was tactical for the strategic pursuit of power. See Bhattarai, *op.cit*, in particular:

> People know only the negative part, but what they forget, or what we have been unable to propagate well since the beginning of the PPW [Protracted People's War], is the new context of world imperialism and the specific geopolitical context of Nepal. In this context, our party decided that we need to adopt some of the features of general insurrection within the strategy of PPW. Therefore the basic strategy will be PPW, but some of the features of general insurrection, which relies on people's movement in the urban areas and leads to the final insurrection in the city, the tactics of the general insurrection, should also be incorporated within that strategy.

This same position was repeated by the Party leadership in November 2010 at the Maoist Sixth Plenum (i.e., party meeting) – well after "peace" agreements had been signed – and in the bitter Central Committee/Politburo debates of January 2012. In the latter, intra-party meetings, agreement was reached to use "revolt" (i.e., urban insurrection) if the stubborn rival parties did not give in to Maoist demands concerning the shape of the "new order." It is alleged betrayal by the mainstream faction of this planned strategy that fueled the intra-party split to be discussed in the text that follows.

40 With the attack upon the army in November 2001 at Ghorahi (Dang district), the Maoists entered the phase of the people's war approach wherein they sought to neutralize the military and sweep into power. Hence they added "main force warfare" of the People's Liberation Army (PLA) to their terrorism and guerrilla warfare. That is, they fielded regular units designed to match those of the government in size, training, and weaponry. It was these units that executed major military operations. Their strength at November 2006 was best estimated at 7,500 personnel, but 22,000–23,000 individuals showed up at the regroupment camps. Verification by observers established the reality of composition but was ignored by the UN Mission (UNMIN). Peace commitments called for PLA combatants to be integrated into existing Nepali security forces or demobilized. The Maoists, determined to take over the army from within, demanded that their combatant units be absorbed intact, and that key positions in the chain of command be given to them (with the topmost positions occupied by officers whom they had either approved or appointed). Contingent events and the intervention of Indian intelligence prevented this from happening. Acting on orders from the Maoist chain of command, most combatants refused integration, which was finally completed in August 2013. Only 1,460 of the entire group opted for incorporation (19,602 had been deemed "eligible" for this action by UNMIN). A larger number have migrated to the breakaway factions (discussed above in the text), which have moved to reconstitute the PLA, armed militia, and paramilitary organizations.

41 From my "The Anatomy of a Maoist Extortion – 'Time to Think of Protecting Ourselves'." Publication included security camera photos.

42 See e.g., "Teacher Kidnapped From Amchok," *Sandakpur Daily*, January 11, 2012 (read in translation). The episode – a January 2012 attack and kidnapping of a teacher in Ilam in eastern Nepal, witnessed by an entire school and reported on the front page of the Nepali language local paper – still apparently failed to be reflected in the main Kathmandu dailies (which appear in English), hence in the relevant compilations of incidents.

43 This could account in large part for the seeming contradiction between official pronouncements, such as the annual statements from the U.S. State Department with

respect to human rights, and media and academic reporting, particularly that from Nepali journalists. Different levels of analysis are in play. At the *macro* (e.g., national) level, the situation has improved in hopeful ways; at the *micro* (e.g., local) level, there has continued considerable organized, directed political violence (i.e., terrorism), exacerbated by opportunistic criminality that has itself been given impetus by continued revolutionary taxation by the Maoists to raise necessary operating funds.

44 No study exists that explores the full range of criminal activities with which Shakti Khor became connected in order to continue the cash-flow necessary for Maoist operations. Evidence strongly indicates all Maoist regroupment camps engaged in such activity. Cutting closer to the bone, portions of the various allowances and support funds intended for the Maoist combatants in the regroupment camps were siphoned off to the Party by the leadership. Nepali media have placed this figure alone at NPR 3.08 billion or US$30,800,000 (at an average 2013 exchange rate of US$1 = NPR 100). For detailed figures used to construct total used here, see e.g., Manandhar, "Impeachment vs Appointment." (It is useful to note that even if the reported figure is off by a factor of 10, the amount embezzled is still substantial: in excess of US$3 million.)

45 In all fairness, it must be stated that the police suffered badly in the conflict. In fact, at key points during the *overt* struggle, they comprised a majority of all dead on the government side. They thus in the period under discussion had every reason to steer clear of further entanglements with their recent foes.

46 Tape with inserted English subtitles available at: www.youtube.com/watch? v=6EoQYZ2oa6M (accessed January 20, 2018). Translated transcript is available at: http://bengalunderattack.blogspot.com/2009/05/maoist-prachanda-red-over-controversial. html (accessed January 20, 2018).

47 Ibid. (transcript).

48 For the original October 2008 *Nepali Sarokar* column, translated and reprinted after her death, see Uma Singh, "A Minister Out of Control"; for assessment, "Press Post Mortem," *Nepali Times*, March 20, 2009; available at: http://nepalitimes.com/news. php?id=15776#.WXYgkaoUkzs (accessed January 20, 2018); for discussion of the Maoist role, especially as it related to seizure of land and its redistribution for political purposes, see "IFJ Demands Justice for Uma Singh and an End to Impunity in Nepal," International Federation of Journalists (IFJ).

49 Though more than a dozen assailants were apparently involved, just two convictions were gained in a 2011 trial and another conviction in a 2015 trial. In both trials, the political nature of the crime was sidestepped, and those ultimately responsible in the Maoist chain of command were not called to account. Conviction, albeit on criminal charges, was unusual enough; but the results speak to the impunity that continues to be the norm in such cases.

50 Most recently, he has rejoined the mainstream Maoist movement and again is in a leadership position.

51 This situation continues.

52 Unlike Western political parties, those adhering to Leninist precepts exercise complete party discipline. Policy is determined by the party rather than by the individual holding a government position – he merely executes the party position and may be replaced at the will of the party for violating its directions. Thus even as prime minister, Dr. Bhattarai neither ruled as such nor claimed to do so. He held his position as a result of party approval and referred all major issues to the party central leadership for debate and consensus on the position he was to take. Similarly, candidates and party members may not engage in revolts from within, such as dominated the recent U.S. presidential election. Under democratic centralism, the parties have the ultimate say in who stands for election and who actually occupies a seat won by a party. To this it may be added that in Nepal representatives do not have to live in the constituency represented.

53 For conviction see "Journalist Dekendra Thapa's Murderers Convicted," *Setopati*, December 7, 2014; available at: http://setopati.net/politics/4482/ (accessed January 20, 2018).
54 As a definitional matter: the overturning of the economic, social, and political systems of social stratification. Revolution is thus an end-state, while the process of "making a revolution" is engaging in revolutionary war. Maoists use the word, though, as both ends and ways; e.g., "I was in the revolution" (which would be the revolutionary war) but "We have not yet achieved the revolution" (which would be a complete overturning of what was in favor of what is to be).
55 Evidence illuminating these positions regularly becomes available due to the constant infighting within Maoist ranks and the profound disillusionment of many cadre, to include a number who once held important positions. Through interviews and examination of translated documents, a record of the Party's strategic thinking has emerged. It can only be labeled "Machiavellian."
56 See: Baidhya, http://revolutionaryfrontlines.wordpress.com/2012/01/03/nepal-revolutionary-maoists-document-to-the-central-committee-ucpnm/#more-21046 (accessed January 20, 2018). This challenge posed a serious threat to the party, because the "radical" faction, as per one media count, controlled 48 of the 148 Central Committee votes, or approximately one-third (the Central Committee being the key validating body for all Party decisions). Of the 16 Standing Committee members (i.e., the Party Politburo, which actually provides continuous leadership), five were members of the dissident faction; of the seven office bearers (i.e., the top Party officials), three were dissidents. Of 236 Maoist lawmakers in the dual legislative/constitutional drafting body, 80 of 236 were from the dissident faction.
57 Ibid.
58 See Gautam et al., "Unarmed Citizens."
59 Kathmandu: Himal Books, 2011.
60 A compilation of *bandas* (also rendered as *bandhs*) for all of 2013 (the relevant point in our discussion) is available at: www.un.org.np/sites/default/files/BandhBlockades_Jan-Dec-2013_A4_6Jan2014_v01.pdf (accessed January 20, 2018).
61 See Office of Foreign Assets Control of the Department of the Treasury "Specially Designated Nationals and Blocked Persons" compendium published by the Office of Foreign Assets Control of the Department of the Treasury; unlike the February 23, 2012 edition, the March 11, 2013 edition omits the Communist Party of Nepal (Maoist). The current list (dated June 21, 2017) is available at: www.treasury.gov/resource-center/sanctions/SDN-List/Pages/default.aspx (accessed January 20, 2018). The assessment as to motive is my own gained through discussions with individuals involved in the decision-making of the process.
62 New Delhi's historically fraught relationship with Kathmandu was driven throughout by calculations of how best to achieve its own security through a stable Nepal, responsive to India's needs. This had led to its playing a very mixed role during the insurgency, backing at various times either or both sides. Having seen its hopes of a soft post-conflict landing dashed by Maoist intransigence, accompanied as it was by intensely nationalistic, anti-Indian rhetoric, India made the decision that it needed to support the traditional parliamentary political parties.
63 As is the case with all Marxist-Leninist organizations, the Nepali Maoists (of all factions) see elections conducted in normal, "one-man/one-vote" fashion as undemocratic and as designed only to perpetuate the domination of the status quo. Though they are careful to conceal their demands for a dictatorship of the proletariat, all Nepali Maoist parties have claimed that parliamentary democracy, the political form of government in Nepal since 1990, is illegitimate and not representative of popular will (i.e., as represented most prominently, they further state, by the Maoists). Their opposition extended to the Constituent Assembly election(s).

64 146 were "live," while 294 proved inert. Figures provided through correspondence with Nepal Army, December 9, 2013.
65 Police figure obtained December 25, 2013.
66 See Citizen's [*sic*] Campaign for Clean Election, *Monitoring Report: Constituent Assembly Election 2013* (Kathmandu: INSEC, May 2014); available at: www.insec. org.np/pics/publication/1401772233.pdf (accessed October 27, 2017, retained in hard copy). Significantly, such reports appear to have been removed from the INSEC site as part of a general cleansing of the record (unsuccessful access was last attempted on February 15, 2018, with earlier attempts producing the same results).
67 Figures from Ibid., esp. Annexes 2 and 5.
68 For translation, see Admin, November 18, 2013, "Nepal Elections: The Kirtipur Audiotape!"; available at: www.kanakmanidixit.com/?p=903 (accessed January 20, 2018).
69 Subedi, "Maoist Victim Banks on Maoists for Polls."
70 Figures from *Monitoring Report: Constituent Assembly Election 2013*, 91–92; assessment as to causation is my own. The *Monitoring Report* does note that 77 ballots were cast for proportional representation, but other information demonstrates that these were votes of the officials sent to conduct the balloting.
71 One of the first actions of the new order was to abolish the monarchy and the official status of Hinduism in what for centuries has been the world's only Hindu kingdom. Destroying the religious and organizational basis for society further disoriented the population, with the Maoists (while in control of the government) going to the extent of trying to name particularly important religious figures, thus to neutralize any possible resistance. This is the same methodology utilized by China in its well-publicized efforts to destroy the viability of Tibetan Buddhism as an independent cultural force; see e.g., Powers, *The Buddha Party: How the People's Republic of China Works to Define and Control Tibetan Buddhism*.
72 In my fieldwork interviews, July–September 2016, I found the Chand strategy expressed in various ways by semi-underground members, especially as "build people's revolt on the foundation of people's war." Unwittingly, then, the Chand group is reinventing the wheel, for it is groping towards what the Vietnamese communists (who were not Maoists but used the same strategic approach) called the "war of overlapping" or "all forms of struggle." That is, Chand seeks to move beyond what in 1996–2006 was an overwhelmingly rural-based, guerrilla mode of warfare – "people's war" – to a mix of people's war with urban action ("people's revolt") to create "unified rebellion." In one sense, this is a misunderstanding of the people's war strategy that the unified Maoist movement actually waged, because urban action was an important part of that effort – thus "unified rebellion." In another sense, his efforts at modification do correctly grasp that previous Maoist urban action was always tactical (e.g., assassinations) and did not support mass organizing in the cities that would lead to "people's revolt." As concerns potential targets (i.e., victims), the distinction hardly matters; but in the longer-term, "people's revolt" would be far more violent than the strategy being followed now by the mainstream. It was the goal of the Vietnamese communists in their now-legendary "Tet of '68" offensive (January–February 1968) in South Vietnam. For an understanding of the Vietnamese strategy the Chand group is stumbling toward, see O'Dowd, " 'What Kind of War is This?' "; for my discussion of the strategic issues of what Mr. Chand is calling "people's revolt," especially as they have impacted people's war groups in recent history, see "Urban Insurgency."
73 See e.g., Singh, "Chand-led Maoist Announce Formation of People's Govt in Bajhung"; reprint with better display of accompanying visuals is at: http://dazi-baorojo08.blogspot.com/2016/01/nepal-chand-led-maoist-announce.html (accessed January 3, 2018). At the inauguration of the Chand party district committee in Bajhang, which is the subject of these sources, YCL/combatants performed martial

skits using dummy weapons but accompanied by martial agitprop promising to oust class enemies and carry the revolution through to success. Violent agitprop is integral to all Chand faction gatherings.

74 Bhattarai's New Force Party, *Naya Shakti Nepal*, was formally launched on June 12, 2016.

75 For excellent summary of issues see Lal, "Rotary of Revolutions."

76 For India's geopolitical position, refer to previous n.62. When UML also appeared to be embracing Nepali nationalism, India worked to undermine its administration by the unofficial blockade (the third such instance of that tactic by independent India).

77 Transitional justice as both a formal category and a human rights challenge refers to the measures to be taken in the post-conflict present to address abuses committed during the conflict past. A constant tension is present between reconciliation and punishment, particularly where crimes involve actions of no, little, or indeterminate military utility (e.g., in the case of the first of this list, rape). As related in detail in the preceding text, the Maoist strategic approach necessarily involved the instrumental use of terrorism against civilians and officials who stood in the way of insurgent popular mobilization, with the actual choice of targets and techniques largely decentralized. This resulted in serious abuses. For their part, though not acting pursuant to policy as was the case with individual insurgents, government security forces also committed widespread legal and human rights abuses. Useful works discussing the issues with specific reference to Nepal are: Sajjad, *Transitional Justice in South Asia: A Study of Afghanistan and Nepal*; and Nepal, "Nepal's Botched Truth and Reconciliation Program."

78 For a poignant but typical example, see Chaudhary, "But Not Living Either."

79 A comprehensive discussion of legal issues and incidents is United Nations Office of the High Commissioner for Human Rights, *Nepal Conflict Report 2012* (Geneva: UNOHCHR, October 2012); available at: www.ohchr.org/Documents/Countries/NP/ OHCHR_Nepal_Conflict_Report2012.pdf (accessed January 20, 2018).

80 Discussion at "Maoists Want SC Verdict Corrected," ICTJ, (April 1, 2015) available at: www.ictj.org/news/maoists-want-sc-verdict-corrected (accessed January 20, 2018).

81 Best evidence supports a position that a healthy proportion of the Maoist weapons in use when the November 2006 peace accords were signed were *not* turned in to the United Nations inspectors. Though approximately 20,000 combatants were accepted for entry into the regroupment camps (see n.40), only some 3,500 weapons were turned in. In particular, many of the most lethal weapons vanished. Though weapons turned in were inventoried, there has to date not been a comparison of those stocks with weapons lost by the security forces during the conflict.

82 See photos in *The Himalayan Times* at: www.google.com/search?q=nepal+bombs+2 013&tbm=isch&source=iu&pf=m&ictx=1&fir=SRDPUJQkgaGSxM%253A%252C q6JRgBYdq9hG5M%252C_&usg=__Ip6AZ_O2Mqr8st_MHCMJC6ZQW7k%3D& sa=X&ved=0ahUKEwjs97-xv5HXAhUr5oMKHXM5D4cQ9QEIPjAE#imgrc=Hqh 7ppvNYzCc8M: (accessed January 20, 2018). Details (with further photos) at "Chand Maoists Torch 10 Ncell Base Towers,"; "Bombs Explode at Two Schools; Scare in Five Others." Media identification of the perpetrators was confirmed by a well-informed former director of military intelligence in a personal communication, October 14, 2016.

83 Media cited government sources in identifying the detained individuals as belonging to the Chand group. They were in possession of 205 5.56 mm rounds and nine loaded magazines for the Indian-made INSAS 5.56 mm semiautomatic rifle, a weapon which had been used by the government during the period of overt hostilities. A number that had been captured do not appear to have been turned in during demobilization. See "Chand Maoist Cadres Held With Huge Cache of Ammunition," *The Kathmandu Post*, December 7, 2016; available at: http://kathmandupost.ekantipur.

com/news/2016-12-07/chand-maoist-cadres-held-with-huge-cache-of-ammunition.
html (accessed January 20, 2018).
84 See "Chand Maoists' First Conclave From 'Feb. 12–21."
85 Nalsing Gad is the site of a massive Asian Development Bank (ADB)-funded (approximately US$10 million) hydroelectric project which is reported to be a source of illicit funds for both mainstream and radical/ultra-radical Maoist groups.
86 See "Chand-led Maoist Cadres Arrested With Explosives."
87 "Biplav Maoist cadres detained with explosives from Rautahat."
88 For cursory details, "Suspected 'Bomb' Defused in Lazimpat," *The Himalayan Times*, September 26, 2017; available at: https://thehimalayantimes.com/kathmandu/suspected-bomb-defused-lazimpat/ (accessed January 20, 2018); for a Twitter photo, www.google.com/search?q=bomb+in+shangri+la+hotel+kathmandu&tbm=isch&source=iu&pf=m&ictx=1&fir=aOqvXXEJcUvG9M%253A%252CIFV20JMR1IIQFM%252C_&usg=__NFE2JmmjHct0jKP2fxmB7NKn4Iw%3D&sa=X&ved=0ahUKEwiotaHX65DXAhUFYiYKHfAdAoAQ9QEINDAC#imgrc=aOqvXXEJcUvG9M: (accessed January 20, 2018). Perpetrators were subsequently identified as Chand Maoists.
89 Seized explosives have been overwhelmingly from mining stocks manufactured in India.
90 Interview, September 4, 2016, Kathmandu. At the time, the member was semi-underground, and the meeting was arranged by an intermediary.
91 The incident, with particulars identifying the Maoists as the perpetrators, is discussed in depth, together with consideration of the background, at "Abduction by Maoist [*sic*] Before Local Election."
92 This is because the Maoists locked up their FPTP seats in virtually impregnable areas electorally, such as Rukum (where they took all seats), the opposition having been eliminated through violence and menace in the previous two decades.
93 See e.g., "Bomb Found at NC Candidate's House."
94 See e.g., Khatiwada, "NC Candidate Among Seven Hurt in Udaypur IED Blast,"; Khadka, "Lost Her Leg, but Not her Zeal."
95 The single attempt of which I am aware, by *The Record* (Kathmandu), has a well-done interactive feature that attempts to detail the most significant IED attacks nationwide. The display, with photos, is but a fraction of the actual IED total; available at: www.recordnepal.com/wire/bomb-blasts-leading-up-to-the-polls/ (accessed January 20, 2018).
96 Video available at: www.youtube.com/watch?v=a9VBn3661EE (accessed January 20, 2018).
97 In Nepali as translated; obtained during April 2017 fieldwork in Nepal.
98 Rather than a direct quote, the reporting used "has said." There is little doubt, though, that the words were uttered. See "Fundamentals of Marxism Cannot be Ignored: Dahal."
99 Guna Raj Luitel, "Nepalese Journalist Defiant After Razor Slashing."
100 Ibid.

Bibliography

"12-Point Understanding Reached Between the Seven Political Parties and Nepal Communist Party (Maoists)"; available at: www.satp.org/satporgtp/countries/nepal/document/papers/12_Point_Understanding.pdf (accessed January 20, 2018).
"Abduction by Maoist [*sic*] Before Local Election," *Goraksha National Daily*, (April 29, 2017) (read in translation); available in Nepali at: http://gorakshadaily.com/ (accessed October 27, 2017, retained in hard copy; domain name expired December 10, 2017).

Admin, November 18, 2013, "Nepal Elections: The Kirtipur Audiotape!"; available at: www.kanakmanidixit.com/?p=903 (accessed January 20, 2018).

Aryal, Mallika. "Open Wounds," *Nepali Times*, (February 5, 2010); available at: http://nepalitimes.com/news.php?id=16781#.VNtgLT8cRZs (accessed January 20, 2018).

Baidhya. Central Committee document, (December 27, 2011) document; available at: http://revolutionaryfrontlines.wordpress.com/2012/01/03/nepal-revolutionary-maoists-document-to-the-central-committee-ucpnm/#more-21046 (accessed January 20, 2018).

Bandas (also *bandhs*) Compilation 2013; available at; www.un.org.np/sites/default/files/BandhBlockades_Jan-Dec-2013_A4_6Jan2014_v01.pdf (accessed January 20, 2018).

Baral, Biswas and Mahabir Paudyal. "Weekly Interview [Pradeep Gyawali]: Maoist Leaders Blackmailed Over the Alleged War Crimes." *Republica*, (July 27, 2016), 7, available at www.myrepublica.com/news/2767 (accessed August 31, 2016).

Basnet, Devendra and B. K. Ganesh "A Conflict-era 'Bourgeois' Teacher Victim." *Republica*, (August 15, 2016), 4; available at: www.myrepublica.com/news/3842 (accessed August 31, 2016).

Bell, Thomas. *Kathmandu*. London: Random House, 2014.

Billingsley, Krista. "Conflict Over Transitional Justice in Nepal." *Anthropology News*, (July 27, 2016); available at: www.anthropology-news.org/index.php/2016/07/27/conflict-over-transitional-justice-in-nepal/ (accessed August 31, 2016).

"Biplav Maoist Cadres Detained with Explosives from Rautahat," *Khabar* (English ed.), (June 8, 2017); available at: http://english.onlinekhabar.com/2017/06/08/401731 (accessed January 20, 2018).

"Bomb Found at NC Candidate's House," *The Kathmandu Post*, (November 30, 2017, 11:49am) available at http://kathmandupost.ekantipur.com/news/2017-11-30/bomb-found-at-nc-candidates-house.html (accessed January 20, 2018).

Briscoe, Ivan and Pamela Kalkman. "Illicit Networks: Rethinking the Systemic Risk in Latin America." *Prism* 5 (4) (undated but 2015–16) 153 (in 151–171); available at: www.clingendael.nl/sites/default/files/Illicit%20Networks%20the%20systemic%20risk%20in%20Latin%20America.pdf (accessed March 3, 2016).

"Chand-led Maoist Cadres Arrested With Explosives," *The Himalayan Times*, (May 10, 2017); available at: https://thehimalayantimes.com/nepal/netra-bikram-chand-led-cpn-maoist-cadres-arrested-explosives/ (accessed January 20, 2018).

"Chand Maoists' First Conclave From 'Feb. 12–21," *The Kathmandu Post*, (February 9, 2017); available at: http://kathmandupost.ekantipur.com/news/2017-02-09/chand-maoists-first-conclave-from-feb-12–21.html (accessed January 20, 2018).

Chaudhary, Gaya Prasad. "But Not Living Either," *The Record*, (August 25, 2015) video op-ed; available at: www.recordnepal.com/perspective/but-not-living-either/ (accessed January 20, 2018).

Della Porta, Donatella. *Clandestine Political Violence*. NY: Cambridge University Press, 2013.

Devkota, Krishna Jawala, Jhalak Subedi and Umesh Chauhan. "*Yudda Hoina Haami ley Garney Kranti Ho*" ([Baidhya] It's Not War we Want but Revolution), *Naya Patrika*, (June 28, 2016); available at: www.enayapatrika.com/2016/06/71334 (accessed August 31, 2016); in Nepali.

Dixit, Kanak Mani. *Peace Politics of Nepal*. Kathmandu: Himal Books, 2011.

Dixit, Kanak Mani. "Revolutionary Comrade." *The Kathmandu Post*, (December 21, 2012); available at: http://kathmandupost.ekantipur.com/news/2012-12-21/revolutionary-comrade.html (accessed August 12, 2016).

Donini, Antonio and Jeevan Raj Sharma. *Humanitarian Agenda 2015: Nepal Country Case Study*. Medford, MA: Feinstein International Center, (August 2008); available at: www.research.ed.ac.uk/portal/files/12595908/Humanitarian_Agenda_2015_Nepal_Country_Case_Study.pdf (accessed January 20, 2018).

"Face to Face: 'Nepal Army Cannot Go Against the People'." *New Spotlight*, December 11, 2015, 16–18; available at: http://spotlightnepal.com/News/Article/Nepal-Army-can-not-go-against-people (accessed August 31, 2016).

Film: *Sicario*. Black Label, 2015; available through Amazon.

"Fundamentals of Marxism Cannot be Ignored: Dahal," *The Kathmandu Post*, (December 27, 2016); available at: http://kathmandupost.ekantipur.com/news/2016-12-27/fundamentals-of-marxism-cannot-be-ignored-dahal.html (accessed January 20, 2018).

Gautam, Krishna Prasad. "Endless Pain for War Victim." *Kathmandu Post*, (March 18, 2012) available at: www.ekantipur.com/2012/03/18/national/endless-pain-for-war-victim/350745.html (accessed January 20, 2018).

Gautam, Kul Chandra, Subodh Raj Pyakurel, Kanak Mani Dixit. "Unarmed Citizens." *Kathmandu Post*, (February 28, 2012); available at: www.kanakmanidixit.com/unarmed-citizens/ (accessed August 12, 2016).

Human Rights Watch. *Children in the Ranks: The Maoists' Use of Child Soldiers in Nepal*. NY: Human Rights Watch, (February 2007); available at: www.hrw.org/sites/default/files/reports/nepal0207webwcover.pdf (accessed January 20, 2018).

"IFJ Demands Justice for Uma Singh and an End to Impunity in Nepal," International Federation of Journalists (IFJ), (March 12, 2009); available at: www.ifj.org/nc/news-single-view/browse/217/backpid/51/article/ifj-demands-justice-for-uma-singh-and-an-end-to-impunity-in-nepal/ (accessed January 20, 2018).

Jha, Prashant. "India Confused by Nepal's Confusion." *Nepali Times*, (July 17, 2009); available at: http://nepalitimes.com/news.php?id=16130#.V619T4-cFjo (accessed August 12, 2016).

"Journalist Dekendra Thapa's Murderers Convicted." *Setopati*, (December 7, 2014); available at: http://setopati.net/politics/4482/ (accessed January 20, 2018).

Keon, Michael. *The Tiger in Summer*. NY: Harper, 1952.

Keon, Michael. *The Durian Tree*. NY: Simon and Schuster, 1960.

Khadka, Basanta. "Lost her leg, but not her Zeal," *Nagarik*, (December 9, 2017) translated and reprinted in *Nepali Times*, (December 15, 2017); available at: http://nepalitimes.com/article/from-nepali-press/Lost-her-leg,4088 (accessed January 20, 2018).

Khatiwada, Dilliram. "NC Candidate Among Seven Hurt in Udaypur IED Blast," *The Kathmandu Post*, (November 30, 2017); available at: http://kathmandupost.ekantipur.com/news/2017-11-30/nc-candidate-among-seven-hurt-in-udaypur-ied-blast.html (accessed January 20, 2018).

Kiernan, Ben. *The Pol Pot Regime: Race, Power, and Genocide in Cambodia Under the Khmer Rouge, 1975–1979*, 3rd ed. New Haven, CT: Yale University Press, 2008.

Lal, C. K. "Rotary of Revolutions," *Republica*, (January 16, 2017); available at: www.myrepublica.com/news/13133 (accessed January 20, 2018).

Lecomte-Tilouine, Marie. "Terror in a Maoist Model Village, Mid-Western Nepal," in Alpa Shah and Judith Pettigrew, eds. *Windows Into a Revolution: Ethnographies of Maoism in India and Nepal*. Delhi: Social Sciences Press, 2011, 207–232; available at: www.academia.edu/9759951/Terror_in_a_Maoist_model_village_mid-western_Nepal_._In_Windows_into_a_Revolution._Ethnographies_of_Maoism_in_India_and_Nepal_A._Shah_and_J._Pettigrew_eds._Social_Science_Press_Delhi_2011._P._207-232?auto=download (accessed August 31, 2016).

Luitel, Guna Raj. "Nepalese Journalist Defiant After Razor Slashing," *Committee to Protect Journalists Blog*, (January 12, 2010); available at: https://cpj.org/blog/2010/01/tika-bista-heard-the-word.php (accessed January 20, 2018).

Manandhar, Narayan. "Impeachment vs Appointment," *The Kathmandu Post*, (September 25, 2016); available at: http://kathmandupost.ekantipur.com/printedition/news/2016-09-25/impeachment-vs-appointment.html (accessed January 20, 2018).

"Maoists Want SC Verdict Corrected," ICTJ, (April 1, 2015); available at: www.ictj.org/news/maoists-want-sc-verdict-corrected (accessed January 20, 2018).

Marks, Thomas A. *Maoist Insurgency Since Vietnam*. London: Frank Cass, 1996.

Marks, Thomas A. Counterrevolution in China: Wang Sheng and the Kuomintang. London: Frank Cass, 1998.

Marks, Thomas A. "Urban Insurgency," *Small Wars and Insurgencies* 14 (3) (Autumn 2003), 100–157; available at: www.tandfonline.com/doi/pdf/10.1080/09592310410001676925?needAccess=true (accessed January 20, 2018).

Marks, Thomas A. "The Anatomy of a Maoist Extortion – 'Time to Think of Protecting Ourselves'," *Nepali Perspectives*, (September 19, 2006); available at: http://nepaliperspectives.blogspot.com/ (accessed January 20, 2018).

Marks, Thomas A. "Counterinsurgency in the Age of Globalism." *The Journal of Conflict Studies* 27 (1) (Summer 2007), 22–29; available at: https://journals.lib.unb.ca/index.php/JCS/article/view/5936 (accessed March 4, 2016).

Marks, Thomas A. *Maoist People's War in Post-Vietnam Asia*. Bangkok: White Lotus, 2007.

Marks, Thomas A. "Sri Lanka and the Liberation Tigers of Tamil Eelam," in Robert J. Art and Louise Richardson, eds. *Democracy ND Counterterrorism: Lessons From the Past*. Washington, DC: United States Institute of Peace Press, 2007, 483–530.

Marks, Thomas A. "The Maoists in Nepal: Strategies of Subversion and Subterfuge." *Faultlines* 19 (July 2008), 33–55; available at: www.satp.org/satporgtp/publication/faultlines/volume19/Article%202.pdf (accessed March 4, 2016).

Marks, Thomas A. "Mao Tse-tung and the Search for 21st Century Counterinsurgency," *CTC Sentinel* 2 (10) (October 2009), 17–20; available at: www.ctc.usma.edu/posts/mao-tse-tung-and-the-search-for-21st-century-counterinsurgency (accessed January 20, 2018).

Marks, Thomas A. "Nepal: Culture and New War." *The Journal of Counterterrorism* 17, no. 2 (Summer 2011), 34–37, 48–52.

Marks, Thomas A. "Recent Titles Dealing With Indian and Nepali Maoists." *Small Wars and Insurgencies* 25 (2) (2014), 487–491; available at: www.tandfonline.com/doi/abs/10.1080/09592318.2014.904036?journalCode=fswi20#.Vtyf_I-cHIU (accessed March 6, 2016).

Marks, Thomas A. "'Post-Conflict' Terrorism in Nepal," *The Journal of Counterterrorism* 21 (1) (Spring 2015), 24–31; available at: https://issuu.com/fusteros/docs/iacsp_magazine_v21n1_issuu (accessed August 13, 2016).

Marks, Thomas A. and David Scott Palmer. "Radical Maoist Insurgents and Terrorist Tactics: Comparing Peru and Nepal," *Low Intensity Conflict and Law Enforcement* 13 (2) (Autumn 2005), 91–116; available at: www.tandfonline.com/doi/full/10.1080/0966284050034728 0?scroll=top&needAccess=true (accessed January 20, 2018).

"Matrika Yadav's Communist Party of Nepal (Maoist) Splits," *The Workers Dreadnought*, (September 29, 2011); available at: https://theworkersdreadnought.wordpress.com/2011/09/29/matrika-yadavs-communist-party-of-nepal-maoist-splits/ (accessed August 12, 2016).

Miklaucic, Michael and Jacqueline Brewer, eds. *Convergence: Illicit Networks and National Security in the Age of Globalization*. Washington, DC: NDU Press, 2013 available at: http://ndupress.ndu.edu/Portals/68/Documents/Books/convergence.pdf (accessed March 3, 2016).

Mohanty, Manoranjan. *The Political Philosophy of Mao Zedong*, rev. ed. Delhi: Aakar, 2012.

Nepal, Indu. "Nepal's Botched Truth and Reconciliation Program," *The Record*, (July 29, 2016); available at: www.recordnepal.com/wire/nepals-botched-truth-and-reconciliation-program/ (accessed January 20, 2018).

"Nepal: Interview With Comrade Baburam Bhattarai," (December 12, 2009); available at: http://marxistleninist.wordpress.com/2009/12/12/baburam-bhattarai-on-nepals-social-revolution/#more-4155 (accessed March 4, 2016).

Nayak, Nihar. "The Madhesi Movement in Nepal: Implications for India." *Strategic Analysis* 35 (4) (2011), 640–660; available at: www.tandfonline.com/doi/pdf/10.1080/0970 0161.2011.576099?needAccess=true (accessed August 13, 2016).

O'Dowd, Edward C. "'What Kind of War is This?'," *The Journal of Strategic Studies* 37 (6–7) (2014), 1027–1049; available at: www.tandfonline.com/doi/pdf/10.1080/014023 90.2014.891982?needAccess=true (accessed January 20, 2018).

Office of Foreign Assets Control of the Department of the Treasury. "Specially Designated Nationals and Blocked Persons," link; available at: www.treasury.gov/resource-center/sanctions/SDN-List/Pages/default.aspx (accessed January 20, 2018).

Ogura, Kiyoko. "Realities and Images of Nepal's Maoists After the Attack on Beni." *European Bulletin of Himalayan Research* 27 (2004), 67–125; available at: http://himalaya.socanth.cam.ac.uk/collections/journals/ebhr/pdf/EBHR_27_04.pdf (accessed August 31, 2016).

Ogura, Kiyoko. *Seeking State Power: The Communist Party of Nepal (Maoist)*. Berlin: Berghof Research Center for Constructive Conflict Management, (2008); available at: www.berghof-foundation.org/fileadmin/redaktion/Publications/Papers/Transitions_Series/transitions_cpnm.pdf (accessed August 31, 2016).

Pettigrew, Judith. "Living Between the Maoists and the Army in Rural Nepal." *Himalaya* 13 (1) (2003), 9–20; available at: www.researchgate.net/publication/254616410_Living_Between_the_Maoists_and_the_Army_in_Rural_Nepal (accessed August 31, 2016).

Pike, Douglas. *Viet Cong: The Organization and Techniques of the National Liberation Front of South Vietnam*. Cambridge, MA: The M.I.T. Press, 1966.

Porter, Geoff D. "Terrorist Outbidding: The In Amenas Attack." *CTC Sentinel* 8, no. 5 (May 2015), 14–17; available at: www.ctc.usma.edu/v2/wp-content/uploads/2015/05/CTCSentinel-Vol.8Issue54.pdf (accessed March 4, 2016).

Powers, John. *The Buddha Party: How the People's Republic of China Works to Define and Control Tibetan Buddhism*. New York, NY: Oxford University Press, 2016.

Prachanda Tape with inserted English subtitles, available at: www.youtube.com/watch?v=6EoQYZ2oa6M (accessed August 11, 2016); translated transcript available at: http://bengalunderattack.blogspot.com/2009/05/maoist-prachanda-red-over-controversial.html (accessed August 11, 2016).

"Press Post Mortem, Uma Singh Was Killed for her Investigative Reporting: IFJ." *Nepali Times*, (March 20, 2009), 2; available at: http://nepalitimes.com/news.php?id=15776#.WXYgkaoUkzs (accessed August 13, 2016).

Pun, Nanda Kishor *aka* Pasang. *Red Strides of the History: Significant Military Raids of the People's War*, trans. Sushil Bhattarai. Kathmandu: Agnipariksha Janaprakashan Griha, October 2008.

Qiao Liang and Wang Xiangsui. *Unrestricted Warfare.* Beijing: PLA Literature and Arts Publishing House, February 1999; available at: www.oodaloop.com/documents/unrestricted.pdf (accessed March 4, 2016).

Robins, Simon, Ram Kumar Bhandari, and the ex-PLA Research Group. *Poverty, Stigma, and Alienation: Reintegration Challenges of ex-Maoist Combatants in Nepal.* York, UK: CAHR/The University of York, 2016; available at: www.berghof-foundation.org/fileadmin/redaktion/Publications/Grantees_Partners/Final_Report_Ex-PLA_Nepal_May_2016.pdf (accessed August 11, 2016).

Roy, Srila. *Remembering Revolution: Gender, Violence, and Subjectivity in India's Naxalbari Movement.* New Delhi: Oxford, 2012.

Sajjad, Tazreena. *Transitional Justice in South Asia: A Study of Afghanistan and Nepal.* New York, NY: Routledge, 2013.

Singh, Basanta Pratap. "Chand-led Maoist Announce Formation of People's Govt in Bajhung," *The Kathmandu Post,* (January 24, 2016); available at: http://kathmandupost.ekantipur.com/news/2016-01-25/chand-led-maoist-announce-formation-of-peoples-govt-in-bajhang.html (accessed January 3, 2018).

Singh, Uma. "A Minister Out of Control," *Nepali Times,* (January 16, 2009); available at: http://nepalitimes.com/news.php?id=15576#.WXYQHKoUkzs (accessed January 20, 2018).

South Asia Terrorism Portal. "India Maoist Assessment: 2015" available at: www.satp.org/satporgtp/countries/india/terroristoutfits/CPI_M.htm (accessed August 11, 2016), with extended 2016 timeline at www.satp.org/satporgtp/countries/india/terroristoutfits/CPI_M_Timeline16.htm (accessed August 11, 2016).

Spencer, Campbell. "Rebels Target Kathmandu," *New Zealand Herald,* (September 5, 2003); available at: www.nzherald.co.nz/world/news/article.cfm?c_id=2&objectid=3522024 (accessed January 20, 2018).

Subedi, Dinesh. "Maoist Victim Banks on Maoists for Polls," *Republica,* (June 22, 2017); available at: www.myrepublica.com/news/22340/ (accessed January 20, 2018).

"Suspected 'Bomb' Defused in Lazimpat," *The Himalayan Times,* (September 26, 2017); available at: https://thehimalayantimes.com/kathmandu/suspected-bomb-defused-lazimpat/ (accessed January 20, 2018).

Tan Hecheng. *The Killing Wind: A Chinese County's Descent Into Madness During the Cultural Revolution,* trans. Stacy Mosher and Guo Jian. New York, NY: Oxford University Press, 2017.

Thapa, Deepak, Kiyoko Ogura, and Judith Pettigrew. "The Social Fabric of the Jelbang Killings, Nepal," *Dialectical Anthropology* 33 (2009), 461–478; available at: http://connection.ebscohost.com/c/articles/45707679/social-fabric-jelbang-killings-nepal (accessed August 31, 2016).

"The Code of Conduct for Ceasefire Agreed Between the Government of Nepal and the CPN (Maoist) on 25 May 2006 at Gokarna"; available at: http://peacemaker.un.org/sites/peacemaker.un.org/files/NP_060525_25%20Point%20Ceasefire%20Code%20of%20Conduct.pdf (accessed January 20, 2018).

The Committee to Protect Journalists. "Mastermind Convicted in 2009 Murder of Nepali Journalist Uma Singh," April 23, 2015; available at: www.cpj.org/2015/04/mastermind-convicted-in-2009-murder-of-nepali-jour.php (accessed August 13, 2016).

Transparency International (Nepal). *National Integrity System Assessment: Nepal 2014.* Berlin, (2014); available at: www.transparency.org/whatwedo/publication/nepal_nis_2014 (accessed March 4, 2016).

United Nations Office of the High Commissioner for Human Rights. *Nepal Conflict Report 2012* (Geneva: UNOHCHR, October 2012); available at: www.ohchr.org/

Documents/Countries/NP/OHCHR_Nepal_Conflict_Report2012.pdf (accessed January 20, 2018).

Upadhyay, Akhilesh and Tika R. Pradhan. "Interview, Pushpa Kamal Dahal: Views That Maoists are Afraid of Transitional Justice are Flawed." *The Kathmandu Post,* (August 1, 2016), 7; available at: http://kathmandupost.ekantipur.com/news/2016-08-01/views-that-maoists-are-afraid-of-transitional-justice-are-flawed.html (accessed August 31, 2016).

Wieviorka, Michel. *The Making of Terrorism.* Chicago, IL: University of Chicago Press, 2004 [new preface ed.].

7 India's two-track response to the Naxalite movement

Security and development, but no political process

Samir Puri

During the rural uprisings of the 1960s, the first wave of Naxalite insurgents faced defeat. Their remnants and successors, however, scattered and rebuilt the movement that, in recent decades, has become a sustained threat that spans several of India's eastern states. In 2004, a merger occurred between the two main factions, the People's War Group (PWG) and Maoist Communist Centre (MCC). They formed the Communist Party of India–Maoist (CPI–Maoist) and this marked the start of a more unified insurgency. In the years 2005–2017, the conflict claimed 7,711 lives (comprising of 3,069 civilians, 2,702 left-wing extremists and 1,940 security forces personnel).[1]

Recognizing the gravity of the threat, India's Ministry of Home Affairs established a "Naxal Management Division" in 2006, which was renamed the 'Left Wing Extremism Division' in 2014. In its own words:

> The central theme of Maoist ideology is violence … to overwhelm the existing socio-economic and political structures. The People's Guerrilla Army (PLGA), the armed wing of the CPI (Maoist) … aims at creating a vacuum … by killing lower-level government officials, police personnel and workers of mainstream political parties.… After creating a political and governance vacuum, they coerce the local population to join the movement.… In areas under Maoist domination, the absence of governance becomes a self-fulfilling prophecy since the delivery systems are extinguished through killing and intimidation.[2]

The government's stated policy response has been to deal with Left Wing Extremism (LWE) in "a holistic manner."[3] India's strategy has comprised of both security offensives and concurrent economic development programs to assuage rural poverty and benefit communities under the sway of the insurgents. Questions of strategy have therefore focused on sequencing the use of the security forces in relation to unfolding development projects, and over whether to clear and hold Naxalite-affected areas before pouring in civilian investment. Comparatively less attention has been invested in factoring political outreach efforts. Overtures for dialogue with the insurgents have been made by some state authorities and by civil society groups, but these efforts have tended to unfold in an ad-hoc manner groups. The real weight of effort has been placed on security. The Left-Wing Extremist Division explains why this has been so:

> The CPI (Maoist) philosophy of armed insurgency to overthrow the Government is unacceptable under the Indian Constitution and the founding principles of the Indian State. The Government has given a call to the Left Wing Extremists to abjure violence and come for talks. This plea has been rejected by them, since they believe in violence as the means to capture State power. This has resulted in a spiraling (*sic*) cycle of violence in some parts of India.[4]

This chapter reconsiders the underlying logic of this approach in terms of the relative weight of emphasis between different possible approaches. India's response can be likened to a stool that balances on two of its three legs only. Why politics has been neglected in the past, at least relative to other responses, and whether politics may yet play a role in the future, are the questions that are considered here.

This chapter is structured in the following way. It starts by examining the socio-economic roots of the Naxalite movement by drawing on existing historical literature. Next, it examines why the two tracks of India's response – security and development – have received the greatest attention and considers the progress that has been made. The relative lack of progress along the political track is then subjected to consideration, before the chapter ends by reconsidering notions of "holistic strategy," and whether India's present policy mix can be accorded this epithet.

The roots of the Naxalite movement

India's communist political tradition has a long lineage. The Communist Party of India (CPI) formed in 1924, inspired by the Russian revolution of seven years earlier. Banned by British Colonial authorities, the ban was lifted in 1942 under the exigencies of the Second World War, when the British Empire joined with the Soviet Union to fight the Axis powers. After the war, the CPI was frozen out of the Indian independence movement, as nationalist political sentiments eclipsed the communist argument. Nevertheless, the CPI established an enduring strain of Marxist-Leninist political thought in India.

The CPI did not advocate armed struggle, and supported Delhi's polices in India's 1962 war against China. The inspiration of China's Maoist revolutionary model on some of India's communists was considerable, however, and a violent strain of the movement split from the political mainstream. The Naxalite movement originated from an anti-landlord campaign waged in 1967 in Naxalbari, in the state of West Bengal. After it grew violent it was suppressed by the security forces.[5] The location of this uprising gave a name to the movement that grew in its aftermath, as Naxalites continued to agitate for landlord evictions and crop seizures. To wage its envisaged rural revolution, the Communist Party of

India–Marxist/Leninist (CPI–M/L) was founded in 1969. The aforementioned MCC, one of its two main factions, split away in 1975.[6]

The original Naxalite uprising was largely decimated by the government response. Since then, India has encountered successive generations of Naxalites in vastly altered conditions. The early phases of the movement unfolded during the apex of communism's global influence. As documented by Sumanta Banerjee, he describes how the original Naxalites capitalized on the government's failure to assuage rural poverty through its agricultural policies and land reforms.[7] The early wave of Naxalite uprisings was crushed during a State of Emergency that lasted 1975–1977. The remainder of the Naxalite movement consolidated in the state of Andhra Pradesh, where the PWG was formed in 1980 (which, alongside the MCC, was the second of the two main factions). Subsequent decades would see the movement grow under the separate efforts of PWG and the MCC, until their merger in 2004 as the CPI–Maoist.

Although its aim of overthrowing the Indian government is in practical terms fanciful, the early Naxalites gained inspiration from rural insurgencies elsewhere in the world, from China to Vietnam. In more recent years, India's Naxalites received some support from Nepal's Maoists, who, after eroding King Gyanendra's faltering rule, entered government in 2006 after a civil war. With the exception of these outside links, India's Naxalite insurgency has been an overwhelmingly domestic conflict, and one that looks increasingly ideologically isolated as the twenty-first century progresses.

The Naxalite insurgency has also been a very local affair, growing in some but not all states, and building its strength by exploiting local socio-economic and governance weaknesses. In Andhra Pradesh, authorities were successful in containing the Naxalites in the 1970s, but in 1991, the ban on the Naxalites was lifted when the then-Andhra Pradesh chief minister discreetly sought their support in state elections (despite the public Naxalite position being to boycott elections). In Bihar, Naxalites exploited unequal land distribution and caste prejudices. In Orissa and Jharkhand, industrial development projects had alienated native tribal people, displacing some of them to make way for mining and energy projects. The Naxalites capitalized on the ensuing resentment in a manner well attuned to local grievances.[8]

Despite operating on a disparate and regional basis, the Naxalites do have some measure of ideological unity. The CPI–M leadership includes members that have been recruited from many states, and who consider how best to exploit local developments like falling wages, caste and religious tensions, or anger at large industrial projects.[9] While some of its rank and file members have been forcibly recruited, others have joined voluntarily. Debate persists as to what motivates people to become Naxalites, as is summed up by Alpa Shah:

> One the one hand, there are those who argue that the grass roots support for the movement is dependent on those caught between the fires who are often swept into the movement out of fear.... On the other hand, there are those

who want to stress the agency of recruits more, and who emphasize that the revolutionaries speak of a practical ideology....[10]

Motivation will vary on an individual basis, with consequent implications for any attempts to offer amnesties to fragment the movement from below. The ideology ought to be taken seriously – the cause of waging a people's war against an India that, in the eyes of the CPI–Maoist, remains semi-colonial and semi-feudal, has found traction in several states.[11] Indian authorities certainly consider LWE to present a complex security and development challenge, but the extent to which it is also considered to be a political phenomenon is moot.

The strategy of managing multi-tracked responses to insurgency

In *Fighting and Negotiating with Armed Groups: the Difficulty of Securing Strategic Outcomes*, I explained that when confronting an insurgent movement, it is unhelpful to conceptualize of a separate "security strategy," "development strategy" or "political strategy." Each of these strands constitutes an *approach* that brings to bear tools and tactics to the overarching response. The *strategy* binds these separate approaches together toward a purposeful goal. Tensions inevitably arise as the various strands of response unfold concurrently, and as they pull apart from each other. In the worst cases, such unravelling strands can leave the strategy akin to the frayed end of a rope.[12]

Security will always be the first instinctive response of the authorities to outbreaks of non-state violence. There are understandable reasons as to why this might be the case, not least because security is likely to be the most urgent first line of response. Activities like gathering intelligence, interdicting attacks, pacifying areas and neutralizing the militants are tough undertakings. They will collectively constitute the most tangible lines of activity in attempts to stem insurgent violence. However, violence always carries its own logic, in which the reciprocity of action and counter-action between security forces and insurgents takes over, as they try to outwit each other, and seek redress for losses they have suffered. Even carefully targeted security tactics can end up inviting tit-for-tat violence, contributing to the militarization of what might fundamentally be an unresolved political and socio-economic grievance.

The question of whether to use force, reconstruction or diplomacy is often a false one. War, economics and politics are inextricably connected. The strategic art of engaging armed groups sits above the separate logics of security measures, economic responses and political processes. Development assistance will mean little if divorced from the provision of security and the inclusive practice of politics in these areas. Politics, whether it involves bargaining with or bypassing the non-state armed group in question, is of little worth if divorced from the security and development tracks. Moreover, while political processes may well be essential to embed the gains made by security forces in countering and containing the violence, negotiations with insurgents cannot be seen to reward violence.

A truly strategic calculus involves accepting the need for a mixed strategy, and the tensions this involves, and managing the mix over time. Whenever "security tracks," "political tracks," "reconstruction tracks" and others run concurrently, they risk drifting apart, and not least because bureaucratic turf wars can end up compartmentalizing the response. Moreover, is possible for a government to become preoccupied with the tough task of implementing just one or even two tracks of what ought to be a multi-tracked response. Thus, when responding strategically (and therefore holistically) to an insurgent movement, it is normal for states to end up doing several things at once: to fight while at the same time non-violently responding to the armed groups.[13]

This is why the "COIN" literature can in fact lead to unhelpful thinking. COIN doctrine has many virtues, not least in explaining how the countering of insurgency involves breaking the bonds between insurgents and the local populace, and how this is achieved by protecting the populace to facilitate the provision of government service to the region. COIN doctrine can helpfully guide the security forces to deploying in a proportionate and purposeful manner. However, COIN thinking can be overly process-focused and divorced from the wider political currents that inevitably define the strategic possibilities of any given war.[14] Also reductionist is the process of winning "hearts and minds," such as when hope is placed in governance reforms that might steadily undermine the insurgency's appeal. As Hazelton explains, insurgencies are countered by buying-out the leaders and warlords responsible for the violence, while deploying enough coercive force to "break the insurgency's will and capability to fight on."[15]

In sum, the strategist must avoid a partisan approach to the tools being used. Each activity must be considered in its own right *and* in relation to other activities. Comprehending all of the relevant factors, how they interact, the trade-offs between them, and their accumulated impact – these are prerequisites for sound strategic thinking in responding to a complex insurgency.[16]

Based on this logic, it is to the government's credit that India has operated a multi-tracked response and has grasped the need for an economic track. However, while there have been attempts at dialogue, involving central and state authorities, they have failed. Opening and sustaining a political track with insurgents is enormously difficult as many governments around the world have found – but many governments also ultimately find it necessary to talk with the insurgents to end a conflict. (An example that will be revisited in the conclusion is Colombia in relation to the Fuerzas Armadas Revolucionarias de Colombia – Ejército del Pueblo (FARC).) With respect to India's unique context, there appears a general absence of political will to pursue a political settlement. But without sustained political dialogue with the insurgents, to what ultimate end in the years to come are the government's security and development approaches moving toward?

The predominant tracks of India's response: security and development

In 2006, as then-Prime Minister Manmohan Singh explained: "Our strategy has to walk on two legs – to have an effective police response while focusing on reducing the sense of deprivation."[17] The Ministry of Home Affairs elaborated: "The Government's approach is to deal with LWE in a holistic manner, in the areas of security, development, ensuring rights and entitlements of local communities."[18] The essential logic of this approach has persisted in the era of Prime Minister Narendra Modi, elected in 2014. Whether the country is run by Congress or the Bharatiya Janata Party (BJP), India's strategy toward the Naxalites remains fundamentally unchanged. If anything, the BJP is likely to BE less compromising, given the party's grounding in the Rashtriya Swayamsevak Sangh (RSS), a patriotic social movement which in many respects is a polar opposite to the Naxalite ideology.

There has been progress in terms of overall security trends, if progress is measured in terms of reducing the number of annual deaths connected to the LWE conflict. After reaching a nationwide peak of 1,180 deaths in 2010 alone (including civilians, security forces and Naxalites), this has dropped to an annual average of 389 deaths each year spanning 2011–2017. At the time of writing, the downward trend is set to persist into 2018.[19] These statistics are a crude measure for overall trends, notably since the conflict comprises of a series of local theaters of action that should be judged on a state-by-state basis.

Although progress has been uneven, security tactics have improved in some theatres. In Andhra Pradesh, the Greyhounds commando force has operated in small units to pursue the insurgents into forest areas. After a peak of 320 deaths in 2005, recent years have seen around only a dozen deaths per year in Andhra Pradesh.[20] However, official figures point to the prevalence of serious insecurity in Chhattisgarh, with at least 100 deaths related to the conflict every year between 2012 and 2017 (although down from an annual peak of 343 deaths in 2010).[21] In a single attack in April 2010 in Chhattisgarh's Dantewada region, Naxalites killed 76 police personnel, leading Prime Minister Singh to describe LWE as India's gravest security threat. On May 25, 2013, an attack in the Darbha Valley killed senior Congress politicians and a former state minister. In April 2017, Naxalites ambushed and killed 24 Central Reserve Police Force (CRPF) personnel in Sukma, also in Chhattisgarh.

The month after this attack in Sukma attack, Home Minister Rajnath Singh presented a refinement of the security approach. Using the "SAMADHAN" acronym, he explained this to comprise of: smart leadership; aggressive strategy; motivation and training; actionable intelligence; dashboard key performance indicators; harnessing technology; actions plans for each theatre; and taking action to block militant funding sources.[22] The use of unmanned aerial vehicles, fortified police stations and such like, as announced by Singh, would modernize the security effort to embed security gains. Despite dramatic setbacks such as the violence in Chhattisgarh, if one considers official data from the Ministry of

Home Affairs, and figures from the South Asia Terrorism Portal, the overall trend is of *decreasing* deaths year-on-year in LWE-affected states.

It has taken India many years to get to this point, not least because there is no single security force to respond to LWE violence. This is a double-edged sword. On one hand it can be beneficial, since just as the insurgents have varied their tactics to local conditions, the security response has similarly needed to be acutely attuned to local dynamics. On the other hand, with different security arrangements from one state to the next, the security response has contended with coordination issues. While the CRPF is operated by the Ministry of Home Affairs, and is tasked with assisting state authorities, it has had to rely on local police forces for knowledge. In response, the insurgents have been adept at exploiting state boundaries to their advantage and have migrated from one place to another to evade the security response.[23]

Moreover, reported instances of excessive force and ill-discipline have at times hampered the security response. There have been anecdotal accusations by activists that blame members of the security forces for intimidating and coercing those suspected of supporting the Naxalites. In 2015, Bela Bhatia, a human-rights activist, claimed police harassed her as she documented incidences of gang rape committed by the security forces in Sukma and Bastar. In 2016, Soni Sori, another activist and a member of the Aam Aadmi Party, claimed police detained and tortured members of her retinue in Sukma to extract forced confessions, and to incriminate her as a Maoist sympathizer. Whether verifiable and isolated or not, such allegations were serious enough for Chhattisgarh's Director General of Police for Anti-Naxal Operations to reiterate that punishment awaited police officers proven guilty of perpetrating such crimes.[24]

The security effort is only one component of the overall strategy. In the words of a police chief in insurgency-wracked Bastar: "I can fight Naxals but not Naxalism. That has to be done through development."[25] However, economically developing conflict affected areas have been complicated by the lack of state penetration. Security concerns mean that while police can travel to these areas, civilians such as health workers, teachers and construction workers often cannot. The Naxalites have duly made the case that they care more for the people than the state does, fortifying their presence through the impenetrability of the areas in which their influence holds sway.[26]

This brings us to the second facet of India's response. According to official sources, federal and state authorities have been involved in the provision of various economic development projects throughout the "Red Corridor" (as the areas of LWE influence are sometimes referred to). As with the security effort, progress can only be judged on a state-by-state basis. Through its infrastructure development projects, the intention has been to improve accessibility and connections for rural communities in a way that might marginalize Naxalite attempts to establish a presence or simply sway their opinion. For example, in 2017 the federal government announced the start of its "Road Connectivity Project for LWE Affected Areas." Costing Rs.110 billion, roads would be built and upgraded in numerous insurgency affected areas.[27] Other examples include more localized projects, such as the plan by Jharkhand state authorities to invest in a rural bus service staffed by unemployed youths, with the intention of fostering relations between rural communities and the state.[28]

The development approach will take time for its effects to be felt. Obstacles facing it are formidable, given how in some cases local grievances have been exacerbated by large commercial projects involved in natural resource exploitation – an issue that the CPI–Maoist have exploited to build its support amongst tribal peoples. The drive by corporations to access natural resources in tribal areas has led to local protests, dispossessions and clashes. Locals have conveyed their hostility to corporate ventures that they fear will endanger their agricultural livelihoods, and that offer them no compensation for the exploitation of their lands.

In some respects, these episodes are modern manifestations of issues long since present in India in some shape or form. The Indian Constitution, adopted in 1950, suggests that rural tribal areas would be granted a modicum of independence from New Delhi in determining the fate of their traditional homelands. Looking back to the origins of the Naxalite movement, its ideological strength was partly built on how tribal people had fared in modern India. Donovan S. Braud has examined the insurgency in relation to land laws and rural production methods. Although post-independence land laws had attempted to protect the land rights of indigenous peoples, their alienation from the land was exacerbated by India's market reforms of the 1990s. Overall, Braud estimates the "Maoist-inspired, agrarian based Naxalite insurgency" to be partly "a response to this capitalization of tribal lands."[29]

States such as Andhra Pradesh, Chhattisgarh and West Bengal have seen contractors seek to access timber, bauxite and other natural resources. This has given newer generations of Naxalites fresh impetus to explain the relevance of their arguments. For example, CPI–Maoist has championed the anti-mining cause in Andhra Pradesh, Odisha, and Chhattisgarh, and has even killed village officials seen as collaborators with the companies involved in natural resource exploitation. Is it fair to judge, therefore, that the Naxalite movement has rooted itself in these socio-economic grievances? While these issues are of the utmost relevance to understanding the grass roots of the movement, there are clearly evident political dimensions as well. Whether or not to swim in these political currents is a strategic choice facing the authorities, and it is to this issue that this chapter now turns.

Explaining relative the lack of progress along a political track

Let us begin by considering two starkly contrasting views. G. K. Pillai, writing on the LWE phenomenon in 2010, during his tenure as Union Home Secretary, delivered the following observation:

> You really cannot have any discussion with the Maoists because they don't believe in discussion as a matter of any principle. They only believe in

armed struggle to overthrow the Indian state and therefore even a discussion is only a ploy to regroup or buy time because they don't believe in parliamentary democracy.[30]

As Pillai explains, offers to negotiate made by CPI–Maoist should be considered a ruse, and pressuring the groups should remain India's default approach.

At the opposite end of the spectrum of opinion, writer Arundhati Roy polemically observed that, when it comes to CPI–Maoist:

> There's no whisper about "talks" of "negotiations". Odd isn't it that even after the Mumbai attacks of 26/11 the government was prepared to talk to Pakistan? It's prepared to talk to China. But when it comes to waging war against the poor, it's playing hard ball.[31]

Taking in these opposite stances – that the CPI–Maoist ought only to be spoken to if first rendered prone by the security effort, versus India perpetuating the war by refusing to countenance talks – is a useful way to consider why the political track has remained so underdeveloped. There are clearly reasons as to why this has been the case, and this is the first port of call.

There have been isolated and episodic attempts at political overtures, but no sustained negotiations have taken place between the government and CPI–Maoist on a track one basis, and in a manner that could be construed as working toward a peace process. Nor are there any prospects for such talks in the short-term. Since the CPI–Maoist has such maximalist demands – the overthrow of the government and the redistribution of resources – it would need to climb down from these positions for viable talks to occur. For talks to be palatable to the government, the parameters of the constitution would need to be observed. Instead, the government's aspiration is for its security effort to sufficiently pressurize the insurgents, and provoke enough defections and surrenders from the movement so that the insurgency is substantially weakened in advance of any future political engagement. Even if the Naxal problem is a political one, and cannot be solved militarily, the security forces must focus on reducing the tempo of violence, so the logic goes.[32]

Dialogue has occurred in the past but on an ad-hoc basis. Negotiations in Andhra Pradesh in 2004 demonstrated a lack of will on both sides. The Naxalites used the experience to stress their organizational credentials, their right to bear arms, and ultimately to announce that their commitment to armed struggle was greater than their commitment to the talks.[33] A civil society movement, the "Committee of Concerned Citizens" (CCC) was involved in outreach efforts in villages under Maoist sway. The CCC built credibility with the insurgents, but the state authorities were lukewarm to this, even stepping up security operations to deny the Naxalites space to use the CCC consultations to build influence. This stance softened in 2004 when Chief Minister Rajasekhar Reddy allowed the ban on the CPI–Maoist to lapse and appointed a CCC mediator to facilitate a dialogue. However, the government issued just one main demand: for the insurgents to surrender and give up their war. The government was concerned not to legitimize the CPI–Maoist revolutionary creed through a negotiation process. Ultimately, the Greyhounds commando unit used the process to gather intelligence and to eliminate the Naxalites, and turned the tentative dialogue into a military victory.[34]

Another failed overture took place in 2010, when Swami Agnivesh, a noted religious figure, was asked by the Ministry of Home Affairs to reach out to the insurgents. Agnivesh's participation secured the Naxalite's attention. Talks were scheduled. However, a CPI–Maoist spokesperson known as Azad was killed on July 2, 2010 by police. The Naxalites alleged that security forces had used the pretext of prospective talks to trap Azad. To save his own credibility, Swami Agnivesh filed a petition to India's Supreme Court, asking for an independent investigation into the killing.[35] This was interpreted by Arundhati Roy in stark terms, writing that: "the elimination of Azad was an important victory [for the war effort] because it silenced a voice that had begun to sound dangerously reasonable. For the moment, at least, peace talks have been successfully derailed."[36]

This was another instance of security forces choosing to score a tactical victory, with little concern for sustaining any kind of tentative political track. It is questionable whether there is much appetite for talks. Some Naxalites may have wanted to participate in electoral politics, even if only to corrupt the system by joining in it. From the perspective of the government, the focus has remained on security operations. It is in no rush and would rather wait to further strengthen the government's position rather than engage in premature negotiations.[37]

The authorities articulate their general lack of political will for talks in terms of sequencing. Home Minister Singh has stated: "I want to appeal to Naxals to shun violence as there is no place for violence in India. Let's come for talks. We are ready for talks and discuss all issues if you give up violence."[38] The official stance is to implore the Naxalites to join the mainstream:

> civil society and the media [need] to build pressure on the Maoists to eschew violence, join the mainstream and recognise the fact that the socio-economic and political dynamics and aspirations of 21st Century India are far removed from the Maoist world-view.[39]

Amnesty schemes, and encouraging defections, follow on from this logic by weakening the insurgency from below. The uptake of these schemes, as measured through the number of Naxalites surrendering, is part of how the government has judged its own progress in dealing with the threat. At the end of 2016, for example, official statistics counted 1,429 surrenders during the year, a marked increase in the 282 who surrendered in 2013. This is presented as proof that the twin-tracked development and security approach was working,[40]

The lack of will to traverse the political track therefore arises partly from the prevailing instinct of the authorities to further weaken the insurgency. Whether the authorities would only ever countenance a sustained negotiation process with a nearly-defeated insurgency is important to consider. The strategic question, therefore, is whether and how India rebalances the weight of emphasis between the different pillars of its strategy in the future. The words of an Andhra Pradesh activist are worth considering in light of the present policy mix:

> A 'two-pronged strategy' is often talked of as the best one ... tough policing of Naxalite violence and benign welfare measures for the poverty-stricken masses.... Quite apart from the fact that in practice, the first prong has been longer and sharper than the second, the truth is that conscious political violence can never be successfully addressed if it is treated as an unplanned outgrowth of social deprivation. It requires constant dialogue at the level of citizens and the government.[41]

India's prospects for ending the insurgency remain dim for as long as dialogue does not at least try to address the root causes and political dimensions of the insurgency. Hunting Naxalites, forcing amnesties from the remaining insurgents, and investing in local infrastructure, are all necessary measures, but may not be sufficient to push the situation beyond the finishing line in the years to come. It is on this point that this chapter concludes – whether politics may yet have a greater role in a revised policy mix, if the progress made in containing the violence continues.

Conclusions

To take an example from an entirely different continent, Colombia's government has faced the rural Marxist FARC insurgency for five decades. Over this time, its policy mix oscillated between security offensives and stalled negotiations. It has taken Colombia a long time to make meaningful progress along its political track. Even then, Colombia's security effort was crucial in preparing the ground for the talks that were held in a mediated setting in Havana. The next challenge has been implementation of the FARC-government deal that was reached in 2016. The deal itself has involved significant concessions by the government, including allowing former FARC members to run for political office, and has relied on a UN inspection mission to oversee the decommissioning of FARC's weapons. The deal could stall or fall apart – but the important thing is that Colombia's war has seemingly come to an end.[42]

The relevance of this example is to illustrate how a long running war with a Marxist-grounded insurgency can indeed end. It is not to suggest that political tracks offer panaceas. Indeed, India's record in negotiating with secessionist groups in the north-east of the country has had mixed results. There, talks have tended to formalize a stalemate in which the insurgents are nominally confined to camps, but often still have free reign to extort the civilian population. At the same time, rejectionist factions break off to continue their attacks on civilians and the security forces from the bush. In some past cases, state borders were modified to effectively buy out the secessionists. For example, in the 1980s, after an insurgency in Mizoram, the authorities allowed former insurgent leader

Laldenga to become chief minister of a new state that he had fought to create.[43] The creation of new states has been a feature of India's history, with Jharkhand and Chhattisgarh created in November 2000 by redrawing state boundaries.[44] CPI–Maoist is not secessionist, and is motivated by socio-economic inequalities, and by the contested politics of resource allocation. As Sudeep Chakravarti has written, "the Maoists are patriots [who] do not want a separate country. They already have one. It's just not the way they would like it – yet."[45]

The prospects for ending this war presently seem improbable. At the national political level, neither the BJP nor Congress, India's leading political parties, offer a fundamental challenge to the existing counter-LWE strategy. Nor are there any voices in mainstream Indian politics to challenge the dominant model of India's economic growth. Alternative opinions have been pushed to the political margins, and this has become more acute since the rise of the BJP in the 1990s and its associated nationalist Hindutva ideology. In remote tribal areas, the Naxalite ideology has competed for tribal attention with BJP-backed RSS cadres that spread their own nationalist message. While the dogged prevalence of the Naxalites seems out of step with modern India's direction of travel, it is also a violent reflection of the notion that, in effect, "two India's" exist side by side: one is increasingly modern and urban and the other remains rural and poor. India has experienced significant and impressive macroeconomic growth in recent decades. But, as Amratya Sen has written, growth has "coexisted with the continuation and sometimes even intensification of deep social failures ... India's success in reducing poverty and deprivation has been very moderate."[46] The CPI–Maoist insurgency is India's violent reaction to wealth being held in the hands of the few; of social injustice; and of marginalization of the agrarian poor.

The anti-LWE two-track approach has made some progress, but it begs the question as to whether there will ever be a tipping point in which the insurgency is deemed to have been fatally weakened. Total eradication of the movement, root and branch, is hardly a viable prospect. All of which increases the importance exploring the possibility of a political track. Managing a successful strategic mix of approaches means maintaining a watchful eye over when that mix needs to be diversified. Or else, India will still be chasing after the Naxalites at its eightieth, ninetieth and hundredth anniversaries following its independence. Containing and reducing the threat is certainly a sensible short- and medium-term aspiration, but the longer-term goal ought to involve seeking an end to the conflict.

Notes

1 South Asia Terrorism Portal. Accessed March 1, 2018. www.satp.org/satporgtp/countries/india/maoist/data_sheets/fatalitiesnaxal05-11.htm.
2 Government of India, Ministry of Home Affairs, Left Wing Extremism (LWE), Division. Accessed March 1, 2018. https://mha.gov.in/naxal_new.
3 Ibid.
4 Ibid.

5 Chakravarti, *Red Sun: Travels in Naxalite Country*, 97.
6 Chadha, *Low Intensity Conflict in India: An Analysis*, 390–401.
7 Banerjee, *In the Wake of Naxalbari: A History of the Naxalite Movement in India*.
8 Berthet and Kumar, *New States for a New India: Federalism and Decentralization in the States of Jharkhand and Chhattisgarh*, 17–18.
9 Chakravarti, *Red Sun: Travels in Naxalite Country* , 113, 297, 316.
10 Shah, "In search of certainty in revolutionary India," in Shah and Pettigrew (eds.), *Windows into a Revolution: Ethnographies of Maoism in India and Nepal*, 39–40.
11 Ramana, "Profiling India's Maoists: An Overview," 152–155.
12 Puri, *Fighting and Negotiating with Armed Groups: The Difficulty of Securing Strategic Outcomes*, 7–12, 149–162.
13 Ibid., for a full exposition of this thesis.
14 Smith and Jones, *The Political Impossibility of Modern Counterinsurgency*, xvi.
15 Hazelton, "The 'Hearts and Minds' Fallacy: Violence, Coercion, and Success in Counterinsurgency Warfare," 80–113.
16 Freedman, *Strategy: A History*, 242–245.
17 Singh quoted in Chakravarti, *Red Sun: Travels in Naxalite Country*, 195.
18 Op. cit., Government of India,, Left Wing Extremism (LWE), Division.
19 Op. cit., South Asia Terrorism Portal.
20 Ibid.
21 Government of India, Ministry of Home Affairs, "State-wise Extent of LWE Violence during 2010 to 2018 (up to 15.02.2018)." Accessed March 1, 2018. https://mha.gov.in/sites/upload_files/mha/files/LWE_06032018.PDF
22 Sandhu, "12 Takeaways from Centre's New Strategy to Deal with Naxals."
23 Nayak, "Maoist Insurgency in East Indian States: Issues and Mobilization," 178–193.
24 Examples cited in the International Institute for Strategic Studies "Armed Conflict Database," 2016 entries on the Naxalite conflict.
25 Quoted in Chakravarti, *Red Sun: Travels in Naxalite Country*, 375.
26 Ibid., 112.
27 Press Information Bureau Government of India Cabinet Committee on Economic Affairs (CCEA), "Cabinet approves Road Connectivity Project for Left Wing Extremism Affected Areas."
28 Johar News, "Rural Bus Service Will Run at 364 Routes in Jharkhand: CM."
29 Braud, "The Asiatic Mode of Production, Indian Land Law, and the Naxalite Movement," 71–85.
30 Pillai, "Left-Wing Extremism (LWE) in India," 6.
31 Roy, *Broken Republic: Three Essays*, 10.
32 Interview with Colonel Vivek Chadha (retired), Institute for Defence Studies and Analyses, New Delhi, November 2013.
33 Ramana, "Negotiating with the Maoists: lessons from the Andhra experience"; Oetken, "Counterinsurgency Against Naxalites in India," 146.
34 Centre for Humanitarian Dialogue and Delhi Policy Group, "Conflict Resolution: Learning Lessons from Dialogue Processes in India," 8–18.
35 Ibid., p. 18.
36 Roy, *Broken Republic: Three Essays*, 187–188.
37 Op. cit., interview with Chadha.
38 The Indian Express, "Rajnath Singh asks Naxals to Shun Violence, Join Peace Talks."
39 Op. cit. Government of India, Left Wing Extremism (LWE) Division.
40 One India, "Record Naxal Surrenders in 2016: 1,420 and Counting," December 27, 2016. Accessed March 1, 2018. www.oneindia.com/india/record-naxal-surrenders-2016-1-420-counting-2301711.html.
41 Balagopal, an Andhra Pradesh Human Rights activist. Quoted in Centre for Humanitarian Dialogue and Delhi Policy Group (2011), 8.
42 Puri, *Fighting and Negotiating with Armed Groups*, 42–52.

43 Ganguly and Fidler, *India and Counterinsurgency*, 24, 42, 49.
44 Berthet and Kumar, *New States for a New India* , 7, 13.
45 Chakravarti, *Red Sun: Travels in Naxalite Country*, 15.
46 Sen and Dreze, *India: Development and Participation*, vi.

Bibliography

Banerjee, Sumanta. *In the Wake of Naxalbari: A History of the Naxalite Movement in India*. Calcutta: Subarnarekha, 1980.

Berthet, Samuel and Girish Kumar, *New States for a New India: Federalism and Decentralization in the States of Jharkhand and Chhattisgarh*. New Delhi: Manohar, 2011.

Braud, Donovan S. "The Asiatic Mode of Production, Indian Land Law, and the Naxalite Movement," *Perspectives on Global Development and Technology*, no. 14, 2015.

Centre for Humanitarian Dialogue and Delhi Policy Group, "Conflict Resolution: Learning Lessons from Dialogue Processes in India," Geneva, 2011.

Chadha, Vivek, *Low Intensity Conflict in India: An Analysis*. New Delhi: Sage Publications India, 2005.

Chakravarti, Sudeep, *Red Sun: Travels in Naxalite Country*. New York: Penguin, 2008.

Freedman, Lawrence. *Strategy: A History*. London: Oxford University Press, 2013.

Ganguly, Sumit and David P. Fidler (eds.), *India and Counterinsurgency: Lessons Learned*. Oxford: Routledge, 2009.

Government of India, Ministry of Home Affairs, Naxal Management Division, available at https://mha.gov.in/naxal_new (accessed March 1, 2018).

Hazelton, Jacqeline L. "The 'Hearts and Minds' Fallacy: Violence, Coercion, and Success in Counterinsurgency Warfare," *International Security*, 42 (1) (Summer 2017).

Johar News, "Rural Bus Service Will Run at 364 Routes in Jharkhand: CM," (May 14, 2017) available at www.joharnews.in/rural-bus-service-will-run-at-364-routes-in-jharkhand-cm/ (accessed March 1, 2018).

Nayak, Nihar. "Maoist Insurgency in East Indian States: Issues and Mobilization" in Arpita Anant (ed.) *Non-State Armed Groups in South* Asia. New Delhi: Pentagon Security International/Institute for Defence Studies and Analysis, 2012.

Oetken, Jennifer L. "Counterinsurgency Against Naxalites in India." In Sumit Ganguly and David P. Fidler (eds.), *India and Counterinsurgency: Lessons Learned*. Oxford: Routledge, 2009.

One India. "Record Naxal Surrenders in 2016: 1,420 and Counting," (December 27, 2016) available at www.oneindia.com/india/record-naxal-surrenders-2016-1-420-counting-2301711.html (accessed March 1, 2018).

Pillai, G. K. "Left-Wing Extremism (LWE) in India," Institute for Defence Studies and Analyses, *Journal of Defence Studies*, 4 (2) (April 2010).

Press Information Bureau Government of India Cabinet Committee on Economic Affairs (CCEA), "Cabinet approves Road Connectivity Project for Left Wing Extremism Affected Areas," (December 28, 2016) available at http://pib.nic.in/newsite/PrintRelease.aspx?relid=155909 (accessed March 1, 2018).

Puri, Samir, *Fighting and Negotiating with Armed Groups: the Difficulty of Securing Strategic Outcomes*, IISS Adelphi Series No. 459. London: Routledge, 2016.

Ramana, P. V. "Negotiating with the Maoists: Lessons From the Andhra Experience," Institute for Defence Studies and Analysis, (October 2011, 13) available at www.idsa.in/idsacomments/NegotiatingwiththeMaoistsLessonsfromtheAndhraexperience_pvraman_131011 (accessed March 1, 2018).

Ramana, P. V. "Profiling India's Maoists: An Overview," in Arpita Anant (ed.), *Non-State Armed Groups in South Asia*. New Delhi: Pentagon Security International/ Institute for Defence Studies and Analysis, 2012.

Roy, Arundhati, *Broken Republic: Three Essays*. London: Penguin Books, 2011.

Sandhu, Kamaljit Kaur. "12 Takeaways From Centre's New Strategy to Deal with Naxals," *India Today*, (May 8, 2017) available at http://indiatoday.intoday.in/story/naxals-mha-rajnath-singh-narendra-modi-government-strategy/1/949072.html (accessed March 1, 2018).

Sen, Amartya and Jean Dreze. *India: Development and Participation*. Oxford: Oxford University Press, 2002.

Shah, Alpa and Judith Pettigrew (eds.), *Windows into a Revolution: Ethnographies of Maoism in India and Nepal*. New Delhi: Orient Blackswan Ltd., 2012.

Smith, M.L.R. and David Martin Jones. *The Political Impossibility of Modern Counterinsurgency*. New York: Columbia University Press, 2015.

The Economist. 'Sweeping up: A stunning debut for a left-leaning, anti-graft party' print edition, December 14, 2013.

The Indian Express. "Rajnath Singh asks Naxals to Shun Violence, Join Peace Talks," (June 18, 2016) available at http://indianexpress.com/article/india/india-news-india/rajnath-singh-asks-naxals-to-shun-violence-join-peace-talks-2861125/ (accessed March 1, 2018).

8 Delimiting an Indian strategic approach to counterinsurgency

Bibhu Prasad Routray

Introduction

In its lengthy post-independence history of dealing with armed rebellions, India has found itself faced with numerous rebellions, against which it has had some degree of success. In fact, several such challenges have been resolved within the framework of the constitution, while in other cases, prolonged security operations have been able to drive extremists to points of hopeless debacle. At the same time, speaking strategically, there has been a failure to find solutions for a large number of conflicts. Emboldened by narratives of alienation, feeding on actual and perceived exploitation by the state, numerous insurgencies, with terrorism embedded as an integral method, have been able to challenge the might of the state for protracted periods. In diverse parts of the country, therefore, both the writ of the state and its version of development remain unrealized. Large areas remain unsafe. "Although by themselves these conflicts may not be enough to hinder India's rise, they do divert the attention of Indian leaders who prefer to look to the global arena."[1]

Among the reasons for this situation is the state's incapacity to marshal its resources in a coherent, applied fashion. State responses to armed rebellions have remained ad hoc and often divided over fundamental questions: Should insurgents be considered enemies or merely misguided people who can be brought back to the mainstream with some allurement? What, therefore, should assume priority in response, development or security? Should the state pursue conflict resolution or be satisfied with conflict management?

Operationalizing the answers has been haphazard and inconsistent. Security response, which invariably takes the lead, has used multiple forces belonging to the states and the center, often without coordination and further negatively impacted by command and control issues, the result often not allowing for much headway. Similarly, unsatisfactory implementation of the state's version of development – consisting as it normally does of a predictable menu of projects connected with roads, electricity, health, education, poverty alleviation, and employment generation – has too often not only failed to speak to sources of marginalization but has actually further alienated people. Where ceasefire agreements have been reached, the state's capacity to move beyond agreements with individual groups to durable peace has proved limited. Protracted negotiation processes are often used by the latter to indulge in extortion and other illegal activities.

It is with this record in mind that this chapter turns to the universe of success, which resides in just four cases, and interrogates their particulars to discern whether they could have been used to shape a body of doctrine providing what might be termed a formula for victory. Even cursory assessment would seem to hold out promise along these lines, leaving a certain analytical frustration at India's failure to do so. In essence, then, we have proceeded from case to case by reinventing the wheel. A detailed examination of this record of futility, though, is not our topic. Rather, the chapter uses the four notable cases of success secured during the past six decades to highlight that even the most elementary compilation of lessons learned should have produced a viable approach that could be reduced to four essential principles. Instead, diverse policies appear to have been implemented in different theatres. Such variation has been made more acute as different political parties in power have pursued attempting to implement their own COIN model.

To provide an analysis of the four key elements of what can be considered an Indian strategic approach to counterinsurgency, the chapter is divided into five sections. The first section sums up the existing debate on whether India has a counterinsurgency doctrine. The second section analyzes India's successes in four cases of success: Mizoram, Punjab, Tripura, and Andhra Pradesh. The purpose is to derive the shared principles that drove such success so that they may be outlined and discussed in the third section in order to set forth a strategic approach. The fourth section then examines whether their non- or partial-application of the approach in the active conflict theatres of India has contributed to the intractability of these conflicts. Finally, in the fifth section, the findings are summarized and a prediction offered as to whether India will be inclined to develop such a counterinsurgency strategic approach and adhere to it.

Indian counterinsurgency strategic approach

Debate as to whether India actually has a counterinsurgency strategic approach is ongoing. Certainly, there is broad body of material available such as one associates with Western militaries (and even states). Further, as strategy is considered here the unique implementation of strategic approach, it is the latter which is used as the terminology to describe my analysis. It is strategic precisely because it advances principles which must then be operationalized; it is not a strategy per se, because specificity derives from the case particulars.

Much of the academic work on counterinsurgency in India has been limited to examining the experience of the security forces, principally the Indian Army, largely in the internal conflict theatres but also in the lone expeditionary case, that against the Liberation Tigers of Tamil Eelam (LTTE) in Sri Lanka, 1987–1990. Since the Indian Army has either been directly involved in irregular conflicts directly or through its derivatives – such as the Assam Rifles (AR) and the Rashtriya Rifles (RR) in Punjab, Jammu and Kashmir, and India's Northeastern states – its *Doctrine for Sub Conventional Operations*,[2] issued in 2006, broadly explains the context and terms of reference for dealing with what in the West is now normally termed irregular warfare. It is, of course, an army manual, yet India has also deployed its various state police and paramilitary forces, with or without the assistance of the Indian Army. Nonetheless, neither the various state police forces (which are effectively autonomous of central direction; India has 29 states) nor the six types of paramilitary forces which have been employed in counterinsurgency operations have issued any doctrine. Therefore, it can be concluded that to a large extent India's counterinsurgency response is being conducted either without any doctrinal assistance or as framed by the military through the lens of the Army. There is nothing inherently insufficient concerning what is contained in the manual noted above, simply that it does not actually tell anyone what "to do." In one sense a strength, it is also a weakness if one is seeking a theory of victory.

Scholars commenting on India's counterinsurgency efforts can be clubbed into three groups. The first argues that the country does not have a strategic approach per se (the wording is mine). Instead, as detailed by Shapoo, the country has a sense that kinetic and nonkinetic facets of response to challenge should be blended so as to produce results.[3] This does not hold up well, they note, terminologically or conceptually, whether what is at issue is labeled strategy or doctrine or approach. Still, the notion is there that a certain consistency informs the Indian approach to armed rebellions within its territory. To this end, one author goes so far as to note:

> We might even call it "grand strategy" because it is primarily a political approach in which military force plays an essential but ultimately limited role. The essence of this strategy is the willingness to compromise with rebellious sub-nationalities on all issues with one exception: secession is taboo.[4]

That such conceptualization has little to do with grand strategy as the term is normally used may be considered at another time; the point is well put.

The second group of scholars argues that the country indeed has counterinsurgency doctrine, with application being specific to each case. As noted by Singh, referencing the publication above:

> The document can be seen as a logical extension of the conventional warfighting doctrine issued in 2004. The doctrine focuses on the principles and practices best suited for sub-conventional operations, and including counterterrorism and low-intensity conflicts. Prior to this, the doctrinal tenets were addressed through a series of departmental training manuals and publications.[5]

Interestingly, as reflected in a number of other recent publications, he associates the nonkinetic aspects (the term is not used specifically) with "Nehruvian" though, which was opposed to the use of overwhelming force. The precise synthesis of nonkinetic and kinetic, he sees as derived from the British experience in Malaya. The result, another observer comments, has "remained fundamentally conservative and traditional."[6] It is thus not surprising that other authors who follow this line of thought point to Indian counterinsurgency as guided by "primacy of political negotiations, adherence to democratic and legal procedures, the use of minimum force and respect for human rights."[7] The result, then, is not altogether at odds with the first school above, a willingness to compromise on all issues save secession.

The third group of experts opines that setting forth actual counterinsurgency doctrine in a complex conflict environment is neither possible nor desirable. To do so – in view of the limitations posed by the nature of India's democratic polity and pressures exerted by the media, human rights groups, NGOs, and activists in conflict theatres – is an unnecessary exercise. It will limit innovation, a key component of resolving conflicts and even subject state efforts to undesirable scrutiny, thereby limiting their effectiveness. Fidler asserts that India's counterinsurgency experience has been far too rich to be reduced to easy synthesis, especially since New Delhi has managed to find the right combination of military pressure and political intervention to manage its conflicts. This, he concludes, is the very reason India has yet to lose a counterinsurgency campaign, albeit some rough patches occasioned by trial and error.[8]

It can readily be seen that at least part of the problem is the easy manner in which everything from strategy to doctrine ends up conflated in a discussion of approach. Of what is meant, though, there is no doubt: how to approach the challenge, whether within signposts committed to writing or merely as lessons learned and passed on. This is an old saw, with the Americans having become noteworthy in recent years for their virtual catalogs of unread guidance and admonitions that absent such, a military cannot move forward. This hardly seems the case, but neither does absence of something more substantive appear inviting. It is with this thought that we turn to specific Indian cases.

Cases of counterinsurgency success

This section provides a summary of India's four successful counterinsurgency efforts. It is not history per se that we are after but analysis of the key features of counterinsurgency as it emerged in neutralizing the challenge.

Mizoram

In the Northeastern state of Mizoram, insurgency started in the wake of a famine causing the deaths of hundreds. The government of Assam, the state of which Mizoram then a part, failed to act in time. In the absence of official support, the Mizo National Famine Front gained popularity by conducting relief work, and by 1962, it had reflagged as the Mizo National Front (MNF). On February 28, 1966, it declared Mizoram an independent sovereign state. MNF cadres overran lightly manned security posts, as well as major government buildings and instal-lations. To clear the hold of the rebels over capital Aizawl, the Indian Air Force was called upon to strafe the town[9] and clear road blocks to allow reinforce-ments. This forced the MNF cadres to go underground and retreat to their bases in the Chittagong Hill Tracts (CHT) in neighboring East Pakistan (which was to become Bangladesh several years later).

The weak police could not be trusted to operate against the MNF cadres. Mizoram was declared a Disturbed Area, and the Army was deployed as the lead counterinsurgent force under the Armed Forces Special Powers Act (AFSPA). The strength of the MNF, apart from its bases in the CHT, assistance from Paki-stan, and its training facilities in China, was in its ability to find hiding space within the civilian population in the sparsely populated and scattered remote villages.

> Determined to starve the MNF of food and extortion-money and deny it information and shelter from the scattered hamlets that were difficult to reach and secure, the Army devised a strategy which had been tried in Naga-land earlier: grouping of villages.[10]

This was carried out between 1967 and 1970.[11] A total of 466 villages with 236,162 persons representing 82 percent of Mizoram's population was herded into "Protected and progressive villages" under military security. While villages where the MNF was not influential were left alone, in the rest, people were asked to move with what they could carry on their back, and the rest was set afire. This and other forms of human rights violations "further widened the gulf between the local populace and the nation and tended to create a degree of empathy for the militants."[12]

Creation of Bangladesh in 1971 proved to be detrimental to the MNF's inter-ests, in particular, robbing the outfit of the support it enjoyed from Pakistan. A pro-India Bangladesh suddenly became a hostile place for MNF. While the top MNF leadership relocated to West Pakistan, many in the rank and file surfaced and crossed over to Mizoram, ending the first phase of insurgency. MNF then moved its remaining cadres to the Arakan Hills in Burma, which became its base for hit and run attacks against the government functionaries and state collaborators.

A series of high-profile attacks, which included ambush of the Lt. Governor in 1974 and killing of the Inspector General of Police and his two senior most officers in Aizawl in 1975, led to another phase of military campaign. As a con-sequence, a large number of MNF cadres surrendered with arms between 1975 and 1977. A peace accord signed in July 1976 collapsed after MNF cadres indulged in rampant extortion activities and targeting of non-Mizo outsiders in the state. In 1979, the MNF was again banned, and its leader, Pu Laldenga was arrested. The established authorities sought to garner popular support against the outfit by highlighting its unending miseries inflicted on the population.

This was used as the springboard to initiate a process of healing of the wounds inflicted by a bitter COIN operation. A process of *de-grouping* of the villages and re-establishment of old villages was initiated in 1972 and completed by 1980, although the character of communities and institutions had in the interim changed drastically. A host of development initiatives was introduced, with substantial influx of resources to the newly created union territory of Mizoram. This could be seen in the Rupees 1.3 billion located for the 1980–1985 five-year plan versus the Rupees 466 million in 1974–1979.[13] Between 1980 and 1986, a protracted peace process continued, with both sides bargaining hard. A peace agreement in June 1986 led to the resignation of Chief Minister Lal Thanhawla and his replacement by MNF chief Laldenga. Peace ensued and has lasted in Mizoram since.

Punjab

The western state of Punjab witnessed a violent militancy in demand for Khalistan, a separate homeland for the Sikhs, in the 1980s and early 1990s. The militants, with support from Pakistan, which funded and armed their movement apart from providing training and bases, carried out civilian massacres, especially targeting the Hindu minority in the state, and engaged in indiscriminate bombings in crowded places. A number of political leaders were assassinated, eventually to include Prime Minister Indira Gandhi in 1984. She was killed by her Sikh body-guards consequent to her decision to send the Army into the Golden Temple, the most sacred religious institution of the Sikhs in Punjab's Amritsar city. It had become the militants' main base, and Operation Blue Star culminated with the killing of separatist leader Bhindranwalle. Unforeseen and difficult circumstances resulted in severe damage to the complex, to include the Akal Takht, the most sacred of all.[14] The result was explosive, with the security forces unable to contain the mushrooming militancy which paralyzed the state's economy and led to wide-spread extortion. Between 1986 and 1990, 10,736 militancy related deaths were reported in the state, to include 7,014 civilians and 920 security force personnel.[15]

A turning point came with the reappointment of K. P. S. Gill, an Indian Police officer, as the Director General of Police (DGP) of Punjab in 1991. Political interference had curbed his hands during his first tenure as DGP, 1988–1990, but soaring violence and insurgency holding the strategic initiative occasioned his return. Given virtually a free hand, he responded with what came to be called the "Gill Doctrine,"[16] wherein the state police held primacy. Their numbers were increased from 35,000 to 60,000, even as the number of centrally introduced paramilitary companies was reduced from 400 to 260.[17] Gill carefully defined the role for each of the agencies involved in his counterinsurgency: the Army, the paramilitary units (from the Central Reserve Police Force (CRPF) and the Border Security Force (BSF)), the intelligence agencies, and the police themselves. Each component was made aware of its role in the campaign and provided with a sense of participation and ownership, thereby minimizing the scope for friction. Three commando battalions were raised.

A central component was raising capacity and capability, which in many respects boiled down to increasing numbers and raising. Substantial numbers of salaried police auxiliaries, Special Police Officers, were mobilized, with surrendered militants also placed in special targeted killing units. Middle-ranking police officials of the Deputy Superintendent of Police (DSP) rank were promoted to Senior Superintendent of Police (SSP), without raising their salary, but adding to their morale.[18] Substantial bounties were placed on the heads of leading militants:

> Rewards of Rupees 25,000 to Rupees 100,000 were standard, and a leading militant like Gurbachan Singh Manochahal was killed by the police with a three million rupee price on his head, as compared to the standard police salary of about Rs.2,500 per month.[19]

This motivated the police to capture senior militants. The physical elimination of senior militants and leniency for the "small-fry" became a central element of the approach.

Enhanced coordination, joint actions, and unity of effort were achieved through a series of unique experiments demonstrating synergy of operations between security forces in 1991. The CRPF and BSF functioned as part of the "cooperative command" structure along with the Punjab Police and the Indian Army. CRPF officers were made part of the joint interrogation teams to handle prisoners. The command structure ensured sharing of intelligence. The radio network of the state police was interlinked with those of the CRPFs.[20] The Inspector General of Police (Operations) performed dual-hat for the CRPF and the Punjab Police.[21] This minimized turnaround time on intelligence inputs and enhanced their actionability. Prolonged interaction between local and central forces dissipated the initial suspicions that each held of the other.[22] All the forces acted in "complete concert, with a clearly defined institutional structure of cooperation and consultation."[23]

Critics pointed to an alleged large number of human rights abuses, including torture, illegal detention, extrajudicial execution, and enforced disappearance. Evidence is decidedly mixed, though the more telling a point may be that when affairs reach the point they had in Punjab. There is little way forward save what has at times been termed, "the good, the bad, and the ugly." Arguably an inevitable feature of most counterinsurgency efforts, such abuses have been criticized by human rights organizations as avoidable. Human Rights Watch, for example, reported:

> Police abducted young Sikh men on suspicion that they were involved in the militancy, often in the presence of witnesses, yet later denied having them in custody. Most of the victims of such enforced disappearances are believed to have been killed. To hide the evidence of their crimes, security forces secretly disposed of the bodies, usually by cremating them.[24]

The Police department, however, highlighted the criminal activities of the militants[25] to mobilize popular support for its operations and to stall, for a long time, any credible inquiry into these allegations. Regardless, by 1994, the upheaval was over. That year, among the 78 dead in militancy related activities, 76 were definitely militants.[26]

Tripura

Insurgency in India's Northeastern state of Tripura started as a reaction to the large-scale movement of Hindu Bengali people from erstwhile East Bengal, which transformed an essentially tribal-majority state into a Hindu-dominated geographical entity. Migration continued even as East Bengal became a part of Pakistan following the bifurcation of India on the eve of its independence from the British in 1947. As the Bengalis dominated society, politics, bureaucracy, and economy and pushed the indigenous population – the so-called Tribals – into the remotest peripheries of the state, insurgent outfits formed to protect tribal identity and safeguard their interests. By the mid-1990s, two outfits, the National Liberation Front of Tripura (NLFT) and the All Tripura Tiger Force (ATTF), came to represent the face of insurgency and indulged in rampant violence, abductions, and extortion, targeting Bengalis. In the words of a Governor of Tripura:

> What gave punch to the insurgency was striking logistics and monetary prowess acquired from the rough, rugged terrain, and the porous and extensive trans-border corridors with Bangladesh. Safe havens in Bangladesh, logistic support from the then solicitous Bangladesh establishment and the external intelligence agencies based there, and networking with potential insurgent outfits aided it.[27]

Between 1993 and 2003, insurgents killed 2,312 civilians. The impact of such bloodbath led to widespread internal displacement, nearly 19,468 families between March 1, 1998 and February 28, 2003, as reported by one source.[28]

Although the state capital, Agartala, remained free of violence, the interior districts became inaccessible. Even National Highway No. 44, connecting Tripura with the rest of the country, had to be cleared on a daily basis to allow for vehicular movement. Infrastructure projects were stalled as insurgents regularly struck the construction sites, kidnapping laborers and killing them if ransom was not paid. For a geographically compact state, but 10,492 sq km and surrounded on three sides by Bangladesh, the rising insurgency meant complete disruption of normal life. Degradation of the normal police intelligence network meant that response produced collateral damage,[29] further accentuating the divide between the Bengalis and the tribals.

In such unpromising circumstances, a turnaround was achieved through a decisive state police-led counterinsurgency led by two successive DGPs who enjoyed political support and achieved a synergy between the state police department and the paramilitary units, again, units of the CRPF and the BSF. The incumbent Left Front government, which had been a target of one of the insurgent groups, the NLFT, and had lost a large number of its cadres to assassinations, provided a free hand to the police, with the condition that cases of human rights violations would be avoided. As the police took the lead, CRPF and BSF personnel performed support functions, both in actual counterinsurgency operations and along the Indo-Bangladesh border, controlling the movements of the insurgents to and from bases in the latter. Well-run joint operations, sharing of intelligence, and regular coordination meetings between the police and CRPF authorities mediated friction.

Police strategy of response was to dominate the most remote areas in the state and to minimize the reaction time for operations. The interior was blanketed by 386 camps of police and security force personnel. In addition, 2,600 Special Police Officers (SPOs) and another 105 Special Police Pickets (SPPs) were also established. Police facilities were upgraded in terms of arms and response capabilities, placing a dispersed but coordinated force at the command of each District's Superintendent of Police.[30] The SPOs gathered intelligence and kept tabs on the activities and movements of the insurgents, collaborators, and sympathizers. The effort overall was kept under "close observation at the highest level in order to check personnel from going berserk and being ruthless, trigger-happy, oppressive and violative of human rights. This paid off."[31]

Ability of the insurgents to find support among the tribal population was curtailed by the implementation of a village reorganization scheme, similar to the one carried out in Mizoram and reminiscent of counterinsurgency operations in Malaya, Kenya, and Vietnam. Tribals were encouraged to relocate to clusters set up along state highways, which not only shrunk the hiding space for the insurgents but enabled the state administration to extend health and education facilities to populations hitherto located in remote and inaccessible forested areas. Instances of insurgent atrocities initiated a process of disenchantment, which, with the state government's sensitive handling of the operations and development initiatives, led to a palpable shift in popular attitudes. Tribal reaction to the population resettlement was surprisingly smooth and non-coercive. Actual movement of the people was preceded by strong information efforts by the workers of the party in power and relevant government departments.

Final assault on insurgent capacities came through the near-completion of border fencing between Tripura and Bangladesh,[32] making free movement across most of the 856 kilometer-long border difficult. This to a large extent stopped the illegal ingress and egress and curbed "hit and run" attacks, the primary mode of insurgent operations. Diplomatic initiatives between India and Bangladesh resulted in raids on insurgent facilities within Bangladesh, which added to the pressures resulting from covert operations by the Tripura police. All this caused a split within insurgent ranks and led to mass surrenders and arrests. By 2006, the insurgency had greatly diminished.

Andhra Pradesh

The southern Indian state of Andhra Pradesh along with the eastern state of West Bengal had witnessed an upsurge of Maoist insurgency in the 1960s, termed "left-wing extremism" (LWE) or Naxalism (both terms still in use). Even after the decimation of the movement in 1971 through a largely kinetic approach by the Indian state, it had embedded itself in Andhra Pradesh; thus it was able to re-emerge in the 1990s. By 2003, Maoists of the People's War Group (PWG) was operating in all of the state's 23 districts, killing civilians and security forces and carrying out several high profile attacks on politicians, to include the then-Chief Minister, N. Chandrababu Naidu.[33] That year saw 575 incidents and 139 deaths.[34]

In May 2004, though, Chief Minister Rajashekhar Reddy of the Congress Party was able to carry out several rounds of negotiations with the group. Several months into the discussions, PWG announced a merger with the Maoist Communist Centre (MCC), an ideologically similar outfit operating in the states of Bihar and Jharkhand. The merger led to the formation of the Communist Party of India–Maoist (CPI–Maoist). Eight months after its initiation, the peace process was called off in January 2005, with the state government demanding that the Maoists surrender their weapons, and the latter accusing the police of carrying out secret operations against them.[35] Violence re-erupted, and a massive re-launch of suspended police operations led to the collapse of the CPI–Maoist organization in Andhra Pradesh. That year, 167 extremists were killed and a large number arrested; 122 more the next year.[36] In view of such pressures, the Maoist remnants crossed over into neighboring Chhattisgarh. By 2007, the LWE problem in Andhra Pradesh had been decimated.

In the absence of the release of the relevant strategic and operational guidance, the components of this counterinsurgency effort, termed the "Andhra model," have remained a matter of speculation. Many point to the critical role played by the Greyhounds, a commando force raised in 1989, as key to the destruction of the Maoist clandestine infrastructure. This has led to other states clamoring to raise similar forces. It is also opined that the eight-months-long peace talks that allowed Maoist cadres to surface and organize public activities exposed them to the police intelligence network. Such information then served as the foundation of the police effort to target all identified cadre once the peace process collapsed.[37] Experts, however, assert that the "Andhra model" in reality was far more comprehensive and was based, they argue, on an intricate understanding of the operational dynamics of the Maoist movement and its protracted war strategy as assailed by years of capacity- and capability-building at the appropriate levels of the state response, especially the police.[38]

In fact, a series of measures had already been implemented to augment the capabilities of the state police as a fighting force. It was they who had primacy throughout the counterinsurgency. The relevant measures included rigorous training regimen for all the young police recruits from the levels of the Assistant Sub-Inspector of Police (ASI), including the Indian Police Service (IPS) officers; absorption of the former Greyhounds personnel, who had a short tenure of three years, into the district police, thereby augmenting the capacities at the district level; a comprehensive plan for area domination, including fortification of police stations, as well as modernization of weapons, communication, transport and support technologies; and delegation of the decision-making authority, making each police station a center of offensive action against the extremists.[39] To boost their morale, insurance cover was provided for all the 121,343 police personnel with an investment of Rupees 656.9 million. For families of those personnel killed on duty, the legal heir was promised to be paid the last pay drawn till the date of superannuation, allowed to retain government accommodation, and given a government job and a house site, besides a generous cash grant.[40]

Greyhounds operations remained central to neutralizing the Maoists, with operations supported by technical as well as human intelligence.[41] In this regard, popular disenchantment with extremism was important. A steady stream of village-level intelligence through a large network of informers ended Maoist ability to remain anonymous and to mount surprise attacks. The police establishment, in addition, implemented a comprehensive rehabilitation package for all the surrendering insurgents and helped them rebuild their lives through setting up small businesses or providing suitable income generating opportunities.

Parameters of engagement

Four parameters of engagement can be extrapolated from these successful counterinsurgency efforts. First, in each, a lead counter-insurgent force is clearly identifiable. In Mizoram, the Indian army had primacy, while in Punjab, Tripura, and Andhra Pradesh, it was the police. There, the DGP was the leader, conceptualizing and overseeing key decisions. The central forces, mostly belonging to the paramilitary bodies (who in aggregate, nationally, have approximately a million personnel) – and in some instances, the Indian Army – were used in support. The use of the Indian Army in counter-insurgency campaigns has been normally limited to theatres sharing borders with a foreign country or involving insurgents who have either maintained bases in neighboring countries or been supported by them. Thus, Mizoram and Punjab witnessed varying degrees of the Army's involvement. The left-wing extremists in Andhra Pradesh were dealt by the state and central police forces. Tripura, where insurgents maintained bases in Bangladesh and were said to have received some support from foreign sources, however, did not choose to use the Army. The police was trusted as capable enough to deal with the problem.

Second, mobilizing popular support for what sometimes turned out to be quite violent counter-insurgency campaigns was essential. Considered to be a legacy of the British effort against the Malayan Communist Party, Indian strategy has as a foundational element a focus on winning popular support. As a result, a conscious attempt at winning the battle of legitimacy has been integral, often preceding the actual launch of the campaign. While excessive violence perpetrated by the extremists turned the civilian population against them in Punjab and Andhra Pradesh, in Tripura and Mizoram, the counterinsurgency campaigns themselves involved controversial village reorganization schemes, which considerably shrunk the hiding space for the extremists, making them exposed to the security force campaigns. For these to be viable, a continuous effort at explanation and integration of popular concerns was essential.

Third, the mandate to deal with the violence of insurgents – with their terrorism, especially – necessarily involves kinetic response. Yet this was in the four cases tempered by the requirement to act proportionally and in moderation as allowed by circumstances. With the exception of Mizoram, where the Indian Air Force carried out bombing activities, at one point targeting the state capital, Aizawl, heavy weapons use was minimal and carefully chosen. None of the campaigns involved systematic use of artillery or heavy machine guns, the exception being the need for armor and its integral armament in the Blue Star assault. While none of the efforts treated above were prophylactic, with all at one point or another drawing criticism, due procedural care was taken to avoid collateral damage and thereby maintain popular support.

Fourth, security force operations, whether by the Army or the police forces, were used to prepare the ground for a political solution to the conflict. However, in this, a clear distinction was made between what was judged externally sponsored terrorism and home-grown insurgency. While the latter merited a political solution, brute force to decimate terror was considered a necessity in case of the former. As a result, Mizoram witnessed political intervention to secure peace. In Andhra Pradesh, the Maoists were given an option of negotiation and security force operations were used as the last resort. Punjab, where extremists secured foreign support, and Tripura, where the insurgents demonstrated no inclination for a peaceful resolution, saw the security forces allowed to perform the role of final arbiter.

Evidence is lacking as to whether these parameters of engagement were deployed in conscious fashion by participants to shape the counterinsurgency efforts discussed in the four distinct theatres of conflict. Adherence to them, though, obviously created the conditions for success. To the extent that they mirror known principles of war (e.g., unity of command and of effort), our assessment is reinforced.

Explaining festering conflicts

In recent years in India, fatalities due to violent extremism have declined. Guns have fallen silent across large parts of the country's Northeast, where some dozen insurgent groups are engaged in peace processes with the government. The security situation has also improved in Maoist-affected states. Even in the seemingly frozen conflict of Jammu and Kashmir, spikes in violence cannot obscure a downward trend. The resultant decline of extreme violence in all these theatres, however, coincides with a sense of unease, stoked by the prospects of extremist revival. Peace efforts in the Northeast, for instance, have generally not progressed beyond regroupment. A large number of groups (at times, more than 50 have been identified) continue to operate, both inside and also from secure locations in Burma. These include factions of the very groups engaging in negotiations. Similarly, albeit CPI-Maoist's current level of weakness, it continues to maintain its bases and to influence large swaths of the country, successfully carrying out attacks on security forces personnel and killing targeted civilians and local defense personnel. In Jammu and Kashmir, since 2015, a new round of militancy, building on the alienation and grievances of local youth, threatens to explode. While the four case studies of bringing insurgency to a close are indeed success stories, India's ability to contain and manage conflicts continues to be stretched, with no end in sight.

This section argues that it is possible to explain this intractability by examining the extent to which the four parameters of engagement outlined in the previous section have been realized or applied. Moving further, non- or partial-application of this strategic approach will be seen to have contributed to the longevity of the conflicts. In addition, this section attempts to explain whether decisions of application or non-application are shaped by external or internally held factors.

A lead counterinsurgent force

There is a widespread realization in strategic circles that irregular challenges are best dealt with by those closest to the problem, in this case, state police forces. In spite of the fact that three of the four internal wars were pacified with the state police force, India's quest for a lead counter-insurgent force continues to oscillate between employing the Army, the paramilitary units, and the state police. This results from a persistent weakness in state police forces both in terms of strength and capacity, as well as a kinetic bias in official circles when confronted with insurgencies.

Since the 1980s, efforts to improve state police capacity has been sought through a centrally funded police modernization scheme. Generous funds are made available to the states for improving training, arms, transport, communication, and facilities. Implementation of the program, though, which is the responsibility of the states, has been uneven. Only a few states have used the available funds, with any number diverting them to expenditures that do not strictly conform to the program's mandate. A dependency syndrome – that is, the Centre will have to assist us should the situation worsen – yet pervades the mind-set of the governments in most of the conflict affected states.[42] This makes them permanently dependent upon central forces.

Not surprisingly, in 2003, a Group of Ministers' report, *Reforming the National Security System*, recommended that the CRPF, the largest paramilitary force in the country, become the frontline counter-insurgency response nationwide and support police actions in various states.[43] The report sought to bring to an end the use of the Indian Army and the BSF in counter-insurgency and counterterrorism. Even in this, the government has faltered. CRPF's strength has increased from 137 battalions in 2001 to 239 battalions in 2016. Apart from performing a range of duties, such as VIP protection, riot control, routine law and order maintenance, the CRPF battalions are involved in counterinsurgency duties in almost all the conflict theatres. Yet the force is not designated primacy for the duty. Analysts have repeatedly pointed out a range of problems pertaining to training and command – among them, a crippling shortage of officers at the cutting-edge Assistant Commandant level[44] – that afflict CRPF. Over the years, its strength and performance have been affected by a steady stream of desertions, suicides, and personnel seeking early retirement.

With the state police being weak in a majority of the conflict theatres, and CRPF being unable to perform as an able supporting force, COIN responsibilities are being shared by other Central Armed Police Forces (CAPF) units and sometimes the Indian Army. The problem of deploying multiple forces has given rise to the problem of coordination, and command and control. Unified Command Structure under civilian leadership has been set up in all the conflict theatres to deal with the issues of coordination among the forces. However, establishing perfect coordination between state and central forces, and among the central forces has proved to be difficult and continues to affect effectiveness of COIN operations.

Gaining popular support

The Indian Army's *Doctrine for Sub-conventional Operations* states that it realizes application of an "iron fist in a velvet glove" (see Foreword). As published on the last day of 2006, the orientation has been embraced to emphasize the unity of kinetic and nonkinetic action, but in particular the criticality of gaining popular support. This is termed people-centric or population-centric, as practiced in the successful efforts previously considered. Popular engagement goes hand-in-hand with perception management. This gives the security forces a long rope should collateral damage occur, generates human intelligence, and helps strengthen the relationship between the civil and military authority.

"Winning hearts and minds" (WHAM) of the civilian population has remained one of the integral elements of all Indian counterinsurgency efforts, at least conceptually. The key is to use impact actions to provide a bridge between immediate needs and normal governance. Hence, the Indian Army and paramilitary forces carry out a range of activities that include arranging educational as well as recreational activities for students in conflict affected areas, providing medical facilities, distributing grains to farmers, providing artificial limbs, and even building community activity centers in villages. CRPF troops have even plowed fields in the Maoist-affected states in order to establish special bonds with the Tribals. State police, too, have attempted to implement low budget civic action programs under the WHAM imperative. Nonetheless, whereas the Indian Army's *Sadvabhavna* operations have been vigorously implemented since their inception in 1997, with a substantial leap in funding in 2007–2008 over the previous year,[45] paramilitary performance has lagged.

Lack of adequate funding for the WHAM projects is only a part of a series of issues faced by the security forces to gain popular support. First, problems regarding lack of coordination with civil authorities have led to duplication of efforts at the district level.[46] Civil authorities have complained of being stripped of their primary role and have alleged that the security forces end up being effectively positioned as "the government" in conflict-affected areas. Their participation in re-starting governance in areas liberated from extremist control, it is claimed, has been unenthusiastic. Second, lack of professionalism has resulted in shoddy implementation of projects. In Kashmir, for example, journalists highlight tales of corruption, scams, and faulty work under the civic action programs; e.g.:

> Three lengths of underground electrification cable were sold at a crossroads, a two-room primary school was built in Reasi without steps to the verandah, three feet off the ground and in one village, cooking appliances were distributed to every home, withdrawn the next day for a "flaw" to be repaired, and never seen again.[47]

Third, the challenge of positioning one as a strict custodian of law and at the same time as a friendly policeman has often been less than successful. While the Indian Army highlights *Sadbhavna* as a game changer in Kashmir, recurrent incidents of mob attacks on Army facilities, especially in July and August 2016, highlight the hollow nature of such claims and the fragility of progress. Allegations of widespread human rights violations by security forces and the police in the Northeastern and Maoist-affected states make WHAM problematic.

A question of human rights

Ivan Arreguín-Toft argues that barbarism – "a state or non-state actor's deliberate and systematic injury of non-combatants during a conflict … increases the costs and risks of military operations, and poisons chances for peaceful post-war occupation and development."[48] Even as exceptions surface to the Indian state's preference for avoidance of overwhelming force, there has been a heartening consistency of approach. Army operations against the Naga insurgents serve to illustrate. As the Naga tribes raised their banner of revolt, seeking an independent homeland immediately after India's independence from the British, the Indian Army in the mid-1950s sought to use airpower against the insurgents. Prime Minister Jawaharlal Nehru rejected the proposal and emphasized that what was at hand was a political problem. "Though there were rumblings within the army about being forced to fight with one hand tied behind their backs, the army accepted these political limits on the use of force."[49] The use of the Indian Air Force in Mizoram thus remains an exception. Even in Punjab, where the state police have been accused of countless human rights abuses, including pressurizing family members of active militants, barbarism never became an element of the counterinsurgency.

Still, in policy circles, there has in recent years been pressure to use unrestrained and disproportionate force against insurgents. As Home Minister, Mr. Chidambaram not only initiated Operation Green Hunt, which massed a large number of paramilitary battalions with an objective of inflicting a swift and decisive defeat on the Maoists, but also lobbied for using the Indian Army. His proposal was not acted upon as a consequence of the Indian Army's principled opposition to being involved in internal wars where a clear nexus between the insurgent and a foreign country does not exist. Since 2010, though, a shift in counterinsurgency policy appears to have been set in motion. The trend has continued under the Bharatiya Janata Party (BJP)-led government.

Not only has the government taken to describing insurgencies as terrorism, but Home Minister Rajnath Singh reportedly has urged state DGPs not to bother about human rights issues while conducting operations against the Maoists.[50] Under the BJP government, a spate of civilian casualty incidents has resulted from counter-Maoist operations in Chhattisgarh and Odisha. Additionally, the Chhattisgarh police have been repeatedly accused of harassing NGOs, civil society activists, and journalists, accusing them of supporting the insurgents. Indiscriminate use of pellet guns on civilian protesters in Kashmir that resulted in hundreds of injuries in July and August 2016, some quite serious, provides further illustration of unrestrained kinetic application. In spite of protests, the government has defended its hardline approach.

Privileging the requirements of the security forces is similarly seen in the continuance of the controversial AFPSA in some of the Northeastern states and Kashmir despite persistent demands for its withdrawal. Since the Indian Army regards AFSPA as an enabling piece of legislation and has refused to operate without it, the government has disregarded several strictures on the act by commissions of inquiry and has refrained from implementing former Prime Minister Manmohan Singh's assurance that the Act would be softened. Opposition to such tactical drift consequent to policy hardline sentiments is particularly odious given widespread recognition in India that targeting the population is a legacy of the past suitable "only for colonial or interventionist armies fighting in foreign lands; that the internal wars that afflict the developing countries are an altogether different military problem."[51] Obviously, what India is facing is objectively not those fighting foreign occupation – even if some of the alienated see the situation in those terms – but marginalized populations, overwhelmingly "ethnic minorities inhabiting geo-politically sensitive border areas" who seek to "move into the process of negotiating new terms of integration with the central authority," albeit some have apparently been induced to "take up arms at somebody else's bidding and promise of support."[52] As such, a population-centric approach is the only viable, ethical option.

Still, in the formulation just offered, it can be seen that a subtle shift in the terminology and orientation can provide justification for those presently in power to treat insurgency as terrorism, dismiss notions of popular support undergirding such movements, and set aside the need of the state to cultivate popular support. Legitimacy as the object of the exercise gives way to compulsion, as is on full display today in Tibet, to India's north, and, further so, Xinjiang. There, colonial imperatives of population and resources control, conjoined to leadership targeting, have been enabled by technology to the extent that entire populations live in open-air incarceration, with thousands of others confined in the old-fashioned concentration camps of imperial yore.

For a democracy, such orientation is not possible or acceptable. Efforts to use ostensibly innovative short-cut measures to speed along embrace of the political opportunity structure, not surprisingly, have been exploited by insurgent movements which have invested in doing away with fixed centers of gravity and set bases, thus demonstrating appreciable powers of withstanding new counterinsurgency assaults. Moreover, use of force without human intelligence has led to a rising level of collateral damage, thereby widening the divide between the civilian population and the government. One thinks immediately of the extraordinarily misdirected effort in West Bengal aimed to cow Nepali estrangement in Darjeeling not through incorporation but kinetic repression.

Preparing ground for a political solution

Such an example leads naturally to a final and fundamental consideration: Whether a force-centric counterinsurgency approach is an end unto itself or an instrument that prepares the ground for a political solution remains as much a question for Indian policymakers as it has been historically for all authorities faced with armed resistance. Rajesh Rajagopalan writes that until 1980s, there was little indication in military writings of any doubt that insurgencies could be *militarily* defeated. "Despite both the recognition of the political roots of the problem of insurgency and the need for close involvement of the civil administration in counterinsurgency, the faith in a military victory remained."[53] By the mid-1980s, this had changed, and writings no longer displayed such faith in kinetics. "Army officers now appeared convinced that only political solutions could end insurgencies,"[54] he argued.

Yet enter a political class that embraces the fallacies of the Raj and hence evinces a deep faith in the viability of the "force first approach," and the implications become clear. First, it must be noted that with the exception of Mizoram, the other counterinsurgency efforts examined – Punjab, Tripura, and Andhra Pradesh – did not have intrusive political intervention. Ironically, in spite of the reality that insurgencies are political uprisings, direct interventions consequent to administration policy designs have not had much success in India's counterinsurgency history. Far from spurring caution, this seems now to have spurred a doubling down on force as the preferred choice for seeking to end conflicts. Second, even the most chartable view – that the "force first approach" is seen as a way to weaken insurgent challenge and enable the government to bargain from a position of strength – the reality is the lackadaisical efforts in moving peace processes to their logical conclusions has served to thwart tangible benefit of redoubled force while leaving behind the negative residue. The Northeastern region, where peace processes between the government and insurgents have continued without an end in sight, provide an object illustration. Some of these frustrated outfits in ceasefire have either abandoned the peace process or have continued with their activities of extortion, abductions, and killings even while negotiating with the government. Under these circumstances, peace processes have curiously coexisted with operations by the security forces against these same outfits.

Third, the failure of state police and even the paramilitaries to restore normalcy has been the reason for the use of the Indian Army in counterinsurgency and associated operations, reinforcing the utility of ultimate force. For example, the Indian Army was redeployed in strife-torn Kashmir in September 2016 after the state police and the CRPF failed to restore normalcy.[55] As the state demonstrated signs of lapsing into a dangerous state for the start of religious extremism with the collapse of administration in many districts, the Army was deployed for the first time since 2014. With its superior training and access to sophisticated arms, the Army invariably takes the violent posture of the state to a higher level.

Conclusion

If we examine this discussion, it seems that indeed India has a strategic approach to counterinsurgency – if it desires to examine its cases of success. The parameters of engagement that constitute this strategic, with few aberrations, have resulted from the conscious political decisions of the actors concerned, as shaped by circumstantial factors. Apart from Mizoram, where the creation of Bangladesh in 1971 boosted the counterinsurgency by shrinking sanctuary for the insurgents, successes in Punjab, Tripura, and Andhra Pradesh have been achieved by a scrupulous adherence to the parameters of engagement. Failure to resolve continuing insurgencies is likewise linked to the opposite, to non-adherence to these same parameters of engagement.

Such a failure is indeed strange but is not without reasons. At one level, failure to modernize police forces and consequent reliance upon the Indian Army as the last resort for dealing with the extremists creates administrative as well as political problems. At another level, the failure is rooted in the deep sense of safety that the political class has enjoyed, being far removed from the conflict theatres. With the exception of Kashmir, most conflicts do not directly impinge upon India's security to a significant extent. Even Kashmir has remained a peripheral problem, situated at one far corner of the country – and a very small corner at that. At no point of time could the insurgents wage urban warfare threatening the territorial integrity of India. Similarly, insurgency in the remote Northeast, claiming as it did in its heyday hundreds of the lives and disrupting governance, was never considered a national threat in New Delhi. The CPI–Maoist, which was once assessed to have been a threat in some one-third of the country's territory, with its announced goal of capturing political power by 2050, never constituted a threat to India's political power centers. This has allowed the political class to pursue an ad hoc counterinsurgency approach, with no attempt made to follow a strategic approach. The failure to designate and develop the CRPF as the lead counterinsurgency force is one example of this indifferent approach.

Likewise, the lack of a long history of political successes in resolving insurgencies and addressing extremism has led India to increasingly depend upon force in internal matters. As discussed, the effort in Mizoram prepared the ground for a political solution to the problem. The peace process facilitated a decisive campaign against the CPI–Maoist in Andhra Pradesh. Barring these few examples, failure has been the norm, as illustrated by the stillborn treaties in

Northeastern states where dissatisfied insurgent factions have walked out and continued their efforts against the Indian state. In this, they have not been unlike the unsuccessful attempts to engage the extremists in Kashmir, which has now again descended into a "force first" policy.

In these circumstances, India's even attempting to implement a counterinsurgency strategic approach appears remote. As the violent potential of the various insurgents grows, the shift to treating political uprisings as terrorism is most likely to evince more acute symptoms. This, in turn, will facilitate greater reliance on the security forces to end conflicts. With a range of deficiencies affecting their performance, they will be ill-placed to deliver on the expectations of the political class. Worse still, human rights violations and lack of effective political intervention will keep the fire of disenchantment burning. It is indeed a vicious cycle.

Notes

1 Daid J Karl, "Is India Ready for Prime Time?"
2 Then Indian Army Chief J. J. Singh while unveiling the Doctrine described, "It is our collective wisdom based on the experience of five decades of fighting low-intensity conflict, including proxy war, counter-insurgency and limited conflict." See Integrated Defence Staff, *Doctrine of Sub-Conventional Operations*.
3 Shapoo, "Red Salute: India's Maoist Maelstrom and Evolving Counterinsurgency Doctrines."
4 Rajagopalan, "Insurgency and Counterinsurgency."
5 Singh, *India's Emerging Land Warfare Doctrines and Capabilities.*
6 Jafa, "Counterinsurgency Warfare: The Use and Abuse of Military Force."
7 Waterman, *Beyond Classical Counterinsurgency: Modelling the Indian Experience.*
8 Fidler, "The Indian Doctrine for Sub-Conventional Operations," 207–224.
9 Verghese, *India's Northeast Resurgent: Ethnicity, Insurgency, Governance, Development*, 141.
10 Ibid.
11 The move to set up grouped villages followed the failure of Operation Blanket in which the Army tried to isolate MNF cadres by deploying 20-man pickets outside every village in Mizoram, which were charged with preventing the entry of insurgents into villages, in addition to providing civic assistance to the population and generally showing the flag. The program failed, because there were far too many villages to cover, even with Army's seemingly inexhaustible manpower resources. See Rajagopalan, op. cit., 161.
12 Verghese, *India's Northeast Resurgent*, op. cit., 143.
13 Ibid. 145.
14 Subsequent findings, however, affirm that the terrorists taking shelter within the Golden Temple had themselves damaged the structure and had transformed it "into a large reinforced pillbox with weapons facing all directions." See Gill, *Punjab: The Knights of Falsehood*, 94.
15 South Asia Terrorism Portal, *Punjab: Annual Fatalities in Terrorist Related Violence 1981–2016.*
16 Mahadevan, "The Gill Doctrine: A Model for 21st Century Counter-terrorism?"
17 Gupta and Sandhu, "True Grit."
18 Ibid.
19 Telford, "Counter-Insurgency in India: Observations from Punjab and Kashmir."
20 Mahadevan, "The Gill Doctrine: A Model for 21st Century Counter-terrorism?," op. cit.

21 Mathur, "Secrets of COIN Success: Lessons from the Punjab Campaign."
22 Mahadevan, "The Gill Doctrine: A Model for 21st Century Counter-terrorism?," op. cit.
23 South Asia Terrorism Portal, *Backgrounder-Punjab.*
24 Human Rights Watch, *Protecting the Killers: A Policy of Impunity in Punjab, India.*
25 Such criminal activities included abduction and rape carried out by the extremists. As the police pressure increased, they sought shelter in the civilian homes across the countryside. Many instances where they demanded not just food and shelter but forced sex with young women in the house emerged. Abductions, rape, and land grabbing by the extremists became common place thereby alienating the extremists from the common people. See Gill, *Punjab: The Knights of Falsehood*, op. cit., 101–103.
26 South Asia Terrorism Portal, Punjab: Annual Fatalities in terrorist related violence 1981–2016, op. cit.
27 Sahaya, "How Tripura Overcame Insurgency."
28 Routray, "Tripura: Creating an Unenviable Record."
29 Ibid.
30 Routray and Sahni, "Against All Odds."
31 Sahaya, "How Tripura Overcame Insurgency," op. cit.
32 By 2012, 730 kilometers of the international border had been fenced.
33 On October 1, 2003, the PWG planted claymore mines on the route of Chief Minister N. Chandrababu Naidu. The Chief Minister's bulletproof car came under the impact and injured him and few other accompanying politicians seriously. Two days later, on October 3, the PWG claimed responsibility for the attack. Devarajan, "Chandrababu Naidu Survives Landmine Blast."
34 Annual Report 2003–2004, Ministry of Home Affairs, Government of India, 41.
35 Farooq, "Why Peace Collapsed in Andhra Pradesh."
36 South Asia Terrorism Portal, *Fatalities in Left-Wing Extremist Violence in Andhra Pradesh: 1968–2016.*
37 Author's interview with a retired senior police official of Andhra Pradesh, New Delhi, August 27, 2016.
38 Sahni, "Fighting The Maoists With Mantras."
39 Ibid.
40 India Today, "Andhra Model."
41 Incidentally, Greyhounds' operations in neighboring states like Odisha have not been successful. A Maoist attack on the Greyhounds personnel in Chitrakonda on June 29, 2008 resulted in the death of 36 Greyhounds personnel.
42 One of the few exceptions in this case is the police of Jammu and Kashmir, which handles a bulk of the COIN duties along with the CAPFs and the Indian Army.
43 Swami, "India's Counter-insurgency Conundrum."
44 Assistant Commandant's typically are responsible for handling a Company or about 125 troops. Since induction has not kept pace with the expansion of the force, most battalions are forced to manage with just half of their sanctioned strength of Assistant Commandants.
45 In 1997, Operation Sadbhavna was formally launched with a fixed budget which was allocated to each formation on the basis of the prevailing security situation. Four core areas initially identified for expenditure were Education, medical facilities, small scale infrastructure, and national integration. Later, women empowerment and human resources development were added to the list. Allocation increased from Rupees 510 million 2006–2007 to Rupees 787 million in 2007–2008.
46 Hasnain, "Sadbhavna is a Humanitarian Exercise, Nothing Else."
47 Devadas, "The Army Problem in Kashmir that No One Wants to Talk About."
48 Arreguín-Toft, *The [F]utility of Barbarism: Assessing the Impact of the Systematic Harm of Non-combatants in War.*
49 Rajagopalan, "Insurgency and Counterinsurgency," op. cit.

50 "Rajnath Singh: As UP CM, I gave free hand to police to tackle Maoists," *India Today*, September 3, 2014, http://indiatoday.intoday.in/story/rajnath-singh-maoists-uttar-pradesh-chief-minister-encounters/1/380692.html. Accessed August 3, 2016.
51 Jafa, "Counterinsurgency Warfare: The Use and Abuse of Military Force," op. cit.
52 Ibid.
53 Rajagopalan, *Fighting Like a Guerrilla: The Indian Army and Counterinsurgency*, op. cit., 165.
54 Ibid.
55 Sreemoy Talukdar, "Kashmir Unrest: Army Deployment a Tacit Admission that India's Writ Over Valley is Slipping Away."

Bibliography

Arreguín-Toft, Ivan. *The [F]utility of Barbarism: Assessing the Impact of the Systematic Harm of Non-combatants in War*, (August 2003) available at http://live.belfercenter. org/publication/17413/futility_of_barbarism.html (accessed August 12, 2016).

Devadas, David. "The Army Problem in Kashmir That No One Wants to Talk About," *The Wire*, (April 21, 2016) available at http://thewire.in/30607/the-army-problem-in-kashmir-that-no-one-wants-to-talk-about/ (accessed July 9, 2016).

Devarajan, A. "Chandrababu Naidu Survives Landmine Blast," *The Hindu*, (October 2, 2003) available at www.thehindu.com/2003/10/02/stories/2003100207560100.htm (accessed August 29, 2016).

Farooq, Omer. "Why Peace Collapsed in Andhra Pradesh," *BBC*, (January 20, 2005) available at http://news.bbc.co.uk/2/hi/south_asia/4183997.stm (accessed August 12, 2016).

Fidler, David P. "The Indian Doctrine for Sub-Conventional Operations." In Sumit Ganguly and David P. Fidler (eds.). *India and Counterinsurgency Lessons Learned*. London: Routledge, 2009.

Gill, K. P. S. *Punjab: The Knights of Falsehood*. New Delhi: Har Anand Publications, 1997.

Gupta, Shekhar and Kanwar Sandhu. "True Grit," *India Today*, (April 15, 1993) available at http://indiatoday.intoday.in/story/police-chief-k.p.s.-gill-turns-the-tide-in-punjab-with-controversial-and-ruthless-methods/1/302060.html (accessed July 12, 2016).

Hasnain, Lt. Gen. Syed Ata. "Sadbhavna is a Humanitarian Exercise, Nothing Else," *Daily Excelsior*, (December 22, 2015) available at www.dailyexcelsior.com/sadbhavna-is-a-humanitarian-exercise-nothing-else/ (accessed July 2, 2016).

Human Rights Watch. *Protecting the Killers: A Policy of Impunity in Punjab*, (October 17, 2007) available at www.hrw.org/report/2007/10/17/protecting-killers/policy-impunity-punjab-india (accessed July 23, 2016).

India Today, "Andhra Model," (April 8, 2010) available at http://indiatoday.intoday.in/story/Andhra+model/1/91900.html (accessed August 12, 2016).

India Today, "Rajnath Singh: As UP CM, I Gave Free Hand to Police to Tackle Maoists," (September 3, 2014) available at http://indiatoday.intoday.in/story/rajnath-singh-maoists-uttar-pradesh-chief-minister-encounters/1/380692.html (accessed August 3, 2016).

Integrated Defence Staff. *Doctrine of Sub-Conventional Operations*, Government of India, (December 2006) available at http://ids.nic.in/Indian%20Army%20Doctrine/doctrine%20sub%20conv%20ow.pdf (accessed August 12, 2016).

Jafa, Virendra Singh. Counterinsurgency Warfare: The Use and Abuse of Military Force," *Faultlines: Writings on Conflict and Resolution*, 3, www.satp.org/satporgtp/publica tion/faultlines/volume3/Fault3-JafaF.htm. (accessed August 12, 2016).

Karl, Daid J. "Is India Ready for Prime Time?," *Asia Policy*, 12, The National Bureau of Asian Research, Washington, DC, (July 2011) available at www.nbr.org/publications/asia_policy/free/ap12_h_indiabre.pdf (accessed July 23, 2016).

Mahadevan, Prem. "The Gill Doctrine: A Model for 21st Century Counter-terrorism?," *Faultlines*, (April 2008) available at www.satp.org/satporgtp/publication/faultlines/volume19/Article1.htm (accessed July 27, 2016).

Mathur, Anant. "Secrets of COIN Success: Lessons from the Punjab Campaign," *Faultlines: Writings on Conflict and Resolution*, 20, (January 2011) available at www.satp.org/satporgtp/publication/faultlines/volume20/Article2.htm (accessed July 31, 2016).

Ministry of Home Affairs, *Annual Report 2003–04*. Government of India.

Rajagopalan, Rajesh. *Fighting Like a Guerrilla: The Indian Army and Counterinsurgency*. New Delhi: Routledge, 2008.

Rajagopalan, Rajesh. "Insurgency and Counterinsurgency," *Seminar*, no. 599, (July 2009) available at www.india-seminar.com/2009/599/599_rajesh_rajagopalan.htm (accessed September 1, 2016).

Routray, Bibhu Prasad. "Tripura: Creating an Unenviable Record," *South Asia Intelligence Review*, 2 (14), (October 20, 2003) available at www.satp.org/satporgtp/sair/Archives/2_14.htm (accessed July 23, 2016).

Routray, Bibhu Prasad and Sahni, Ajai. "Against All Odds," *Outlook*, (August 31, 2005) available at www.outlookindia.com/website/story/against-all-odds/228448 (accessed July 20, 2016).

Sahaya, D. N. "How Tripura Overcame Insurgency," *The Hindu*, (September 22, 2011) available at www.thehindu.com/opinion/lead/how-tripura-overcame-insurgency/article2465348.ece (accessed July 27, 2016).

Sahni, Ajai. "Fighting The Maoists With Mantras," *Outlook*, (July 25, 2008) available at www.outlookindia.com/website/story/fighting-the-maoists-with-mantras/237994 (accessed August 13, 2016).

Shapoo, Sajid Farid. "Red Salute: India's Maoist Maelstrom and Evolving Counterinsurgency Doctrines," *Small Wars Journal*, (July 8, 2016) available at http://smallwarsjournal.com/printpdf/47347 (accessed September 1, 2016).

Singh, Harinder. *India's Emerging Land Warfare Doctrines and Capabilities*, RSIS Working paper, (October 13, 2010) available at www.idsa.in/system/files/WP210.pdf (accessed August 11, 2016).

South Asia Terrorism Portal. *Backgrounder-Punjab*, www.satp.org/satporgtp/countries/india/states/punjab/backgrounder/index.html (accessed July 23, 2016).

South Asia Terrorism Portal. *Punjab: Annual Fatalities in Terrorist Related Violence 1981–2016*, www.satp.org/satporgtp/countries/india/states/punjab/data_sheets/annual_casualties.htm (accessed July 23, 2016).

South Asia Terrorism Portal, *Fatalities in Left-Wing Extremist Violence in Andhra Pradesh: 1968–2016*, www.satp.org/satporgtp/countries/india/states/andhra/data_sheets/annual_casualties.asp (accessed August 13, 2016).

Swami, Praveen. "India's Counter-insurgency Conundrum," *The Hindu*, (July 23, 2010) available at www.thehindu.com/opinion/lead/indias-counterinsurgency-conundrum/article528762.ece (accessed July 21, 2016).

Talukdar, Sreemoy. "Kashmir Unrest: Army Deployment a Tacit Admission that India's Writ Over Valley is Slipping Away," *First Post*, (September 10, 2016) available at www.firstpost.com/india/kashmir-unrest-army-deployment-is-sign-that-centre-recognises-the-states-authority-2999002.html (accessed September 12, 2016).

Telford, Hamish. "Counter-Insurgency in India: Observations from Punjab and Kashmir," *The Journal of Conflict Studies*, 21 (1), (Spring 2001), https://journals.lib.unb.ca/index. php/jcs/article/view/4293/4888 (accessed July 24, 2016).

Verghese, B. G. *India's Northeast Resurgent: Ethnicity, Insurgency, Governance, Development*. New Delhi: Konark Publishers, 1997.

Waterman, Alex. *Beyond Classical Counterinsurgency: Modelling the Indian Experience*, www.polis.leeds.ac.uk/assets/files/students/student-journal/Winter-2014/Waterman-Beyond-classical-counterinsurgency.pdf (accessed July 23, 2016).

9 Countering violent extremism

The Singapore experience

Mohamed Bin Ali

Introduction

Terrorism and violent extremism have affected Southeast Asia considerably in the past decade. The region faced the threat of Jemaah Islamiyah or JI[1] in the late 1990s. In 2001, the Singaporean authorities uncovered a plot by JI members operating in the country and carried out a first wave of JI arrests in December of that year. JI members planned to conduct several terrorist attacks in the city. The plans included assaults on local critical infrastructure, the transportation system, and several diplomatic targets to include the U.S. Embassy.[2] With this arrest, the plan to bomb these targets was disrupted.

In Singapore, the threat of terrorism and violence is not new. The country had endured racial riots, confrontations and even international terrorism in the past.[3] Unlike previous threats however, the violence from JI is religiously motivated, and this has posed a new challenge for the government and its society. To counter it, the government developed a holistic and comprehensive approach that included rehabilitation of those arrested. Apart from direct initiatives, the authorities also initiated programs that aimed to strengthen the relationship and trust between the different religious communities in Singapore. This was critical, as terrorism and violent extremism have the potential to disrupt social order through exploiting racial and religious faultlines. Thus, ever since the plot of JI network was discovered, the government and community have focused on maintaining social stability amongst its people from different religions and races.

Another important element was the government's endorsement of community-based initiatives to co-exist alongside more traditional counter-terrorism measures. This came from the realization that the affected community would be in the best position to locate the local sources of misunderstanding or grievances, thus facilitating targeted solutions. This gave rise to local community-based initiatives such as rehabilitation conducted by the Religious Rehabilitation Group (RRG) and social services extended by the Aftercare Group (ACG) during the early days of the JI threat. RRG comprises a group of local Muslim clerics who provide religious counseling to JI detainees, while ACG was formed to assist the families of those arrested. The Islamic Religious Council of Singapore, a governmental body responsible for Islamic and Muslim affairs along with local mosques also initiated programs which aimed to counter radical ideologies and prevent radicalization in the community.

Since 2001, the threat of extremism from violent Islamist groups in Singapore has evolved from JI to the phenomenon of self-radicalization where individuals have come under the influence of ISIS narratives and propaganda. This has spurred the Singapore government and its community to continue to develop new strategies to ensure the security of the nation and to prevent such radicalization.

This chapter provides a brief overview of the JI threat in Singapore. It analyzes the efforts taken by the Singaporean government and community to counter the threat. The chapter focuses on the efforts and contribution of RRG members in their counter ideology and rehabilitation work. The chapter also highlights the social rehabilitation and counter extremism initiatives of other community groups such as ACG and MUIS. The chapter concludes by sharing some important lessons learnt from Singapore's experience in CVE.

The threat of JI in Singapore

Singapore's global profile and its affiliation with Western countries such as the U.S. have made it a target among for extremists and terrorists in the region. The existence of JI in Southeast Asia came to the attention of the governments in the region after September 11, 2001, when Singapore's Internal Security Department (ISD) disrupted a JI plot against American, British, Australian, and Israeli diplomats in December 2001, as well as against other targets. As a consequence, Singapore began collaborating with law enforcement, security, and intelligence agencies of Malaysia, the Philippines, and Australia to dismantle the regional JI network.

JI was known to be an al-Qaeda associate. It was construed that al-Qaeda had funded the JI attack on the region's well known Indonesian tourist resort, Bali, in October 2002. The attack killed 202 people, including 88 Australians. Despite being targeted by the Indonesian authorities, however, JI survived, and its subsequent years saw continued, intermittent attacks.

The arrest of JI members in Singapore revealed the need for an improvised counter-terrorism strategy as we were dealing with ideologically motivated movements. JI members portrayed a sense of exclusiveness and used Islamic concepts such as *Jihad* (struggle or fighting) to justify their violent acts under the banner of Islam. As such, the threats and acts of terrorism in Singapore were more than just the potential destruction of lives and properties. They could also endanger the harmonious, multi-racial fabric of society, a cornerstone of the nation's progress, and lead to possible distrust and discord among the multireligious population. Such challenge could not be managed by law alone. The long-term elimination of the threat required elimination of extreme and radical ideologies from the community.

Responding to JI: operational and strategic approaches

Terrorism needs to be dealt with at two levels – the hard edge of operations and the soft edge to address generators of marginalization and alienation. The first involves the use of law enforcement agencies and personnel such as the police, army, and intelligence. The second, equally important, aims to respond to the long-term threat. This includes countering of terrorist ideology and the disruption of terrorist financing. Both approaches have been adopted in Singapore. Operationally, the authorities adopted a new integrated national approach structured around the prevention, protection, and response domains in Singapore's national security program. On the soft side, countering ideological support to terrorism (CIST) has become crucial, because the ideology and extremist messages articulated by groups such as al-Qaeda, ISIS, and JI are widely and effectively disseminated through all forms of media. This includes the Internet, which has a powerful appeal in much of the Muslim world.

The beliefs and motivations of these movements must be eroded to effectively counter the extremist ideology. This can be achieved by cutting off the supply of recruits and the elimination of their financial support networks. A campaign is also required to prevent the metastasizing of terrorism to new regions. This demands the need for an ideological counterstrategy, as well as concerted efforts by the government and the community to respond to the pervasive, ever-spreading ideology of terrorists.

The present threat embraces a strategy that seeks to bring about a global transformation consequent to their ideological struggle against their adversaries. Using religion as both tool and disguise, they actively target and indoctrinate the wider Muslim community using all methods and platforms available. These include the use of lectures and pamphlets, together with the utilization of mosques, educational institutions, and private residences. Emphasis is placed on Information Technology, especially the internet and social media, to spread their vision. Relatively cheap, largely unregulated, and able to reach millions of people, the internet serves as an ideal instrument for disseminating their ideological themes, providing moral inspirations and recruiting new supporters. Singapore faced a similar phenomenon with the arrival of extremist groups such as JI. The government's pro-activeness and its co-enlistment of the Muslim community led to the formation of an ideological counterstrategy to deal with extremist ideological threat.

Despite its appeals to the larger Muslim world, the extremist ideology runs contrary to Islam's mainstream traditions. A majority of the global Muslim population rejects violent extremism in the name of Islam. Violent Extremism movements have badly tarnished Islamic concepts such as *Jihad*, often contextualizing it incorrectly to justify criminal behaviors, even as they seek to position themselves as contemporary manifestations of Islam's heroic past. Informing and educating those not yet affected by these ideologies is the best long-term strategy in order to sustain the fight against extremist ideology in the long run. Additionally, rehabilitation and reformation can help those potentially affected by these dangerous ideologies to create a new path for their lives.

Singapore's unique counter-ideological program

Terrorism occurs when ideological motivation meets operational capability. The way a terrorist group shapes its radical worldview and its publicly disseminated messages plays an important role in the public interface between the group and its target audience. A group can successfully indoctrinate the public to become sympathizers, mobilize supporters, and recruit members through its methods of propaganda.

A multi-pronged approach is needed to counter terrorism effectively. The ideological counter in the "war on terrorism" should include not only a "shooting war" or law enforcement operation but a "war of ideas," as well. The response needs to disrupt and degrade a terrorist group's military and economic infrastructure and target the organization's political apparatus. If left unchecked, this apparatus will continue to mobilize political support and logistical assistance, eventually generating new recruits.[4]

At the Committee of Supply Debate on the Ministry of Home Affairs on March 3, 2005, former Minister for Home Affairs Wong Kan Seng was questioned on the ideological efforts in confronting terrorism. He said that the government's approach has been to encourage the Muslim community to police itself against ideological attacks by terrorist groups such as JI as the community is in the best position to ensure that its own members are ideologically inoculated against incorrect and dangerous religious teachings.

There is also a dedicated group of Islamic leaders and teachers who provide religious counseling for JI and Moro Islamic Liberation Front (MILF) members detained or under International Security Act's (ISA) Restriction Orders. It is part of the Internal Security Department (ISD)'s overall religious rehabilitation program instituted for JI detainees and their families. This group, together with a few local mosques, has also extended contributions to the welfare of detainee families. Apart from house visits, educational assistance and upgrading courses are also offered to the families to ensure they are able to cope with the situation emotionally, socially and economically.

The four types of rehabilitation that are carried out for Singapore's detainees include:

1 *Family Support* – By keeping their families informed and in contact with them. This is also done through family integration and emotional support.
2 *Financial Support* – For the spouses and children of the detained JI members
3 *Psychological Counselling* – For individual wellbeing.
4 *Religious Counselling* – This is aimed at explaining the true concept of *Jihad* (struggle), *Ummah* (Islamic Community), *Bai'ah* (Pledge of Allegiance), *Daulah Islamiah* (Islamic State), and other Islamic concepts manipulated by JI.

Religious counseling for detained JI members

Religious counseling sessions in Singapore were conducted for three groups of people: detained JI members, JI members placed under Restriction Order (ROs), and their family members. These religious counselling sessions were conducted by members of the RRG, who provided the detainees with a clear understanding of the religious concepts previously misinterpreted. The counselling sessions helped those detained to realize their earlier destructive ways and assisted them in overcoming their feelings of betrayal of fellow JI members during ISD investigations.

These investigations showed that JI's terrorist plans have been based on religious ideology. *Jihad* is actively promoted as a religious duty for the purpose of committing acts of terror, and Islam was used as a tool for terrorism. JI's misinterpretation and misunderstanding of certain religious concepts led it to believe that it was on the crimes it committed against individuals and society. These misunderstandings reached a dangerous level and required definition of the actual meaning of *Jihad* to place the concept in proper perspective.

Singapore's religious counselling assisted JI detainees in uncovering misinterpretations and incorrect beliefs in their religious doctrine. This was achieved by providing them with a better perspective and understanding of Islam. The detainees had to be inoculated, because failure to correct their misconceptions – especially their belief that *Jihad* is war and that it is *Fardu Ain* or compulsory for all Muslims – would render them a potential time bomb when released. The JI belief system and the understanding of *Jihad* had to be addressed and defined to ensure they did not return to their violent ways, posing a future threat to Islam and Singapore.

The religious rehabilitation group (RRG)

Unlike the use of hard counter-terrorism approaches, Singaporean leaders knew that the Singapore Muslim population was facing the threat of radicalization by violent extremist groups based outside the country and that it was essential to partner with the Muslim community to reach out to the vulnerable. While only a very small number of Singaporean Muslims were detained for terrorist-related activities, they could not be held indefinitely, so the country had to develop approaches to meet the contemporary challenges of ideological extremism.

Prominent Islamic scholars[5] were invited to an initial dialogue with the JI members by the local security agencies. These scholars assessed and concluded that the religio-ideological component of the JI movement needed to be dealt with in order to achieve a longer term strategic response to this form of terrorist threat. According to them, the grave danger of JI's religio-ideological inclination needed to be treated as a concern to Singapore's security, thus addressed. This led to the formation of the RRG.

RRG was officially inaugurated in April 2003. It originally had 11 members and is now over 46 members strong. These members consist of mainly *asatizah* (religious teachers) drawn from diverse age groups, careers, and educational backgrounds. RRG also includes a secretariat made up of members from the *asatizah* and other non-religious backgrounds. It functions as administrative support.

RRG's main and initial task is provide religious counseling to the JI detainees and their family members. Today, counseling efforts by the RRG have been extended to include self-radicalized individuals, those influenced with ISIS narratives, and anyone deemed to possess radical and extremist views. The group's other objective is to serve as an expert resource panel in assisting the government and the community's understanding of Islam.

Counselling sessions discuss concepts pertaining to *Jihad, Hijrah, Bai'ah, Ummah,* and *Daulah Islamiyah,* and refute JI's conceptualization of the terms.[6] Apart from examining these traditions, RRG members also have a duty in educating the detainees on the peaceful and moderate message of Islam and the universal Islamic and moral values, topics that are deemed important in understanding the religion. The counseling of the detainees is a long-term process that requires perseverance. While counselling efforts were ongoing, RRG broadened its activities to include educational and social programs. These programs aim at proactively educating the wider public on the dangers of extremist ideologies and preventing Muslim youths from being lured by deviant teachings.

Today, the concept of religious rehabilitation, particularly for Islamist militants, has gained wide acceptance both locally and internationally. Many governments have realized that religious rehabilitation is an important component to formulate a more effective counter-extremism and -terrorism strategy. This can be seen in countries such as Saudi Arabia, Yemen, Pakistan, Egypt, Bangladesh, Malaysia, and Indonesia.[7]

Singapore's approach to religious rehabilitation, in particular, has accrued interest of many governments and scholars. Due to the importance and interest of such efforts, RRG members have traveled widely to share Singapore's approach of detainee rehabilitation with the authorities in many countries including Malaysia, Indonesia, Saudi Arabia, United Kingdom, United States, Germany, Belgium, Austria and Denmark. Bruce Hoffman of Georgetown University (Washington, DC) states:

> The path-breaking work of Singapore's Religious Rehabilitation Group (RRG) provides a model and inspiration for counter-radicalisation efforts everywhere. The RRG's outreach efforts not only to radicals but to their families are a seminal example of the most innovative and novel approaches to addressing this phenomenon. Most importantly, it proves that there is no war on Islam, as the radicals often claim, and that communities can indeed co-exist peacefully and harmoniously.[8]

Responding to the ideological threat of ISIS

Although the threat of JI has been lessened in the face of counter, RRG continues to play its role in curbing extremist ideologies, particularly those of ISIS by strong use of technology and especially social media to spread its ideological influence. The Singaporean security agency has arrested several young Singaporeans who have been swayed by this ideology. These arrests demonstrate the vulnerability of Singaporean youths to influence and targeting by extremist propaganda.

One of the arrested individual revealed that he intended to carry out violent attacks in Singapore. The vulnerability of Singaporean youth is further heightened by their heavy reliance on Internet sources for religious guidance since they are quick, easy to use, and highly accessible. Without a strong foundation in religious knowledge, these youth are unable to discern correct Islamic teachings drawn from traditional sources with perverted messages disseminated by the ISIS.

Like JI, ISIS justifies its radical narrative by using and mis-contextualizing Islamic teachings and concepts. It claims its actions are in line with Islamic teachings and calls upon Muslims to migrate to the so-called "Islamic Caliphate" or "Islamic State" established in Iraq and Syria. ISIS has also called on Muslims to pledge allegiance to their self-proclaimed leader, Abu Bakar Al-Baghdadi. They extended their territory to include the *Wilayat* in the Philippines. This shows the strong support given by ISIS to like-minded militants in Southeast Asia.[9]

In July 2014, RRG established the RRG Resource and Counselling Centre or (RCC) to centralize its efforts in counter-ideological work. The RCC aims to chronicle the work of RRG since its establishment and to enhance its capability and capacity to meet the evolving challenges of radicalism and terrorism. It also provides training and resource materials to its counsellors, *asatizah*, and others who are interested to do research on issues related to religious and violent extremism. RCC is seen as an important addition to Singapore's overall efforts in countering radical and extremist ideologies. It further augments RRG's position as an authority in the field, as well as promoting greater awareness, knowledge, and assurance to the public that a concerted effort in counter-ideology is underway to prevent Singaporeans from falling prey to violent extremism.

Since 2014, RRG has published a number of pamphlets and booklets as part of its Public Education Series. The first pamphlet, entitled *The Syrian Conflict*,[10] aims to inform the public on the nature of the conflict in Syria. Among other things, it debunks the allegations made that the fight in Syria is an act of *Jihad* that requires the participation of all Muslims. The second pamphlet, *The Fallacies of ISIS Islamic Caliphate: A Brief Introduction*,[11] was published in early 2015 to explain the illegitimacy of the Caliphate. Through the publication series, RRG aims to educate the public on ISIS, its threat and atrocities, and deter them from being influenced by the group's narratives and propaganda.

RRG has also produced short video clips as part of its community outreach against religious extremism and terrorism. The videos feature RRG members speaking on issues related to religious extremism and have been uploaded on YouTube and Facebook. These initiatives aim to delineate any misunderstandings of Islamic teachings and help the community to better understand the threat of extremism and terrorism. RRG has produced a manual that is used by its members to counter ISIS ideology. The manual provides arguments that debunk the ISIS notions of the Islamic Caliphate, obligations of *Jihad*, and several other ISIS narratives. It discusses important topics, including Muslims living in secular environment and the need for critical thinking to evaluate religious sources. The manual is used by RRG members as a signpost in re-educating individuals and youth who have been influenced by ISIS narratives and is also used for public education purposes.

Additionally, a mobile application and a helpline have been set up by RRG at its center to facilitate public inquiries on issues regarding radicalization or aspects of religion that can potentially lead one to become radicalized. This initiative aims to provide the public with a legitimate reference point on radicalization matters, eliminating the options of being left without any religious guidance, thus the need to turn to non-credible sources on the Internet. It has also begun conducting short religious talks before the weekly Friday prayers to raise awareness on the threat of radicalization in the community. This is designed to leverage the large congregations who attend mosques on Friday and to spread counter-ideology messages on a broader scale.

The inter-agency aftercare group (ACG)

The holistic approach of the Singapore rehabilitation program includes a component aimed at providing social, motivational, and financial support to the families of the detainees. This social rehabilitation endeavor is helmed by an umbrella organization known as the Inter-Agency Aftercare Group (ACG). Its members include leaders of several local Muslim organizations such as Taman Bacaan Pemuda Pemudi Melayu Singapore (Taman Bacaan), the Association of Muslim Professionals (AMP), Yayasan Mendaki, and Khadijah Mosque, as well as a Muslim community leader, Rhazaly Noentil, a former representative of En-Naeem Mosque. ACG draws on existing networks of social and financial support to ensure that the wives and children of detainees receive the help they need and continue to feel a sense of community spirit and belonging in Singapore. Over the years, ACG has helped the wives receive training to take on jobs, ensure that their children continue their studies free from financial burden, and find work placements for former detainees to reintegrate into society. Overall, ACG's holistic rehabilitation framework plays a role in helping to socially reintegrate the families affected.

Such an approach is important in Singapore's CVE effort due to the country's multi-racial and religious position. It underlines the need for a program of this nature to strengthen the community's social resilience. ACG is an example of a ground-up initiative demonstrating self-help within the community and the willingness of the Muslims in Singapore to lend a hand to those in need. More importantly, the approach seeks to eventually break the local cycle of violence. As former Minister Wong observed:

The work of the ACG has been an integral part of the reintegration of the former JI detainees into mainstream society. The ACG has been quietly helping the families of the detainees for more than a decade and has been a strong pillar of support for these families.[12]

Islamic religious council of Singapore (MUIS)

As the ultimate religious body looking into the affairs of the Muslim community in Singapore, the Islamic Religious Council of Singapore (MUIS) has taken several important steps to ensure the correct and proper propagation of Muslim teachings to the local community. This is in line with the need to address directly the current trend of people turning to the web to supplement their religious understanding. Since the discovery of JI in 2002, MUIS has initiated programs to combat extremist ideas in Singapore.[13]

One of its most significant efforts is the development of the Singapore Islamic Identity (SMI), a project focused on disseminating modern state-centric Islamic ideals to local Muslims.[14] It comprises ten desired attributes for a Singaporean Muslim community of excellence, focusing on knowledge, principle centeredness, progressiveness, and inclusiveness, which are believed to be the identity of Singaporean Muslims today.[15] Thus SMI reinforces the fundamental precept that as active citizens living in a secular state, embracing the modern world by being progressive as well as adaptive does not make Singaporean Muslims any less faithful to Islam.[16] Such a precept is believed to be useful in the efforts to counter extremist characteristic of exclusivism.

Another key initiative of MUIS is the regulation of religious teachers in Singapore. This involves ensuring that all Muslim religious teachers in the country are accredited and that they do not espouse any extremist and exclusivist teachings when they preach and teach Islam here. One of the suggestions by the Singapore government, as stated in the White Paper on the arrest of JI members, was for the Muslim community to develop a comprehensive self-regulatory system to monitor religious education and detect the dissemination of extremist teachings.

MUIS, in collaboration with the Singapore Islamic Scholars and Religious Teachers Association (PERGAS), launched the Asatizah (religious teachers) Recognition Scheme (ARS) in December 2005. The scheme's key objective was to oversee the professional conduct of all *asatizah* in performing their roles as religious teachers, scholars, propagators, and advisors on Islam. The assumption is that such a system will curtail the spread of deviant teachings which mislead the community, create social disharmony, and present security threats to the country.[17] Critically, ARS was established to enhance the standing of *asatizah* and serve as a credible source of reference for the Singapore Muslim community. The recognition which is assessed and approved by the Asatizah Recognition Board (ARB) is granted only to qualified Islamic religious teachers and scholars who have met the minimum standards of qualification and are considered fit and proper to preach and teach Islamic religious knowledge in Singapore.[18]

Other initiatives developed by MUIS include the development of specific websites to counter radicalism and extremism. In 2005, MUIS developed several useful websites (www.iask.com.sg, www.invoke.sg, www.radical.mosque.sg) to provide a correct and relevant understanding of Islam and its practices. These websites are primarily aimed at the internet-savvy young generation, endeavoring to provide tips, guides, and information to differentiate radical and/or extreme ideologies from societal norms, as well as how to live a harmonious and progressive religious life. With the enduring threat of extremism prevalent today compounded by the looming peril posed by ISIS, MUIS continues to inform through publication of a guidebook titled *Resilient Families: Safeguarding against Radicalism*. Produced by the Office of the Mufti at MUIS, this guide aims to ensure the continued endurance of the local Muslim community and offers tools for families to protect their loved ones from radicalization.[19]

MUIS has also initiated various programs for adults and young children, with the objective to counter extremist narratives and ideologies. For instance, the curricula of MUIS aLIVE[20] and Adult Islamic Learning (ADIL)[21] include lessons that help inoculate learners against online radicalization. The aLIVE programs aims to develop children nurtured with *taqwa* (God-consciousness) and good *akhlak* (character), into practicing Muslims who are knowledgeable in Islam and show care and concern toward others, whereas ADIL was developed to enrich the Islamic learning experience for adult Muslims in Singapore. This new program was developed to allow greater choices for Muslim adults to learn Islam in structured, modular settings, providing more opportunities for closer interaction and engagement with the *asatizah*.

Apart from efforts to counter extremism initiated by ACG, MUIS, PERGAS, and other organizations, there are also efforts by individuals. Noticing the urgency to counter radical views that are widespread in the cyberspace, several commendable initiatives have been undertaken by the Muslim community to fight the phenomenon. For example, Muhammad Haniff Hassan, a Research Fellow at the S. Rajaratnam School of International Studies, has developed a blog to refute extremism. The blog http://haniff.sg/en/ seeks to offer responses to radical ideology that underlies violent radical Islamist terrorism, share perspectives on how counter-ideologies can be executed, and provide relevant materials on Islam's position on jihad, terrorism, extremism, and other related issues.

Lessons learnt from the Singapore experience

To some, Singapore is but the little red dot on the map. Although small in size, Singapore's strategic position both in geography and global affairs has made it an important player in the regional fight against terrorism. In fact, Singapore's role and contribution to fighting terrorism has moved beyond Southeast Asia. In the global fight against terrorism, Singapore has joined the multinational coalition force set up to combat ISIS. The Singapore Armed Forces (SAF) team based in Kuwait provided a medical team in Iraq and satellite imagery analysts in Kuwait. SAF also has a KC-135R air-to-air refueling plane, and intelligence and planning officers based in Kuwait. According to Singapore's Defence Minister Ng Eng Hen, these are Singapore's contributions and long-term commitment to the international efforts to counter extremism and terrorism.[22]

At the CVE level, Singapore' appreciation for the importance of countering extremist ideology and rehabilitation of detainees has attracted global attention and recognition. Singapore also hosted the East Asia Summit Symposium on Religious Rehabilitation and Social Reintegration in 2015. The symposium's focus was to map current and emerging threats of ISIS, global implications, and how government and civil society should respond. Prior to that, Singapore hosted the International Conference on terrorist Rehabilitation and Community Resilience in 2013 and the International Conference on Terrorist Rehabilitation in 2009. All these conferences have brought together experts, academics, religious scholars, and practitioners in CVE and rehabilitation from across the globe. The establishment of the Religious Rehabilitation Group Resource and Counselling Centre in Singapore marks another milestone in theses CVE efforts. The Centre has attracted a continuous stream of visitors from local schools and madrasahs, as well as foreign personalities from all over the globe. In May 2018, for instance, the Grand Imam of Al-Azhar, Syaikah Dr Ahmad Al-Tayyeb, the highest authority in the renowned al-Azhar University in Egypt, visited the Centre and recorded his admiration of the importance of RRG's work in countering the dangers of religious misinterpretation.

Singapore's approach to CVE stands out in its singularity and is tailored to the multi-racial and multi-religious local setting. The lessons learnt from the country's experiences can be summarized as follows:

Importance of countering extremist ideology

The ideology of violent radical Islamist groups such as al-Qaeda, JI, and ISIS frames their organizational structure, leadership and membership motivation, recruitment, and support. It also shapes their strategies, tactics and world view. Their threat will persist unless the extremist ideology and its perverted understanding of religion is countered. To do this, their religious orientations must be understood. VRI groups attempt to identify themselves as representative of the authentic and original Islam as practiced by the early Muslims. They advocate strict adherence to *their* understanding of Islamic practices as enjoined by Prophet Muhammad, the final prophet, and subsequently practiced by the early pious Muslims known as the *salaf al-salih*.[23] That by understanding demands, they claim, that certain actions be taken. These actions, as discussed above, are terrorist in their targeting and intent.

VRI groups have managed to radicalize individuals and convince them to take action in their name. An example can be seen in the ISIS *Dabiq* magazine. The articles are written by IS members whose chain of knowledge is unknown, and their religious content can be characterized as having Sunni-versus-Shiite orientation, circular discourses on religious concepts, and the extensive use of eschatological or "End Times"[24] narratives.[25] Extremist ideology and propaganda will continue even if the groups cease to exist. To confront the threat, though, robust tools and religious scholars are needed to steer misguided views and critically invalidate threat ideologies; namely, its questionable religious legitimacy.

Government and community partnership

Responding to the threat posed by violent radical Islamist groups requires a holistic approach that targets both the terrorist organizational and ideological infrastructure, a great challenge for many secular governments due to lack of capabilities and experience. Governments must thus work with religious communities. To counter the perverted understanding of Islam propagated by extremists, the *ulema* or religious scholars need to come forth and assume responsibility for framing religion correctly. Ultimately, they are needed to rehabilitate detainees and to inform the larger Muslim community about the dangers of extremist narratives.

Furthermore, religious scholars must openly and proactively reject violence and intolerance through debates and open dialogues about the nature of religious extremism. This will better assist policymakers in dealing with such issues.[26] The essential reality is that extremists believe their immoral acts of violence are moral and that they are on the right path to God. This is drawn from a long tradition of extreme intolerance that does not distinguish between politics and religion, instead distorting them both. This must be deconstructed.

The story of RRG stands out in this regard. A group of local *ulema* working closely with the local government is a fine example of the importance of a government-community partnership to deal with the threat of terrorism and extremism. The Singaporean government has been amenable about the idea of working with the community for a number of reasons. These include: (1) the threat of terrorism is not the problem of the Muslim community alone but of the nation and hence requires the attention of all; and (2) Participation of religious scholars is needed, because they are the right people to confront extremist ideologies and rehabilitate the detainees.

Role of civil society and non-governmental organizations

As the Singapore experience has demonstrated, Muslim and non-Muslim organizations such as RRG, ACG, and Inter Racial and Religious Confidence Circle (IRCC) have managed to help the government effectively to reduce the threat of terrorism and extremism in the country. They have valuable expertise and experience in addressing conditions conducive to the spread of terrorism.

Conclusion

The discovery of the JI network in Singapore and the arrest of individuals radicalized by ISIS narratives have produced invaluable lessons. Importantly, it has been demonstrated that the government and the Muslim community can work together to produce the ideological counter required to defeat terrorists and extremist groups. This is not a war against Islam but a war against any misrepresentation and misunderstandings of Islam. This is not a clash of civilizations but a clash of ideas that has divided the world into peace and war.

The efforts made by the Singapore government and the community to fight terrorism and mitigate the threat of ideological extremism have been robust. While the threat has been degraded, it is likely that the country will continue to be targeted. Efforts and necessary resources must continue to be directed to support an efficient and effective strategy. This includes the allocation of sustained resources to train manpower, improved infrastructure to ensure greater security, and the support and intervention of the community and religious organizations. Strong leadership is in this respect key and has thus far been demonstrated. Singapore's counter-terrorism success has been a direct result.

Notes

1 Jemaah Islamiyah is a Southeast Asian militant Islamist organization which was responsible for many terrorist attacks in Indonesia from 2001 to 2009 including the Bali incident 2002. JI is dedicated to the establishment of a *Daulah Islamiyah Nusantara* (regional Islamic state) in Southeast Asia incorporating Indonesia, Malaysia, the southern Philippines, Singapore and Brunei.
2 Singapore's Ministry of Home Affairs White Paper *"The Jemaah Islamiyah Arrests and The Threat of Terrorism,"* January 7, 2003 available at www.mha.gov.sg/get_blob.aspx?file_id=252_complete.pdf (accessed August 16, 2017).
3 Individuals associated with foreign terrorist groups such as Hezbollah and Liberation of Tamil Tigers Eelam (LTTE) operating in Singapore and Singapore Airlines plane hijack in 1990 by four Pakistanis in Singapore are examples of international terrorism that happened in Singapore.
4 Muhammad Haniff Hassan, "Key Considerations in Counter-Ideological Work Against Terrorist Ideology," 561–588.
5 They are Ustaz Ali Bin Haji Mohamed who is currently chairman of Khadijah Mosque and Ustaz Mohamad Hasbi Hassan, President of Singapore Islamic Scholars and Religious Teachers Association (PERGAS). Both of them co-founded the RRG.
6 For more on JI's ideology, see Gunaratna and Bin Ali, *"Countering The Ideology of Jemaah Islamiyah – A Point by Point Approach"*; Gunaratna, *The Ideology of Al-Jama'ah Al-Islamiyah*, 68–81; Gunaratna, *Ideology in Terrorism and Counter Terrorism: Lessons from combating Al Qaeda and Al Jemaah Al Islamiyah in Southeast Asia.*
7 See Gunaratna and Bin Ali, *Terrorist Rehabilitation: A New Frontier in Counter-Terrorism.*
8 Professor Bruce Hoffman as quoted from Bin Ali, "The Religious Rehabilitation Group (RRG): A Community-Government Partnership in Fighting Terrorism," 250.
9 See Singh and Ramakrishna, *Islamic State's Wilayah Philippines: Implications for Southeast Asia.*
10 *The Syrian Conflict* published by RRG available at www.rrg.sg/wp-content/uploads/2016/04/Syrian-Conflict-Pamphlet-20140702.pdf (accessed on September 2, 2017).
11 *The Fallacies of ISIS Islamic Caliphate: A Brief Introduction* published by RRG available at www.rrg.sg/wp-content/uploads/2016/04/RRG-ISIS-Booklet-ENG-.pdf (accessed September 3, 2017).

12 Mr Wong Kan Seng quoted from Hussain and Kader, *Inter-Agency Aftercare Group: Fostering Social Reintegration and Building Community Resilience*, 17.
13 See Hassan, *Update: Singapore's Community-Based Counter-Ideology Initiatives.*
14 See Abdul Azeez, *Creating A Modern Singapore Muslim Community*, ISEAS Working Paper 32, (2014) available at www.iseas.edu.sg/images/pdf/iseas_working_papers_2014_2.pdf (accessed September 10, 2017).
15 The explanation of the ten desired attributes of the Singaporean Muslim community can be found in a document entitled "*Risalah for Building a Singapore Muslim Community of Excellence.*"
16 Mohammad Alami Musa, "Singaporean Muslim Identity: Tolerant, Adaptive and Progressive Yet Keeping the Faith."
17 Hassan, *Recognising the Teachers of Religion: Some Food For Thought.*
18 See official website of the Asatizah Recognition Scheme available at www.muis.gov.sg/ARS/About/Background.html (accessed September 11, 2017).
19 See MUIS guide book on *Resilient Families: Safeguarding against Radicalism* available at www.muis.gov.sg/officeofthemufti/documents/Countering%20Radicalism%20version%204%20(Final).pdf (accessed September 10, 2017).
20 See MUIS aLIVE website at www.muis.gov.sg/alive/ (accessed September 12, 2017).
21 See MUIS ADIL website at www.muis.gov.sg/adil/index.html (accessed September 12, 2017).
22 See Chuan, "Singapore's Contributions in International Fight Against Terrorism," *The Straits Times* Singapore, (October 1, 2017) available at www.straitstimes.com/singapore/singapores-contributions-in-international-fight-against-terrorism (accessed May 29, 2018).
23 Bin Ali, *IS Terrorism: How to Win The Ideological Battle*, RSIS Commentary, No. CO16203, (August 11, 2016) available at www.rsis.edu.sg/rsis-publication/rsis/co16203-is-terrorism-how-to-win-the-ideological-battle/#.WcGjzcgjGUk (accessed September 3, 2017).
24 See Bin Ali, *Jihad' in Syria: Fallacies of ISIS' End-Time Prophecies.*
25 Bin Ali and Yussof, "*IS Ideology: Debunking its Pseudo-Religious Character.*"
26 Bin Ali, "*Countering ISIS Ideological Threat: Reclaim Islam's Intellectual Traditions.*"

Bibliography

Abdul Azeez, Rizwana. *Creating A Modern Singapore Muslim Community*, ISEAS Working Paper 32, (2014) available at www.iseas.edu.sg/images/pdf/iseas_working_papers_2014_2.pdf (accessed September 10, 2017).
Bin Ali, Mohamed. *Jihad' in Syria: Fallacies of ISIS' End-Time Prophecies*. RSIS Commentary, No. CO14149, (July 30, 2014) available at www.rsis.edu.sg/rsis-publication/rsis/jihad-in-syria-fallacies-of-isis-end-time-prophecies/#.WcZuDMgjFhE (accessed September 3, 2017).
Bin Ali, Mohamed. "The Religious Rehabilitation Group (RRG): A Community-Government Partnership in Fighting Terrorism." In Zainul Abideen Rasheed and Norshahril Saat (eds.), *Majulah 50 Years of Malay/Muslim Community in Singapore*. Imperial College Press, World Scientific Publishing Company, June 2016.
Bin Ali, Mohamed. *Countering ISIS Ideological Threat: Reclaim Islam's Intellectual Traditions*. RSIS Commentary, No. CO16016, (January 25, 2016) available at www.rsis.edu.sg/rsis-publication/srp/co16016-countering-isis-ideological-threat-reclaim-islams-intellectual-traditions/#.WcHOgcgjGUk (accessed September 3, 2017).
Bin Ali, Mohamed. *IS Terrorism: How to Win The Ideological Battle*. RSIS Commentary, No. CO16203, (August 11, 2016) available at www.rsis.edu.sg/rsis-publication/rsis/

co16203-is-terrorism-how-to-win-the-ideological-battle/#.WcGjzcgjGUk (accessed September 3, 2017).

Bin Ali, Mohamed and NurulHurda Binte Yussof. *IS Ideology: Debunking its Pseudo-Religious Character*, RSIS Commentary, No. CO17170, (September 19, 2017) available at www.rsis.edu.sg/rsis-publication/rsis/co17170-is-ideology-debunking-its-pseudo-religious-character/#.WcHQxMgjGUk (accessed September 20, 2017).

Chuan, Tong Yong. "Singapore's Contributions in International Fight Against Terrorism," *The Straits Times*, Singapore, (October 1, 2017) available at www.straits-times.com/singapore/singapores-contributions-in-international-fight-against-terrorism (accessed May 29, 2018).

Gunaratna, Rohan. *The Ideology of Al-Jama'ah Al-Islamiyah*. Current Trends in Islamist Ideology, Hudson Institute, 1, 68–81.

Gunaratna, Rohan. *Ideology in Terrorism and Counter Terrorism: Lessons from Combating Al Qaeda and Al Jemaah Al Islamiyah in Southeast Asia*. CSRC Discussion Paper 05/42, September 2005.

Gunaratna, Rohan and Mohamed Bin Ali. *Countering The Ideology of Jemaah Islamiyah – A Point by Point Approach*. ICFAI University Press, Hyderabad, April 2007.

Gunaratna, Rohan and Mohamed Bin Ali (eds.). *Terrorist Rehabilitation: A New Frontier in Counter-Terrorism*. Imperial College Press, 2015.

Hassan, Muhammad Haniff. *Recognising the Teachers of Religion: Some Food For Thought*, RSIS Commentary No. CO06009, (February 7, 2006) available at www.rsis.edu.sg/rsis-publication/rsis/764-recognising-the-teachers-of-re/#.WcCR9cgjGUk (accessed September 11, 2017).

Hassan, Muhammad Haniff. "Key Considerations in Counter-Ideological Work Against Terrorist Ideology," *Studies in Conflict and Terrorism* 29 (6), September 2006, 561–588.

Hassan, Muhammad Haniff. *Update: Singapore's Community-Based Counter-Ideology Initiatives*, RSIS Commentary no. CO08006, January 11, 2008

Hussain, Zakir and Abdul Halim Kader. (eds.), *Inter-Agency Aftercare Group: Fostering Social Reintegration and Building Community Resilience*. Taman Bacaan Pemuda Pemudi Melayu Singapura, First Edition 2015.

Musa, Mohammad Alami. "Singaporean Muslim Identity: Tolerant, Adaptive and Progressive Yet Keeping The Faith" in Abdul Halim Kader (compiler) *Fighting Terrorism: The Singapore Perspective*, Taman Bacaan Pemuda Pemudi Melayu Singapura, Fourth Edition 2007.

"Risalah for Building a Singapore Muslim Community of Excellence" MUIS, available at www.muis.gov.sg/officeofthemufti/documents/Risalah-eng-lr.pdf (accessed September 11, 2017).

Singh, Bilveer and Kumar Ramakrishna. *Islamic State's Wilayah Philippines: Implications for Southeast Asia*. RSIS Commentary, No. CO16187, July 21, 2016.

10 Challenges in counter terrorism and counter violent extremism in Malaysia

Andrin J. N. Raj

Current threats

Islamic State (IS) now constitutes the greatest challenge for Malaysia, a threat which includes the virtual sphere and has prompted Malaysian authorities to increase their efforts at counter. Malaysian authorities have identified the "new" home-grown terrorism as the biggest threat component. It is reported that 75 percent of IS supporters in Malaysia were radicalized online in 2015.[1] The threat of IS propaganda through the virtual caliphate has inspired and promoted radicalization, violent extremism and religious-motivated terrorism in Malaysia.

Besides these threats, Malaysia faces external threats from foreign fighters who use Malaysia as a conduit, recruitment and fundraising state for IS. Many of these foreign fighters, both locals and foreign, have been arrested throughout Malaysia over the past three years. However, this has not seemed to curtail the influx as borders are porous in security protection and a high level of corruption exist within some security agencies in Malaysia. The threat comes from poor security measures from air, land, and sea which have made Malaysia an easy access point from which for terrorists to operate.

Since Malaysia lies within three active centers of religious extremism and terrorism threats – the Arakan region in Myanmar, Southern Philippines, and Southern Thailand – it is highly likely that it will encounter further threats of terrorist activities within Malaysia. Toward the east and south, Indonesia plays an active role in supporting religious Islamist groups operating in Malaysia. In an interview with Myanmar security intelligence in 2014, an estimated 50 Rohingya operatives trained by al Qaeda were alleged to have crossed into Malaysia and were scattered among madrasas (religious schools) in the northern states of Malaysia.[2]

Abu Sayyaf Group (ASG) poses another significant threat to Malaysia as it has pledged allegiance to IS, and Filipino ASG members have been arrested over the past couple of years in Malaysia. ASG has close ties with east Malaysian terrorist group Darul Islam Sabah, as well as the Muslim Filipino group, the Knights of the Right Keepers.[3] According to a military source, this group is made up of Filipinos living in Sabah with Malaysian identification cards and has been linked to supporting the kidnap and ransom activities of ASG.[4]

Malaysian authorities have also to deal with the supporters of terrorist groups who provide resources and funding to support the global jihadist war. Terrorism financing is another threat posed by terrorist groups with transnational linkages who provide the supply chain process of money laundering. Numerous non-government organizations which were religious in their operations have been caught funding terrorist groups, as well as involving themselves with local Malaysians. Many of these organizations have been shut down by the authorities. Although the Malaysian authorities are trying their level best to curtail this threat, it remains very difficult to address.

Sympathizers in Malaysia play a leading role within the virtual caliphate, where they operate, share information, and engage in religious radical discussions online. This group poses another threat of a virtual community as conceptualized in Benedict Anderson's sense of an "imagined community." He argues that these sympathisers believe they are a part of the "nation or in IS case, a caliphate."[5] As mentioned earlier, 75 percent of Malaysian IS supporters have since 2015 been reported to be radicalized through the internet, and prior to that, hands-on recruitment was conducted within the local arena using Jemaah Islamiyah platforms, the *shuara* ("places of meeting").

Since 2014, the total number of IS operatives and supporters has grown, with some traveling to Iraq and Syria to join the global Jihad. Supporters based in Malaysia have recruited Muslim youths from higher institutions and infiltrated government agencies such as law enforcement agencies, the civil defense, military, political, and religious institutions. The infiltration of IS and extremists into state-run organizations is now a reality according to local authorities. In August 2017, a total of more than 30 suspected IS operatives were apprehended by the Malaysian authorities. Nationalities included Malaysians, Maldivians, Syrians, Iraqis, Bangladeshis, Indonesians, Palestinians not mentioning earlier arrest of Uighurs from China, Pakistan, Algeria, Morocco, Saudi Arabia, Turkey, Philippines and the list does not seem to end there.[6]

The rise of IS and the support for radical and extremist views are essentially a function of internal political processes where Muslims by and large feel, rightly or wrongly, that they have been deprived of their rightful place in their own societies and thus would like to shape it in accordance with the principles of Islam. The belief that sharia laws will be the proponent for an Islamic state, the struggle and the course for such progress can only be brought by jihad. Although Malaysia is heading toward greater Islamization, the proponents of radicalization remain small, and most by a large margin do not advocate violence or hatred. Still, the reality is that some proponents within the Muslim community, supported by foreign entities from the Middle East, are thriving, driving the demand for radical and full *sharia* to be implemented. Wahhabism and Salafism from Saudi Arabia are impacting Malaysia and must be addressed to avoid moving toward violent extremism as this can be damaging if not tackled with proper educational learnings in the theological framework, approaches and studies within the doctrines and the teachings of the Islamic faith.

Re-emerging threats: Jemaah Islamiyah in Malaysia

The threats of religious terrorism and violent extremism in Malaysia date back to the 1993 formation of Jemaah Islamiyah (JI) in Southeast Asia by Abu Bakar Ba'asyir. Yet it was in Malaysia that Abu Bakar Ba'asyir started the initial process of Jemaah Islamiyah in the late 1970s while in exile from the Suharto secular administration for trying to implement *sharia* in Indonesia.

Mantiqi 1, set up under the four main cells operating in the region, was based in Malaysia and represented Thailand, Malaysia and Singapore. Singapore and Thailand under Mantiqi 1 were considered operational targets for JI except for Malaysia, which remained a non-militia operational cell for a long period of time. Several planned attacks to begin in 2003 in Malaysia by JI were thwarted by the Royal Malaysian Police (RMP), Special Branch Counter Terrorism Unit. These targeted foreign visiting military delegations from the U.S. War College and various foreign dignitaries' visitations.[7]

Malaysia was seen to be cautious about the sensitivities of the Muslim majority as to avoid inciting accusations of religious discrimination against religious organizations and groups spearheading the rights of the Muslim majority and hence overlooking the threats posed by JI. In certain Muslim countries in the Middle East and Africa, JI is not labeled as a terrorist group. These approaches allowed for JI movement, fundraising, money laundering, recruitment, and operational planning within the region and to exploit Muslim sentiments in Malaysia.

The Mantiqi 1 cell actively recruited Malaysians and Indonesians in its early stages. JI recruited mainly from institutions of higher learning in Malaysia as well as within the *Shuara* groups established by JI members. JI's organization was well set up under a fully operational and organized structure as per below.

Further to this, JI Malaysia had transnational linkages with al-Qaeda and through its associated network of al-Qaeda in Pakistan (AOP) and splinter groups within the region such as the Moro Islamic Liberation Front (MILF), the Moro National Liberation Front (MNLF), Abu Sayyaf Group (ASG), Free Aceh Movement (GAM), and the Southern Thailand Patani groups.[8]

Following the setup of JI in Malaysia, many other radical and extremist religious groups were later established, such as the Kumpulan Militant Malaysia (KMN), Al Maunah, Malaysia Mujahedeen Group, Al Arqam, and Darul Islam. Of particular concern, Jemaah Islamiyah and al-Qaeda have regained momentum in Southeast Asia. This was confirmed during a session in Singapore in connection of Milipol conference, October 31, 2017. The Marawi siege (May 23, 2017– October 23, 2017) saw JI operatives assisting the Maute[9] group and Abu Sayaaf during the conflict.

Roots of radicalization in Malaysia

Malaysia falls under the category of "territorial communities" where Islam plays a dominant role within the Muslim population. As such, the Muslim community

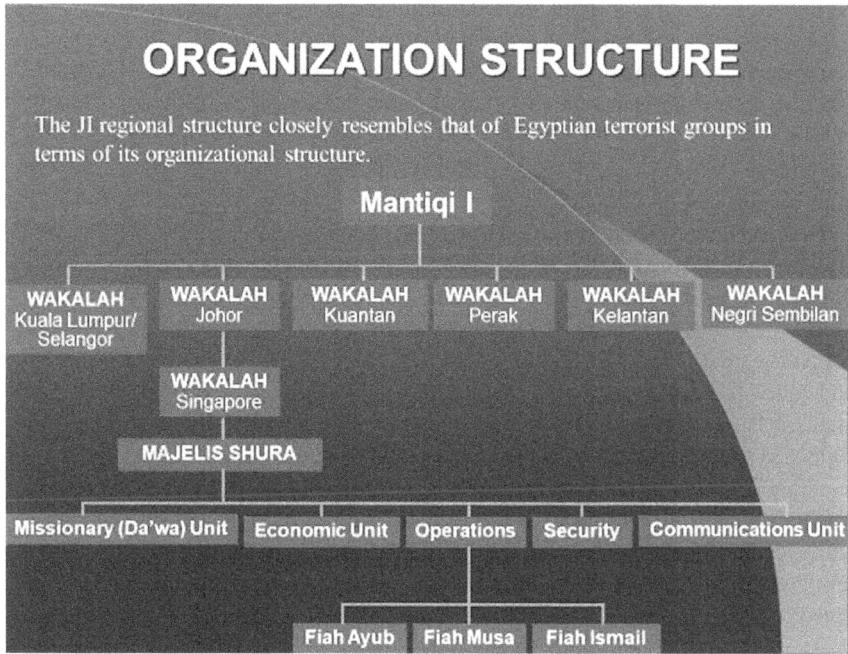

Figure 10.1 JI's organization structure.

Sources: Jemaah Islamiyah Organisation Chart, Malaysia; IACSP SEA data base centre.

has better access to mainstream clerics and their moderate Islamic teachings in comparison to the diaspora or migrant Muslims in the West. While this curtails and mitigates the risk of radicalization and evolving of extremist beliefs to a certain extent, the threat of radicalization and extremism within Malaysia has steadily grown and continues to increase.

The reason for this increase has its roots in Malaysia's foreign policy shift by the Mahathir administration in the early 1980s toward supporting Palestine against the perceived atrocities committed by the Israelis against the Palestinian Muslims. Later, in the early 1990s, the Bosnia-Herzegovina conflict steered more support for the idea of Muslims being attacked in their home states by "Western" powers, and killing of Muslims was labeled "a clash of religious beliefs." According to a former JI member from Malaysia and discussions conducted with some radically inclined believers, they see Islam being threatened by the West. As the Western world is thought to be predominantly Christian, the logic runs to a war of religion, particularly one directed at Islam. The response therefore has been not only against Christianity but against all other religions, for they are considered a threat to Islam.

These issues have not been debated extensively within Malaysia, and one can see why the threat has escalated based on perceptions that Muslims are being persecuted by the "West." Although the country generally maintains peace and harmony, underlying currents from foreign radical and Islamist terrorist groups, political hands, and local chauvinist groups have manipulated the system to cause mainstream Muslims to fall prey to the thinking of "clash of civilisations." Islamist groups in Malaysia also started forging alliances with regional jihadist groups within Asia and the Middle East. Jemaah Islamiyah and local groups with links to al-Qaeda managed to accomplish this feat by inducing local groups to embrace the notion of global jihad.

Radical thought and radicalization was primarily homegrown in Malaysia. However, following the 911 attacks Islamist groups such as the Kumpulan Mujahidin Malaysia advocated and supported violence within the country and encouraged other Islamist groups within Malaysia to apply the same beliefs within the context of Jihad. The role that al-Qaeda played is evident in this transformation, and more than half of the 9/11 attackers met in Malaysia prior to the actual attacks. The meeting was to discuss, prepare, and plan. The attack was initially supposed to originate from Thailand with a flight enroute to New York.[10]

After 9/11, religious radicalization and extremism grew in Malaysia. This naturally caused to surface the canard that some manner of violence was inherent or embedded within Islam, when analysis made clear it was being used to misinterpret and abuse the faith.[11] Of the result, though, there can be no doubt, and (as noted above) any number of local Muslims fell prey of radicalization. Rejecting Western norms of pluralism and individual rights, they accepted the use of violence in inter-community and inter-faith discourse. This process was naturally facilitated by the larger context of emerging transnationalist Islamist movements. The result has been the major shift toward Islamization in the country.

Joseph Liew of Nanyang Technological University has observed that the former leader of Malaysia (now returned to office), Mahathir Mohamad, blended Islam with his modernity and urged Malay-Muslims to undergo a mental revolution and cultural transformation, a "jihad" of the self. He perceived Islam to be a progressive religion and could be embedded into modernity, economic development, and a knowledge economy.[12]

Malaysian authorities have contended that terrorism is not global but independent of state system and has to be countered by various measures besides military action. It therefore requires a strategy of counter built around winning the "hearts and minds" as used during the communist insurgency from the 1960s to the 1980s. This strategy, because it followed the more intense actions required during the emergency, could concentrate on reducing hostility and emphasized intelligence, diplomacy, better policing, and covert actions over hard measures of law enforcement and security agencies.

Islamization in Malaysia

The political elite of Malaysia sought primarily to facilitate the development of civil society and democracy within the context of Islam and the constitution. However, many politicians toward the 1980s began to advance a skewed

"Islamic state" notion as opposed to an "Islamic country," and religion was exploited to seek political gains. The constitution clearly states that Islam is the main religion of the country, which itself is a secular state. The efforts by some politicians to exploit the term "Islamic State" has fostered division. According to Mansor Mohd Noor, studies have shown that ethnic and religious divide does prevail in Malaysia.[13]

In 1969, the New Economic Policy (NEP) marked the beginning of Islamization in Malaysia. The policy formulated after the racial riots of May 13, 1969 to empower and improve the socio-economic positions of the majority Malay race, *Bumiputra* (prince of the soil, meaning aboriginal). This policy was basically a radical affirmative action program that allowed many Islamic-centric institutions to be established and to propagate Islamization within the country.[14] With the passage of time, the growth of strict religious adherence and radical views grew within the spectrum of discussion concerning the development of Malaysia.

These institutions as well as radical institutions continued gathering momentum without scrutiny during the first Mahathir administration, 1981–2003. Mahathir also introduced "Islamic policies" that included "inculcation of Islamic values" within the government.[15] Some of these policies included segmenting the banking sector to cater for Islamic financing, *Bumiputra* privileges such as priority for businesses, purchasing of assets at a lower value, and religious institutions to foster Islamic values. Government agencies adopted Islamic jurisprudence of values enshrined in *Sharia*. As these policies thrived so did radical institutions; and within the pretext of these platforms, the threat of radicalization and extremism also grew in tandem.

As time passed, Malay Islamic groups from the far right placed political pressure on the Badawi administration, 2003–2006, the successor of Mahathir, and he reluctantly acceded to more Islamic measures in competition with his opponents from the opposition. These political approaches began to form platforms for a more radical and extremist stand to take precedents within the ruling administration and for other such institutions to be formed openly within the country.

As a consequence, stricter and more conservative religious doctrines were adopted within the Muslim community, both in civil society and government. Today, the threat of radicalization and prevalence of extreme views are playing a dominant role within Malaysian politics, shaping the economic and financial landscape with "tendencies" to incorporating *sharia* rulings discreetly in the development of the country. Thus moderate Muslims are becoming fearful of threats of instability to a multi-religious and diverse society.

According to Farish Noor, the rise of religious power by the Islamic party, Malaysian Islamic Party (Parti Islam Se Malaysia, PAS) in the late 90s was directly linked to the arrest of then-former Deputy Prime Minister Anwar Ibrahim.[16] Though the details remain contested, the reality was that the Malay Muslim community was torn in its interpretations of the faith, and faith and politics served to distort each other to the detriment of security.

Several higher learning institutions, both public and private, and religious schools have been detected by the local authorities to be breeding grounds for radicalization, but little is said about this.[17] According to a former Jemaah Islamiyah operative radicalized in 2000 in a local university, the radicalization was growing rapidly. Recruitment into Islamist groups, especially JI, was totaling initially about 500 a year from the year 2000 in one particular university. The operative was recruited by Dr. Azahari bin Husin, who was the main recruiter for JI in Malaysia and key funder and technical mastermind of the Bali bombing in 2002.

An extract taken from a section of a published document from the University Malaya Press of 2003 and adapted by the author of this chapter indicates a shift toward radicalization and extremism:

> A former President of a major Islamic university in Malaysia, describes himself as being approached by some jihadist professors and lecturers from India, Pakistan and Sudan informing him that he should not report to duty at the University even after he had an official letter from the University and patron of the University who was a Sultan (Royalty).
>
> At his testimony in his speech paper published by University Malaya Press section of a conference held in December 13–14, 2011, Seminar on Entrepreneurship in Higher Education, he terminates them within three days of his appointment and soon after that he was approached by three jihadist students to assess his commitment to Islam and Jihad. He was also informed that some 30 jihadist students were on campus.

The group of G25 made up of prominent Malays in Malaysia, in 2017 stated that Malaysians are becoming increasingly concerned that "the rise of extremism and intolerance by various groups is a result of a perceived ambivalent attitude, if not tacit countenance, under the Government."[18] In May of 2018, the G25 again claimed that a local think tank, Malaysian Islamic Strategic Research Institute (IKSIM) was labeling other Muslims as apostates and liberals. A G25 member, Noor Faridah Mohd Ariffin said that "We were very concerned when we heard they had been given a contract to brainwash students and government officers to their way of extremist thinking," she added.[19]

This clearly indicates that the rise of radicalization and extremism within institutions in Malaysia is evolving and alarming. Radicalization in educational institutions and religious schools in Malaysia is related to the threats of jihadist terrorism as these institutions have had foreign clerics from the Middle East, Africa, and the Southeast Asia region, most of them being radical-thought believers and vectors for spreading radical Islamic ideologies within institutions. According to the Inspector-General of Police, 389 persons have been arrested since 2013 for suspected terrorism-related activities.[20] However, there is no proof of these individuals coming from these institutions.

Other reasons related to radicalization and extremism are the unregulated setting up of *Takfirs* schools (Islamic religious schools), which are not religiously governed or supervised by the proper authorities. This allows clerics who have no religious or theological backgrounds to disseminate radical teachings and extremist views to the students. Malaysia has seen several of these schools sprout with no proper licenses or directive from the Ministry of Education and the relevant Islamic authority. According to Mohd Mizan Aslam of the Islamic University of Perlis:

> The Malaysian Government has not made improvements in the Islamic education curriculum yet ... to avoid occurrence of additional number of homegrown terrorists, the government must provide more missionaries or preachers, who are not only certified but also possess vast knowledge of Islam and the Jihad concept.[21]

Promoting moderation

Malaysia, being a multi-ethnic and multi-religious society, has witnessed religion play an important role in the promotion of integrity of the individual, society, and the nation as a whole. Inculcating the masses with values in the management spheres of the society such as education, the media, economy, politics, and administration of the country has been emphasized to create a harmonious and moderate society.[22] As such, religious institutions and religious diversity were needed to be managed in the best possible means available to the government. The 9/11 attacks created an adverse effect for the Muslims in Malaysia as well as globally as the image of Islam and being Muslim was tarnished. The attacks also complicated relations between Muslims and other religions within Malaysia and globally. The Malaysian government drew lessons from these circumstances.

Since religious diversity is present within Malaysia, the government addressed the practice of mutual respect, understanding, and cooperation among religions and further strengthened the structural platform already in existence. Values which were universal in all religions were to be highlighted with a view of promoting unity and mutual understanding with all religious faiths' followers within the country. Within the integrity plans for religious institutions, the agenda adopted best practices for the religious sector that had the following objectives: promoting cooperation, understanding, and mutual respect among followers of different religions; the upholding and practice of noble religious values; and the upholding and practice of Islam as a religion of progress. Religious organizations within Malaysia have also established the Interfaith Dialogue group, supported by the Malaysian government, hence addressing the growing trends of intolerance.

State responses

A dilemma ensues since Jihad-as-seeking is enshrined within Muslim teachings. A small percentage of believers, as is by now well known, have interpreted the quest for purification as a violent endeavor to be directed at others. Such ideological belief is promulgated by the IS and has been disseminated to the wider

Muslim community. Nevertheless, it lives outside the moderate majority of the Muslim world and alienates itself from the four main schools of Islam, primarily the Sunni school of jurisprudence.[23] The move from moderation into the predisposition toward violent extremism is into the Salafi-Takfiri = Jihad equation. Hanbali can be a precursor toward Salafi-Takfiri ideology. The Hanbali school of thought is the strictest and smallest form of the Sunni jurisprudence, which it is mainly practiced in Saudi Arabia but also in Malaysia by radical groups today. The Salafi ideology is to follow the *Salaf*, "the Way," which is the teachings of the first three generations of Islam. The Salafi is not a threat to Islam if the Salafi is without an action arm of the Takfiri. This is where the point of Jihad actually lies.[24] The tandem of radicalization of a Salafi can be determined by both the arms joining together, and if the Salafi decides to take action, it then becomes Takfiri. Takfiris are those who are Salafist believers and commit to action which can be related to a jihad.

Many Takfiris believe that those outside the belief of Salafi/Takfiri are infidels and believe that action must be taken against them in order to adhere to the teachings of Islam. Sharia plays an important role within the Salafi-Takfiri ideology, because all who are non-believers and infidels must be targeted. The Salafi-Takfiri teachings are derived from Sayed Qutub and other radical and extremist writers. Qutub's teachings have been central to the actions of many religious terrorist groups such as al Qaeda, Boko Haram, JI including IS.

The current threat thus relates to the war of ideas and has required Malaysia to deconstruct radical ideology and work with moderate Muslims to minimize the appeal of radicalization and extremism. Malaysian authorities have initiated a two-pronged approach which involves ideological and counter-terrorism components.

Counter-terrorism laws

Malaysian authorities have taken various steps to curtail the threats of terrorism arising from religious radicalization and extremism. Policies that derived from the legislative aspects of countering terrorism in Malaysia were based under the Directive No. 18 (Revised) under the National Security Council (NSC) of Malaysia, which was to denounce any form of terrorist acts, protection of hostage's life and property, preference for negotiated solutions, no exchange of hostage to resolve the crisis and strike action as a last resort following failure of negotiations.

National security threats within Malaysia are complex. The 1970s saw numerous threats from ultra-conservative religious organizations that caused mistrust amongst the populace as well as threats to democratic, secular government. These stemmed from misinterpretations of Islam by interest groups, some of which operated within the national boundaries and others that developed extensive external networks (though the two were certainly not mutually exclusive).[25] Malaysia identified Islamic extremism and militancy as a threat to its national security since the 1970s and concomitant executive actions taken under the Internal Security Act (ISA) of 1960 (Act 82) This act was "non-law abiding" as it gave full powers to the government to hold a person under detention without any trial or court proceedings. ISA was controversial right up to its repeal in 2011, when its place was taken by the Security Offences (Special Measures) Act (SOSMA) 2012 (ACT 747).

This new law related to security offences under the Penal Code of Chapter VI and VIA of the various criminal acts committed against state and terrorism-related offences.[26] With the increased threat of terrorism in Malaysia by religious violent extremist groups such as IS, the Prevention of Terrorism Act (POTA) 2015 and Terrorism in Foreign Countries Act (SMATA) 2015 were enacted. In line with these laws, Malaysia was also part of the Anti-Money Laundering and Anti-Terrorism Financing Act 2001 (AMLATFA), which came into force globally in 2001 and was initiated by the United Nations from its earlier Suppression of Financing of Terrorism law in 1999. These laws provided means and procedures to address the threats of terrorism, radicalization and extremism and allowed for detention and prevention of the said threats within Malaysia and globally.

Malaysia was also instrumental in engaging in regional and international cooperation to counter the threats of terrorism at the Association of South East Asian Nations (ASEAN) and the United Nations International Conventions and Protocols on combating terrorism.[27] In the UN General Assembly on September 24, 2016, Malaysia declared that the fight against terrorism could not be fought by force or punitive measures alone. It stressed that countering the threats through de-radicalization and rehabilitation program had been proved effective in addressing the evolving threats within Malaysia and should be applied globally. Following the UN speech by then Deputy Prime Minister Zahid Hamidi, Malaysia pushed forward its efforts to curtail the threats of radicalization and extremism through its initiated programs.

De-radicalization initiatives were considered an important facet of the overall approach. Various government agencies were tasked to rehabilitate and subsequently reintegrate former radicals back into society. The Royal Malaysian Police (RMP), Department of the Islamic Development of Malaysia, Malaysian Prisons Department, Ministry of Education, and higher education institutions in Malaysia were included in the program. Recently, Malaysia also introduced a Counter Messaging Centre (CMC), which is under the jurisdiction of the RMP, and subsequently, the Digital Strategic Communication Division under the Ministry of Foreign Affairs Counter Terrorism Centre. However, to-date, this Centre has been repelled by the new government of Malaysia.

Malaysian authorities claim to have rehabilitated and de-radicalized at least 90 percent of subjects. A former senior director of the Malaysian prisons department, who was involved in many of these rehabilitation efforts and de-radicalization processes, indicated the limitations of the program. According to him, it only addressed ideological concepts and did not identify re-integration of these subjects, which would have allowed them to assimilate back into a racially and religious diversified setting. These would allow the subjects to address other religious groups and to respect the other races who live within a larger Muslim community by adopting civics into their rehabilitation program.

Based on analysis of experience within the program of countering violence extremism (CVE), the Malaysian authorities are far from achieving 100 percent effectiveness. Religious supervision is also lacking over the authorities who employ clerics for CVE programs. There are no structured criteria to employ these experts by the relevant authorities. Some of these experts have been found to have radical and extremist views in delivering lectures in the Malaysian de-radicalization program.[28]

Ideological approaches and concerns

Although the approaches set forth within the context of countering terrorism, radicalization, and extremism from theological and ideological views are seen as the only defining approach to curtailing the threats, there is a need to carefully examine if these really address the threat or simply allow it to evolve further, giving extremists the leverage to act. The international community, which includes government leaders, scholars, and religious institutions, addresses counter-terrorism policies from an ideological approach rather than actionable law enforcement capabilities on the ground. Countering terrorism from an ideological perspective is seen as the most appropriate counter-measure by experts and counter-terrorism analysts.[29]

Ideological approaches also claim to address the "root" causes rather than dealing with the consequences. The experts claim it presents a more effective tool in comparison to police and intelligence actions designed to apprehend the terrorists. It also indicates a promise of limiting the radicalization process and preventing violent extremism.[30] Further, what is being advocated is not either/or but rather a balancing of what has been termed kinetic and nonkinetic response. Many governments who advocate an ideological approach are simultaneously engaged in combating terrorism with military force.

Identifying an ideological approach can be considered in two broad lines: one to counter the contested ideology and the other to promote an alternative, a more preferred one. Many of these approaches do not highlight and specify how this is to be done rather than only describing the goal to confronting and defeating the ideology.[31] In respect to the approaches from an ideological perspective in countering terrorism and violent extremism, the term "ideology" can be described as being vague as it leads to different interpretations. What Malaysia identifies as an ideological perspective can be different to other Southeast Asian countries.

This holds true for other theatres, as well. Prime Minister Abdullah Badawi addressed the threat of an ideological approach to the root causes of resolving the Palestinian and Iraq issues, which he identified as the factors driving terrorism. He claimed that the injustice, oppression, and marginalization of peoples in the countries concerned were an immediate driver for extremism and terrorism. Abdullah stated, "So long as we continue to deny or to confuse the issues in order to avert attention from the real issues, insurgency and terrorism will persist."[32] Left unstated, of course, was the stark reality that all issues do not lend themselves to compromise when interests as elemental as homeland and religion are at stake.

This highlights the need, in addressing roots of conflict, to include in analysis not only those in the threat group but those who have chosen to embrace altern-ative routes. A futile search for attributes must look instead to the manner in which pathways unfold in particular contexts. Put very simply, terrorists' beliefs and goals are also shared by the larger Muslim community who are not radical-ized but in various ways are willing to support the terrorists for their jihad toward the infidels and suppressors. The challenge for counter hence becomes substantial.

Policing and law enforcement

Attacks worldwide, especially in Southeast Asia and in particular Malaysia, have made clear that to counter ongoing threats in the near-term requires more than addressing root causes. A holistic multi-pronged strategy to prevent a terrorist attack should include political will and public support and cooperation. High-quality intelligence must be collected to curtail the evolving threat. International intelligence sharing and overall counter terrorism cooperation should be enhanced by integrating the agencies approach to counter the ideological and other threats. ASEAN Regional Forum (ARF) is a major platform for ASEAN governments to address collaboration and information sharing. In Malaysia, the police have established a special unit to monitor terrorism-related activities – the Counter Terrorism Unit, which has resulted in arrests through monitoring and investigations. It collaborates with the Malaysia Islamic Development Depart-ment (JAKIM), State Islamic Religious Departments (JAIN), Department of Prisons, and other relevant organizations.

Likewise, in the battle of narratives, long-term efforts to win the "war of ideas," with the assistance of community leaders and educators, should be inter-twined with counter-terrorism law enforcement operations. Other counter-terrorism protocols such as adoption of biometric, forgery-proof passports and tighter border controls and comprehensive and rigorous counter-proliferation measures to prevent acquisition of chemical, biological, radioactive and nuclear (CBRNE) weapons materials by terrorists must also be adopted. In 2017, an international syndicate was crippled by the Malaysian authorities, involving forgery of foreign passports, travel visas, and work permits.[33] This syndicate was involved in several related crimes involving illegal migrants and criminals acquiring residential and visas to live in Malaysia. According to a reliable source, there were indications of possible terrorists acquiring new identities to live within the region.

Rigorous implementation of international conventions and agreements must be pursued and addressed by law enforcement agencies so as to mitigate the risk factor. Law enforcement agencies have a lead role to play in reducing the risk of, and responding to the threats of, terrorism domestically. Law enforcement agen-cies must engage former surrendered terrorists to adopt strategies in countering the threat and this is vital for the curtailing of radicalization and violent extremism globally.

Hence, the current posture on terrorism by law enforcement agencies needs to place particular emphasis on preparedness, capability planning, maritime security, land transport security, critical infrastructure protection, threat assessment, and regional security. The need for cross-government approach such as responses by multi-agencies, comprehensive risk management that requires information, and intelligence sharing, coordination, and cooperation are vital to defeating the threats of terrorism. Industries and the public must be involved in mitigating the risk factors.

Civil society engagement

The Najib administration, which had to deal with security and religious threats of its predecessor, was unable to stop the hemorrhage that was Islamization in Malaysia, and so had to re-address it with new laws, such as POTA, SOSMA, and others.[34] It also founded in 2010 *the Global Movement of Moderates Foundation (GMMF)*, an institution to study and prevent radicalization and extremism among the Muslim community within Malaysia as well as sharing the idea and platform with global Muslim communities. Though positive steps, these were accompanied by some major mistakes. These included allowing a radical preacher from India, implicated in terrorist activities there and in Bangladesh, to remain in Malaysia even though a warrant of arrest was issued by New Delhi. Another error was to embrace radical clerics in the conduct of de-radicalization segments within the government-approved programs, the intent being to manipulate-while-appeasing some quarters of the ultra-conservative Muslim communities to retain political power, the objective being electoral gain.

With the April 9, 2018 change of government, Malaysia has a new administration, headed for the second time by Mahathir Mohamad. By June 2018, the government was in the process of assessing the laws pertaining to national security of Malaysia, to include those relating to terrorism and extremism. The steps it will take necessarily remain a matter of urgent concern, particularly given Mahathir's past missteps.

Conclusion

Countering the threats of terrorism and violent extremism requires a balanced approach and collaboration with law enforcement and communities. Addressing the roots and framing effective responses must be balanced with actual steps directed at prevention, reaction, resiliency, and consequences management. This is especially true in the Malaysian case, as the current situation demonstrates that threats have moved beyond long-term remediation. In fact, it is what the authorities have done, as they have moved to counter both threat message and operational efforts.

Threats such as the ASG, the IS, al-Qaeda, and JI loom large over Malaysia. The apprehension of ASG and IS operatives in August–October 2017 by the

Malaysian Counter-Terrorism Unit of the RMP is applauded but at the same time worrisome as these included a number of foreign nationals from within the region and globally. In addition to these are the home-grown terrorists who have been recruited and indoctrinated by Indonesian, Malay-speaking IS arm Katibah Nusantra, as revealed in its videos released from 2015 to 2017. Social media is also gaining a large number of followers within Malaysia and inspiring home-grown terrorists to act within the country. The Movida attack, in June 2016, the first IS-inspired attack – on a restaurant-cum-bar on the outskirts of the Kuala Lumpur, where two hand grenades were thrown into the restaurant – was carried out by local Malaysians supported and financed by Russel Salic, a Filipino.[35] No direct intervention was found in the inspiration chain. Such radicalization and extremism has grown since the days of JI being active in Malaysia. The new extremists are now prepared to carry out violent attacks on the nation's people and property in the name of jihad.[36]

The International Association for Counterterrorism and Security Professionals-Centre for Security Studies conducted an interview with a former rehabilitated JI operative member, who claims that Malaysia is not prepared for what is to come. The interviewee claims a growing number of radical youth recruited as early as 2000 have not been identified by any security agencies and that radical jihadists now in their early thirties to forties are within ultra-conservative religious and political groups, government, and non-government organizations.

The Malaysian government has declared that IS as a major threat to Malaysia and its people. Consequently, law enforcement agencies must collaborate with public and the communities. Communities are the first line of defense as has been proven by the actions already taken by the relevant authorities in Malaysia. This is recognized, and local support is the foundation of efforts to address root causes and propagate a counter-narrative. Their assistance allows the authorities to identify, infiltrate, and intervene to either thwart possible attacks or reverse the path of radicalization. Former convicted JI members who have been de-radicalized have also been mobilized to assist and provide information, intelligence, and expertise.

One of the key reasons why the Counter Terrorism Unit has been able to work in Muslim communities is because many of the personnel involved in these operations are themselves Malays, with easy accessibility and community acceptability. Being embedded in a receptive population makes it easier to curtail and identify the threat based on assessing individuals' radicalization stages. Such an approach interfaces with conventional counter-terrorism measures such as intelligence work to identify, monitor, interrupt, and neutralize the actions of these groups prior to planning or engaging in any act of violence. This being true, sometimes law enforcement agencies may not monitor an individual further if he or she is not seen to be progressing in the radicalization process. This can be risky as illustrated by a former JI member whose wife was not being monitored as she operated a food stall within the Malaysian High Courts.

Over the past years, RMP has managed to disrupt and capture a large number of terrorist suspects, totaling more than 300, and has been able to thwart many planned attacks. It seems clear that the preferred strategy for engaging terrorism-related activities is basically police work and intelligence sharing, locally and globally. To this end, Malaysia actively engages with its Southeast Asian neighbors in sharing intelligence and combating terrorism domestically and internationally.

As urgent and intense the task is, CVE strategies involves all relevant government agencies. The Counter-Terrorism Unit of the RMP normally takes the lead, followed by agencies tasked by the Ministry of Home Affairs. These also include the religious agencies and the prisons department. Community policing for CVE is playing a vital role as it involves engaging and building relationships with the community to fighting crime as well as curtailing extremism.[37] Key elements in countering terrorism and violent extremism include engaging security agencies and applying four key elements: reduction of the threat, preparedness, response with appropriate resources, and recovery in order to restore normal conditions. Finally, the authorities must focus on social cohesion, prevention, engagement, and specific intervention programs, as well as continually assessing and monitoring for vulnerabilities in an evolving threat environment.

Notes

1 "Islamic State: Defeating the Virtual Caliphate." *The Express Tribune*, October 6, 2017, http://tribune.com.pk/story/1524180/Islamic-state-defeating-virtual-caliphate.
2 Classified Source Myanmar: Interview by Author.
3 Classified Source Philippines: Interview by Author.
4 Ibid.
5 Anderson, *Imagined Communities: Reflection on The Origin and Spread of Nationalism.*
6 Source: IACSP data base centre, 2017.
7 Source: U.S. Military Intelligence; "In personal discussion with Andrin Raj" 2007.
8 Ibid.
9 Maut Brothers; a group linked to the Islamic State.
10 Source: IACSP database center; 2017.
11 Noor, "Crossing Ethnic Borders in Malaysia: Measuring the Fluidity of Ethnic Boundary and Group Formation,": 61–82.
12 Transnational Islamist Movement in Asia; Networks, Structure and Threat Assessment, conference jointly organized by the International Centre for Political Violence and Terrorism Research (ICPVTR) and Centre for Eurasian Policy, The Houston Institute, United States; September 19–20, 2006, Sentosa, Singapore.
13 Noor, "Crossing Ethnic Borders in Malaysia: Measuring the Fluidity of Ethnic Boundary and Group Formation," 61–82.
14 Ibid.
15 Abuza, "Militant Islam in Southeast Asia: Crucible of Terror," 53.
16 Abuza, "Militant Islam in Southeast Asia; Crucible of Terror," 55.
17 Lt. Col. Sani Royan; former Royal Malay Regiment and Intelligence Officer of the Royal Malaysian Armed Forces and current Deputy Director, International association for Counterterrorism and Security Professionals-Centre for Security Studies, Kuala Lumpur, Malaysia. Personal Interview 2012.
18 www.thestar.com.my/news/nation/2017/09/28/g25-religious-extremism-must-be-curbed/#yXDjROD37bBy80em.99. This statement was issued during the Najib's administration.
19 www.themalaysianinsight.com/s/51282/.

20 www.nst.com.my/news/nation/2018/03/349273/389-people-arrested-suspected-terrorism-related-activities-2013.

21 *International Journal of Management and Applied Science*, ISSN:2394-7926; 3 (5), May 2017; Radicalisation in Malaysia: Number of Homegrown Terrorist is Rising; Mohd. Mizam Aslam.

22 National Integrity Plan; Government of Malaysia; Strategy 5: Integrity Agenda for the Religious Institution page 44; Integrity Institute of Malaysia, April 23, 2004.

23 Lt. Col. Sani Royan; former Royal Malay Regiment and Intelligence Officer of the Royal Malaysian Armed Forces and current Deputy Director, International association for Counterterrorism and Security Professionals-Centre for Security Studies, Kuala Lumpur, Malaysia. Personal Interview 2012.

24 Ibid.

25 *Journal of Public Security and Safety*, 6 (2) (2016); Malaysia's Policy on Counter Terrorism and Deradicalization Strategy; Dato' Seri Dr. Ahmad Zahid Hamidi, Deputy Prime Minister of Malaysia.

26 Ibid.

27 Ibid.

28 "Controversial Malaysian preacher attracts fresh flak, but Home minister calls him and 'asset'": November 10, 2017; *The Straits Times*, Singapore.

29 The Southeast Asian Approach to Counter-Terrorism, Learning from Indonesia and Malaysia; Kumar Ramakrishna, Nanyang Technological University, Singapore; *The Journal of Conflict Studies*, 25 (1) (2005); http://journals.lib.unb.ca/index.php/JCS/article.view/189/333;A,A.

30 Counter-Terrorism in Southeast Asia: The Dangers of an "Ideological" Approach by Natasha Hamilton-Hart; Southeast Asian Studies Programme, National University of Singapore 2002.

31 Ibid.

32 Abdullah Ahmad Badawi, 2006. Speech at the launch of the book "The Malayan Emergency Revisited 1948–1960," Kuala Lumpur, May 9. www.pmo.gov.my.

33 www.thestar.com.my/news/nation/2018/03/07/bangladeshhi-forger-made-millions-in-msia/.

34 Hamidi "Malaysia's Policy on Counter Terrorism and Deradicalisation Strategy," *Journal of Public Security and Safety*, 6 (2) (2016), 4–6.

35 Bukit Aman verifying FBI report on financier of Movida blast, October 9, 2017, www.malaysiakini.com.

36 Raj, "Fighting the Enemy Within," September 11, 2017, www.nst.com.my.

37 Ibid.

Bibliography

Abuza, Zachary. "Militant Islam in Southeast Asia: Crucible of Terror," 2003.

Abdullah Ahmad Badawi. Speech at the Launch of the Book "The Malayan Emergency Revisited 1948–1960," Kuala Lumpur, May 9, 2006. www.pmo.gov.my.

Anderson, Benedict, *Imagined Communities: Reflection on The Origin and Spread of Nationalism*. London: Verso, 2004.

Chandra Muzzaffar. *Islamic Resurgence in Malaysia*. Petaling Jaya: Pernerbit Fajar Bakti, 1987.

Department of Military Intelligence Malaysia. Personal Interview August/September 2017.

Hamidi, Ahmad Zahid. "Malaysia's Policy on Counter Terrorism and Deradicalisation Strategy," *Journal of Public Security and Safety*, 6 (2) (2016), 4–6.

Hussin, Mutalib. *Islam and Ethnicity in Malay Politics*. Singapore: Oxford University Press, 1989.

International Conference on Global Movement of Moderates; Conference Proceedings, January 17–19, 2012, Kuala Lumpur Convention Centre, Malaysia.

Jemaah Islamiyah Deradicalized Former Operative (Personal Interview August/September 2017).

Lt. Col. Sani Royan, former Royal Malay Regiment and Intelligence Officer of the Royal Malaysian Armed Forces and current Deputy Director, International association for Counterterrorism and Security Professionals-Centre for Security Studies, Kuala Lumpur, Malaysia. Personal Interview 2012.

Lt. Col. Sani Royan; former Royal Malay Regiment and Intelligence Officer of the Royal Malaysian Armed Forces and current Deputy Director, International association for Counterterrorism and Security Professionals-Centre for Security Studies, Kuala Lumpur, Malaysia. In-person discussion August/September 2017.

National Integrity Plan. Government of Malaysia; Strategy 5: Integrity Agenda for the Religious Institution. Integrity Institute of Malaysia, April 23, 2004: 44.

Noor, Mansor Mohd. "Crossing Ethnic Borders in Malaysia: Measuring the Fluidity of Ethnic Boundary and Group Formation," *Akademika* 55 (Julai 1999).

Raj, Andrin. U.S. Military Intelligence: Interview based on confidentiality, 2007.

Raj, Andrin. *Militant Islam in Malaysia: Synergy between Regional and Global Jihadi Groups*. Middle East Institute, January 16, 2015.

Raj, Andrin. Columnist News Straits Times Malaysia: National Security 2001 present, available at www.nst.com.my.

Raj, Andrin. Confidential and Restricted documents based on trainings provided to governments globally and within the Southeast Asian region.

Raj, Andrin. "Fighting the Enemy Within," (September 11, 2017) available at www.nst.com.my.

Royal Malaysian Police (Personal Interview August/September 2017).

Southeast Asian Regional Military and Security Agencies Personnel, Personal Interview based on confidentiality, August/September 2017.

Transnational Islamist Movement in Asia; Networks, Structure and Threat Assessment, conference jointly organized by the International Centre for Political Violence and Terrorism Research (ICPVTR) and Centre for Eurasian Policy, The Houston Institute, United States; September 19–20, 2006. Sentosa, Singapore.

Transnational Islamism and Its Impact in Malaysia And Indonesia; IDC Herzliya, Rubin Centre, Research in International Affairs, August 29, 2011.

Part III
Quagmires

11 The counterinsurgency quandary in post-2001 Afghanistan

Antonio Giustozzi

1 Introduction

Afghanistan showed the initial signs of a developing insurgency as early as 2002. By 2005, a full blown insurgency was evident and was generally acknowledged as growing fast until at least 2009. During 2010–2014, greater commitments by Western powers and the growth of the Afghan national security forces appeared to have contained the expansion of the insurgency, even if by other accounts the insurgency was still large-scale. During this period, several thousand advisers from mostly Western countries advised the Afghan security forces, down to the army's company level and police at many of their local stations. From 2015 onwards, the scale of the advisory effort was dramatically reduced, while the insurgency was rebounding. The combination of a protracted threat and of a long-term advisory effort (15 years altogether) did not lead to a coherent, let alone fully adequate counter-insurgency effort, enabling the Afghan government to pursue the fight on its own. During 2017 the Trump Administration in the U.S. was expanding its commitment again, even if not to the levels seen in 2006–2013.

This chapter argues that a coherent Afghan counter-insurgency policy only started emerging in 2017 and tries to explain why this is the case. The chapter draws from the ever-growing literature on the Afghan post-2001 insurgency and on first-hand observation and participation in the development of the Afghan security forces. First of all the target of a counterinsurgency policy, that is, the insurgency of the Taliban is briefly assessed. The Taliban as an organization and a movement is particularly complex to analyze because of its decentralized character. But the purpose of this article is not to produce a particularly innovative analysis of the insurgency; it rather aims at delivering to the reader a short synthesis of its main features and sketch of the main characteristics of its evolution.[1]

This chapter analyzes the various factors that contributed to the failure to develop a coherent Afghan counterinsurgency policy after 2001. In this regard it is necessary to briefly discuss the 1980s counterinsurgency effort as the immediate historical precedent to the current conflict. Many Afghan officers and policy-makers lived through the 1980s, fighting either on the government side, or against it, and therefore whether consciously or not were influenced by the policies adopted in that period (Paragraph 3.1). It is also necessary to briefly discuss the impact of several years of extensive advising efforts and partnering by Western military and police forces with the Afghan security forces. Inevitably Afghan policymakers and implementers were influenced by this effort, even when they did not particularly agree with its content. Finally the longest section of this chapter (3.3) is dedicated to the post-2014 period, when the Afghan government is expected to fully take responsibility for the counterinsurgency effort.

2 The challenge

2.1 Insurgent structure, organization, linkages

After 2001, Afghanistan has been affected by multiple insurgencies, but by far the largest one has been the one mounted by the Taliban. It accounted at any given time for 80–90 percent of the insurgency if measured in terms of armed fighters, and probably an even larger percentage in terms of violent attacks. Several independent non-Afghan insurgent organizations have supported the Taliban (or part thereof) in their struggle, but from 2005 onwards they have become increasingly marginal to the conflict, due to the growth of the Taliban.[2] Between 2002–2016, Hizb-i Islami also operated as an insurgent groups in Afghanistan, but with relatively marginal impact alongside the Taliban. Hizb-i Islami reconciled with the Afghan government in 2016–2017.[3] Finally, the Islamic State (locally known as Daesh), which started its activities in 2014 and its insurgency rapidly reached significant proportions, even if never approaching the power of the Taliban.[4]

Because of the overwhelming weight of the Taliban within the insurgency, this section will be focused on the Taliban, leaving aside the smaller groups. Militarily, the Taliban were initially (2003–2004) lacking any organization; any Taliban cadre active in the insurgency would try to recruit as many men as possible, based on his personal connections to obtain funding and supplies. Some components of the Taliban, such as the Haqqani network, started to impose some discipline early on, appointing "representatives" in the provinces and districts. The Rahbari Shura, where the political leadership of the Taliban was concentrated, from 2004 onwards started sending "governors" to the provinces and then to the district, asking the combat groups loyal to the Taliban to acknowledge their authority.

In their effort to spread their reach nationwide, the Taliban started co-opting individuals and networks that had not previously been associated with the Taliban. This accentuated the polycentric character of the Taliban. Top-level decision making among the Taliban has been a complex matter of negotiating and trading horses; at the tactical level complex arrangements had to be made over who was responsible for which areas and about how to cooperate together. These complex mechanisms did not always work and often tension emerged among the Taliban over territorial control, recruitment and funding.[5]

External funding to the Taliban grew steadily over the years until 2014, enabling them to mobilize larger and larger forces. By 2012, the Taliban were able to deploy as many as 40,000 mobile fighters inside Afghanistan, with substantial numbers resting or in reserve in Pakistan and Iran, in addition to tens of thousands of part-time fighters and support elements. The growing ability of the Taliban to deploy mobile fighters, who could be deployed strategically, was perhaps the most important development of the insurgency (see also 2.3 below).[6]

The polycentric character of the Taliban did not impact dramatically on their operations until 2014. Fighting a guerrilla war against a far superior enemy (the US armed forces and their allies) as they were, the Taliban did not need to coordinate much among themselves. Quite the contrary, their convolute and redundant organization added resiliency, allowing them to take heavy casualties while remaining operational. With the withdrawal of Western combat troops at the end of 2014, the weak ability of the Taliban to coordinate militarily became a problem. Still the Taliban had by then developed the ability to deploy small armies at the regional level (at least in the south and east), fighting competently and effectively.[7]

2.2 Factors aiding the insurgency

The Taliban insurgency has been driven by several factors. First and foremost, revenge taking by members of the Taliban in 2002–2003, which drove many of them back toward the battlefield. The lopsided social, political and economic developments taking place in Afghanistan after 2001 also seem to have fueled mobilization into the Taliban, the marginalization of the clergy (which had by contrast been courted by the Taliban emirate), the economic and social marginalization of remote and poorly connected parts of the countryside, which did not benefit from the strong economic growth of post-2001 Afghanistan, and the marginalization of much of the rural youth.[8]

The geopolitical situation, with its multiple rivalries, also contributed to fueling the Taliban's insurgency. Disgruntled by Kabul's cozy relationship with India, the Pakistani authorities encouraged the Taliban's tapping into the large Afghan community living in Pakistan for recruitment and support. Indeed, the 4 million or so Afghans in Pakistan turned into a major source of recruits for the Taliban. Another important factor was the US intervention in Afghanistan and Iraq, and, in particular, the association of these interventions with an agenda of reshaping the region, introducing "democracy" and asserting American hegemony. After the intervention in Iraq it became much easier for the Taliban to raise funds on a large scale, and correspondingly expand their operations. Not only Pakistan continued to offer a safe haven to the Taliban along the Afghan border but also from 2005 onwards Iran started establishing relations with the Taliban, and by 2011 it was ready to open a safe haven for them on its territory as well. Funding from the Arab Gulf monarchies increased as well. The funding and the safe havens contributed decisively to the resilience of the Taliban as an insurgency, allowing them to train, rest, recuperate and establish a command and control center where their enemies could not threaten them.[9]

Nonetheless the remarkable success of the Taliban in courting support from different and even mutually hostile sponsors came at a price. It made the Taliban dependent to various degrees from foreign sponsors, and exposed to their whims. The heterogeneous character of the Taliban's external supporters started pulling different groups of Taliban in different directions particularly from 2014 onwards, as the regional conflict between Iran and Saudi Arabia was intensifying. The Taliban's internal cohesiveness, already problematic as discussed above, was further compromised.[10] Originally almost exclusively composed of Pashtuns, as the Taliban expanded geographically, they also started attracting other ethnicities, mostly Tajiks and Uzbeks. While this gave the Taliban a more solid foothold in some northern provinces, it also created tensions locally between Pashtun and other Taliban over appointments and the distribution of resources.[11] With the botched announcement of the death of Mullah Omar in July 2015, the Taliban's internal divisions became public, and their debates became more acrimonious than ever. Infighting among Taliban also intensified.[12]

2.3 Changing character of the insurgency

Originally a movement aimed at capitalizing on grievances against the new regime established in Kabul after 2001 and its external supporters, the Taliban gradually evolved in a counter-state that tried to control territory and population. Among the various commissions and offices the Taliban set up were some dedicated at administering specific aspects of life inside Afghanistan, first and foremost the judicial commission was tasked to administer justice through a network of Taliban courts.[13]

The Taliban military also became gradually more sophisticated. Initially the Taliban had little idea of guerrilla tactics and would often mass against government outposts; from 2011 however the leadership imposed the adoption of more professional guerrilla tactics, often offering training courses in Pakistan and later Iran. The Taliban became increasingly apt at laying ambushes and most importantly at laying IEDs, which soon became their weapon of choice. Although mostly not sophisticated technically, the Taliban's IED were mass produced and deployed so widely to represent a very serious challenge even for the most advanced armies of the world.[14]

Harder to notice initially, the Taliban also started developing the ability to conduct relatively complex maneuvers on the battlefield, even when hundreds or sometimes thousands of fighters were massed. The Taliban started experimenting with their new ability to conduct large-scale operations in 2014, but applied their capacity seriously only from 2015 onwards. This effectively represented a transition away from guerrilla warfare, toward "hybrid warfare" at least, if not conventional war. The new Taliban units retained the ability to disperse their men as well as to concentering them, and could fight with guerrilla tactics whenever convenient.[15]

The Taliban also developed their own "expeditionary" logistics, that is, the ability to carry out large-scale operations far away from their safe havens, where the bulk of their supplies are usually kept. The campaign leading to the capture of Kunduz city in September 2015, for example, was planned a year in advance and implied positioning supplies and organizing local support for thousands of fighters over a period of seven months in the middle of enemy territory – with supply lines crossing some old feuds of the anti-Taliban opposition of the 1990s – like Badakhshan and Takhar. The Taliban mixed their own centrally controlled logistics with procurement systems run by local Taliban (in part through black market actors) and with the services provided on a commercial basis by smugglers. The Taliban also improved on their initially non-existent medical support for their fighting units, establishing hospitals in Pakistan and deploying medics to the battlefield.[16]

3 Afghan counterinsurgency approach

3.1 The legacy of previous Afghan governments

The on-going insurgency in Afghanistan is not the first large-scale insurgency experienced by the Afghan state in recent history. In 1978–1992 a pro-Soviet government fought with direct Soviet support (1979–1989) against an array of insurgent groups, ranging from tribal to conservative Islamic, to Islamic fundamentalist, Muslim Brotherhood-type and Khomeinist. In the early days of that insurgency, the Afghan state did not have anything resembling a counterinsurgency strategy, beyond reacting to challenges on an ad hoc basis with extreme (but ineffectively targeted) violence, trying to pre-emptively exterminate the potential leadership of opposition groups, and mobilizing local communities against other communities deemed to be involved in the insurgency.

Under Soviet patronage and following Soviet advice, a new pro-Soviet leadership from early 1980 onwards tried to hammer together a more sophisticated counterinsurgency approach. Following the model of Soviet Union, primacy in developing and running the counterinsurgency campaign was given to the security services (KhAD), which by 1986 were upgraded to full ministerial status. With abundant funding and a major investment in training new professional cadres in the Soviet Union, KhAD organized tens of counterinsurgency battalions and a wide array of local militias, which proved much more effective in fighting the insurgency than the army and at least allowed the regime to survive the withdrawal of the Soviet troops in 1989, if not win the war outright.[17]

The Afghan army has remained focused on conventional warfare, at its best being able to protect key assets such as cities and strategic locations, or clear specific areas from insurgent presence by deploying massive firepower, but never being able to effectively engage with the insurgents, unless these were willing to stand and fight (which happened rarely). Even when the insurgents decided to stand and fight, the mountainous environment made it difficult for the conventionally trained army to score tactically. The police forces throughout this period limited their role to protecting the cities and the roads, playing a modest combat role only.[18]

The counterinsurgency of the 1980s can be construed an effort to develop the tools for targeted repression (non-existent in 1979 as the new regime had purged the state apparatus of most experienced officials), combined with a KhAD-driven effort to co-opt local opposition commanders (but without trying reconciliation with the higher leadership of the opposition). Only toward the very end of the regime (1991–1992), as Soviet support was clearly waning, President Najibullah also tried to engage with some moderate elements of the Pakistan-based opposition, but his leverage declined quickly as it became clear that he was losing his only source of international patronage.[19]

Many officers who had served in the 1980s within the ranks of the pro-Soviet regime made their way into serving for the post-2001 governments. By 2015 some of them had risen to senior positions, including three who served at different times as ministers of interior. They carried with them at least some legacy of the 1980s experience in counterinsurgency.

3.2 The ISAF legacy

From 2002–2014 *military* counterinsurgency efforts in Afghanistan have been largely driven by the leadership of the international contingent deployed to protect the new regime. The ISAF mission was not involved in counterinsurgency initially, but started doing so as the insurgency started picking up. By 2006 it was deployed to a number of provinces outside Kabul. Since the Americans were paying the bills of the ANA and European and other donors were paying the bulk of the bills of the Ministry of Interior, in addition to providing equipment, they a had a huge say over the formulation of the military counterinsurgency policy. In addition, their Afghan counterparts did not have clear ideas about the formulation of a counterinsurgency strategy, making it even easier for westerners to impose much of their agenda.[20]

This overwhelming Western influence might at least have speeded up the formulation of a clear (if not necessarily adequate) counterinsurgency strategy, if only the westerners themselves had had clear ideas in this regard. In reality each country contingent contributing to ISAF had its own views about the right counterinsurgency approach, and some of the national contingents changed their approach several times over the time span of ISAF. This was largely true of the Americans, whose influence was largely predominant. The result was constantly shifting counterinsurgency approaches, resulting in confusion and lack of clarity among the Afghans about what counterinsurgency approach the patrons and advisers were advocating. Another factor of confusion was the inherent discrepancy between what Western advisers and military leaders recommended and their own practice. For example, they would recommend that Afghans applied the rule of law to prisoners as well, but they were found to have tortured Afghan prisoners in Bagram. The Abu Ghraib scandal in Iraq also continued to undermine Western recommendations in this regard. Western advisers would advise against the use of massive firepower in inhabited areas, but often ended up bombarding Afghan villages. They recommended tight rules of engagement, but failed to consistently punish their own forces when they blatantly failed to observe such rules. They recommended that the Afghans deploy a lightly equipped infantry against the insurgents, but were unwilling to be deployed in battle without close air support, armor and artillery support. And so forth.[21]

The legacy of ISAF's advice resulted in misperceptions among their Afghan advisees. At the drawdown of Western troops in 2014, the legacy of ISAF in terms of the perception of their Afghan advisees could be described as consisting of the following main points:

1 the "modern way" of fighting is relying on air power to do most of the job;
2 the Taliban insurgency was assessed as well past its peak and weakened to an extent that it could not mount a strategic threat;
3 the political and military dimension of counterinsurgency are two separate tracks;
4 governance and social issues are dealt in yet another separate track (development);
5 in addition, ISAF imposed its own views about how the Afghan security establishment should be structured;
6 maintain a weakly mobile army to locally support a first line of defense constituted of police units and local militias;
7 focus on "useful Afghanistan," that is cities, heavily populated rural areas and the highways;
8 heavily garrisoned key population areas in the southern provinces of Kandahar and Helmand, in the assumption that the Taliban will be hurt by a direct challenge in their traditional turf.[22]

A major point of consensus had emerged after 2009 – completely crushing the insurgency in the presence of the Pakistani safe haven was not a realistic expectation and that the main aim of the Afghan security forces should be holding the ground while regional engagement leads toward some political settlement of the conflict.[23]

Points 3 and 4 above explain why in this paragraph what has been discussed so far has been described as "military counterinsurgency." It was assumed that well-behaved and disciplined forces (itself an ambitious target) could win the war even in a governance and political vacuum. The politics of territorial and population control remained throughout the conflict. Attempts were made by some of the contributing nations (the Britons, the Dutch …) to influence Kabul's appointments in terms of governors and chiefs of police. But while it proved possible after some tug of war with Kabul to get some people sacked, it never proved possible to get Kabul to appoint and properly support replacements that would meet the job descriptions as defined by Western advisers and diplomats.

As a result another aspect of ISAF's and, more generally, the Western countries' legacy in the counterinsurgency effort was a mix of appointees to senior provincial and district positions with good credentials meeting the standards recommended by Western advisers, and political appointees chosen by Kabul to further the political interests of the ruling coalition. The differences between these two groups and their coexistence in the governance and security apparatus contributed to make it more dysfunctional. The two groups promoted different approaches to counterinsurgency (among other things), contributing to make the Afghan government's approach even more incoherent.[24]

3.3 Post-ISAF Afghan counterinsurgency

3.3.1 The view from the Palace

The Kabul government took full formal control over security in Afghanistan by 2014; there was at this point no excuse for not developing a fully-fledged counterinsurgency strategy. The 2014 presidential elections and the political crisis that followed delayed the start of the formulation of such a strategy until the autumn of that year. More importantly, having started developing a counterinsurgency approach did not mean that progress came easily or quickly. The composite character of the new National Unity Government and the attendant differences among its members did not help to speed up the process. Once the process started, the Afghan National Unity Government had to deal with the mixed legacy of ISAF and reconcile it with a variety of views held by Afghan politicians and security officers, often irreconcilable with each other. Finally, ISAF was succeeded by an advisory mission named Resolute Support (RS) that despite having a much more modest footprint, could still rely on the fact almost all the bills related to maintaining the security sector were being paid by Western powers (mostly Americans). The then newly elected President Ghani was keenly aware of continuing Western influence over Afghanistan and indeed considered it a main asset for himself, being widely seen the Washington's favorite politician in Kabul. Being Washington's key partner in Kabul gave Ghani a power that he would not have been wielding otherwise. Ghani was happy to display American support by having the head of Resolute Support, Gen. Campbell, and a few other American officers and officials attend the meetings of the National Security Council on a regular basis.[25]

In the absence of anything resembling a coherent counterinsurgency strategy, during 2014–2015 Afghan counterinsurgency *practice* in Kabul can be described as follows:

1   hold on to not just key assets as defined by ISAF in 2013–2014 (see above), but also to all administrative centers, including those in remotely located districts;
2   adopt a largely defensive and reactive strategic posturing, while pursuing political negotiations brokered by the Pakistani authorities (hosts to the Taliban's leadership) and by the Chinese government;
3   rely primarily on the police in a counterinsurgency role, with the army mostly deployed in a back-up role and for occasional "clean-up operations";
4   rely primarily on special forces for offensive operations;
5   rely on local governance structures and the security services for co-opting individual opposition commanders (a legacy of the 1980s);

6   expand official militias (ALP) to fix selected areas affected by the insurgency.[26]

However, there was very little consensus within the NUG, or between it and the leadership of RS on these issues. For example, RS continued to advocate the ISAF line of thinking that it was not realistic to try and hold all the district centers. During 2015 in Kunduz, the Taliban bled the Afghan army and police in battles over mountain districts, attacking the district centers and then forcing the security forces to send reinforcements and supplies on poorly maintained mountain roads. The Taliban were able to force the security forces to fight on their ground, inflicting heavy casualties and wearing the enemy down. Still the NUG insisted that it would have been politically unbearable to abandon any area of the country to the armed opposition.[27]

The role of militias, official and unofficial, was also controversial: the human rights lobby strongly opposed them, and even within the NUG views varied greatly. Similarly, negotiations with the political leadership of the Taliban were supported by Kabul's Western allies, but viewed with strong suspicion by the supporters of the Chief Executive, Dr. Abdullah, and that of the first vice-president, Abdul Rashid Dostum, and even by some close collaborators of President Ghani.

Another political controversy concerned the inability or unwillingness of the army to take a more proactive role in counterinsurgency. The police bitterly criticized what they considered the laid-back approach of the army, while the police were taking much higher casualty rates than the army. Fighting in Helmand during 2015 and early 2016 demonstrated that the ability of the army to do more was highly questionable, but the criticism nonetheless continued.[28]

As the end of its first year, the occupation of Kunduz city by the Taliban for a few weeks in September–October 2015 prompted the NUG to reconsider its counterinsurgency policies. The option of "regularizing" hundreds of private militias connected somehow to pro-government politicians and local strongmen was considered, even though no final decision had been taken in this regard.[29] In January 2016 the decision was also taken to increase the powers of the provincial governors and task them to coordinate all the security forces within their area of authority.[30] These discussions however illustrated also how the NUG was lagging behind in addressing the developments on the ground. The options discussed might have been useful in fighting an insurgency until 2013 and to some extent 2014, but by 2015 the Taliban were switching full gear to civil war mode, essentially deploying relatively well organized and trained fighters to contest government control over the main arteries of communication, the most heavily populated portions of the countryside and the cities.

New appointments were made during 2015 to the army and police, trying to bring more competent commanders to the fore. These new appointments had the potential of being more relevant to addressing the emerging threat of Taliban "armies" entering cities like Kunduz, Lashkargah, Girishk, Maimana, Pul-i Khumri and even raiding Kabul city itself. These appointments were however in line with the merely reactive attitude that had characterized decision-making in security matters in post-2013 Afghanistan. In some cases, it might have been too late, given that in Helmand army corps 215 had been virtually destroyed in the fighting.[31]

In practice until 2017, President Ghani did not have the kind of control over the security apparatus, that he needed to impose his own approach toward reform. Only with the appointment of Ghani loyalists with a strong anti-corruption reputation at the top of Interior and Defence during 2017 (Barhami at the Ministry of Defence and Barmak at the Ministry of Interior) was Ghani able to see large-scale appointments of technocrats taking place. Ghani clearly thought that his priority should be stamping corruption out, as opposed to putting in charge an experienced general, able to handle the on-going conflict.[32] Perhaps Ghani's assumption was that with US support reaffirmed in August 2017, he had time to cleanse MoD and MoI of corruption, before focusing on developing a battlefield strategy.

At the same time under advice from Resolute Support, Ghani and his team developed, out of the *practice* described above, a more coherent strategy, which also featured a major restructuration of the security sector starting from 2017. This involved the transfer of the Border Police and of the gendarmerie (ANCOP) from the MoI to the MoD, the doubling in size of the Special Forces and the expansion of the Air Force. A decision was also taken to establish a new militia called Afghanistan Territorial Army (ATA) under the MoD, although implementation plans were delayed after some political backlash.

Of all these changes, the most important one was the planned expansion of the Special Forces. Given that such expansion would have to come at the expense of the regular army (with officers and NCOs being taken away from it to staff the new units), the decision implicitly reneged on the idea of turning the Afghan National Army into an effective armed force. Offensive operations were the preserve of the Special Forces after 2014, and so will be in the foreseeable future. While effective, the Special Forces do not have the manpower to secure the ground after their raids, which can mean one of two things:

a   Kabul gives up on control of the rural areas and content itself with disrupting the Taliban as much as possible, delaying if not preventing their offensive operations against government-held cities.
b   Ground is secured by other forces i.e., militias.

The ATA concept is therefore functional at the new special forces-centered emerging strategy. It remains to be see seen as to how the ATA would be managed and integrated with the rest of the security apparatus.[33]

3.3.2 The view from the field

Whatever the thinking about counterinsurgency among the policymakers in Kabul might have been, what was going on in the provinces could differ substantially, depending on whether any strong personality was running the effort locally and on the local alignments of factions and groups. It is interesting to note for example that the legacy of ISAF's efforts led to completely different outcomes in Helmand and Kandahar, two provinces that had been affected in very similar ways by ISAF's surge in 2009–2012. As mentioned above, ISAF's counterinsurgency at its peak had been very much focused on these two provinces, believed to be (rightly or wrongly) key to the effort to break the Taliban's morale. A thick net of outposts manned primarily by ANP and ALP was built up under ISAF guardianship from 2009 onwards, extending government control and influence to many rural areas previously controlled by the Taliban. Helmand was considered important because of its large opium production – preventing the Taliban from tapping on this source of revenue was believed to be another key tool of pressure that could have tipped the balance of the conflict against them. Under ISAF advice Helmand was assigned an army corps (215) just for itself and for the neighboring province of Nimruz (where hardly any Afghan National Army units was deployed anyway), while Kandahar shared another with the provinces of Uruzgan, Zabul and Daikundi. On paper, Helmand had 18,000 army troops, plus a few thousand National Police and Local Police, which during 2015 were reinforced with units sent from most other army corps.[34]

Despite being more strongly garrisoned than Kandahar, the Afghan government deployment to Helmand nearly collapsed under Taliban pressure during 2015–2016. Only the deployment of over 300 US and British troops and several hundred Afghan special forces averted the collapse. From October 2015, the main cities of Lashkargah and Girishk were under direct threat.[35]

By contrast during 2015–2016, the Taliban hardly made any significant gains in Kandahar. There could be several explanations for these dramatically different outcomes. Certainly, dynamics internal to the Taliban led to Helmand being considered a priority target compared to Kandahar, but the performance of the government's forces in Helmand fell short. Regardless of the actual causes of the dramatically different outcomes, what is interesting to note in this context is the explanations provided by Afghan police and army officers for such different outcomes. These explanations essentially boiled down to leadership. The chief of police of Kandahar province, Gen. Raziq, was almost universally seen as a charismatic and able field commander who was able not only to fully control ANP and ALP, but even de facto lead the security forces as a whole and kept the Taliban at bay.[36] Gen. Raziq was not the kind of military professional that Western advisers and policymakers were trying to promote after 2001: with no military education or indeed any education, a former militia commander with a reputation for smuggling, Raziq boasted about executing Taliban prisoners and was often accused of arbitrary executions of suspects as well.[37]

The failure in Helmand was attributed to a lack of capable leadership and lack of coordination among the different security forces. The Kandahar–Helmand debate accentuates the contrast between a Western-driven push to professionalize the security forces, and the political instinct of Afghan politicians to rely on charismatic "strongmen" who could deliver control over their regions of influence if patronage was extended to them and not too many questions asked about how the goods were being delivered.[38]

Under Karzai it was fairly common to appoint some powerbrokers of questionable reputation to key positions in the security apparatus. One such man was Amanullah Guzar who was chosen in 2015 as the new police chief for the northern region. Other examples include the appointment of Gen. Daud as head of the northern police zone in 2010; Matiullah Khan as chief of police of Uruzgan in 2011; Gul Agha Shirzai as governor of Nangarhar in 2005 and of several lesser powerbrokers in the ALP in various locations and at various times. During 2015–2017, however support for Raziq within the security establishment of Kabul grew steadily, even among professionally trained officers, as he appeared to be the only serving senior officer able to handle the Taliban. A common refrain was "we need one like him for our province."[39] Among the top politicians, such a line of action (empowering or re-empowering charismatic powerbrokers to lead the counterinsurgency effort locally) was also receiving growing support after 2014. Vice-president Dostum openly advocated a role for himself of this kind, while in Balkh one of the most important allies of Chief Executive Dr. Abdullah, governor Atta Mohammad Noor, claimed a key role for himself in fighting the Taliban, without asking for anybody's authorization.

4 Conclusion

The Afghan political and security establishment struggled to develop a coherent counterinsurgency strategy after 2001 for a number of reasons. Until 2005, there was little sense that a counterinsurgency strategy was needed, as the Taliban were seen as a mere nuisance. From 2005–2013, it was largely left to the ISAF command to determine which counterinsurgency strategy was most suitable. At least until 2011 few believed the Americans would leave any time soon and until the formal security transition took place in 2012–2013. It is not surprising therefore that until 2013 few in Kabul felt the need to start developing a coherent counterinsurgency approach. During 2013 and throughout 2014, there was a lot of uncertainty due to the long political transition; the political debate was not about how to defend the country and contain or defeat the Taliban, but about who should succeed him and which coalition could propel him to power. Likewise, the NUG in 2014 assumed the Taliban would not be able to mount a strategic threat and had instead faith in the capacity of the Afghan army to deal successfully with any military challenge coming from that side. The real turning point in the formulation of a coherent Afghan counterinsurgency strategy was instead the occupation of the city of Kunduz in the Taliban at the end of September 2015. The shock waves caused by the episode were then reinforced by the virtual disintegration of army corps 215 in Helmand, which had been under constant Taliban pressure from February 2015 onwards and was on the verge of disappearing altogether by November 2015. At this point the Afghan political and security elite finally started serious discussions about how to confront the threat.

By 2017, President Ghani and his closest collaborators were able to impose their strategic approach on the NUG: secure cities and highways, raid Taliban-held rural areas and occupy some of them with militias, force the Taliban into a stalemate and wait for their internal contradiction to explode, while at the same time hoping that the Kabul authorities' own internal contradictions would not burst first. This approach makes limited (and decreasing) use of strongmen and powerbrokers, and relies on the on-going reform of the security forces to contain corruption and improve performance. It is too early to assess the outcome of this strategy, but as of early 2018 there were signals that intra-Taliban contradictions were indeed worsening. The adoption of an American model of army organization continues to impose, however, a considerable burden on the Afghan security establishment; similarly, efforts by European advisers to reshape the police in the image of European police forces were never very realistic. Competing as they were with a burgeoning private sector, the ministries of defense and interior were never able to attract the kind of educated personnel who could run the relatively sophisticated machinery that Western advisers were helping to build.

Detractors of various leanings opposed Ghani's line, but the main opposing "party" argued that in an internal conflict what is needed is charismatic leaders who know their local environment well, can mobilize local support and enjoy credibility as effective leaders because they have a record of delivering success. An obvious weakness of this approach is that the ability of the security forces to function ends up depending on a limited number of individuals, linked personally to some key members of the ruling elite and liable to be marginalized in any political transition at the top. The same individuals are also liable to be targeted by insurgents and political rivals for assassination – Daud and Matiullah were killed alongside scores of less well-known local powerbrokers, while attempts were made on the lives of Raziq, Shirzai and Noor. The ability of the powerbrokers to project their influence beyond their own turf might invoke the jealousy of local rivals. Atta Mohammad Noor's efforts to extend his influence westwards caused for example a bitter conflict with Abdul Rashid Dostum, Uzbek powerbroker of north-western Afghanistan.[40]

Under Karzai the two approaches had been mixed together, with the effect of demoralizing the security establishment. Professional officers resented the promotion of many unqualified powerbrokers, and the powerbrokers-turned-security-officers feared the prospect of future professionalization. President Ghani instead had campaigned for election on a program of professionalization, reform and institution building.[41] He eventually proceeded to develop a more coherent counterinsurgency, but only time will tell whether his approach is the right one or not.

Notes

1 A sample of the project's early findings can be found in Farrell and Giustozzi, "Taliban at War."
2 Interviews carried out with Taliban cadres in 2012–2015 in virtually all provinces of Afghanistan showed that foreign fighters always represented a small proportion of all

fighters associated with the Taliban, although there was a high degree of variation from province to province.

3 See Tahir, "Gulbuddin Hekmatyar"; Marzban, "Gulbuddin Hekmatyar"; Ruttig, "Bomb and Ballot."
4 For a short discussion of the Islamic State in Afghanistan and of its strength, see Giustozzi, "Islamic State in Khorasan"; Rassler, "Situating the Emergence."
5 For a more detailed discussion, see Franco and Giustozzi, "Revolution in the Counter-revolution."
6 Interviews with Taliban leaders and cadres, various locations, 2012–2015.
7 Giustozzi, "Military Adaptation"; Giustozzi and Mangal, "Assessing the Taliban's Capture"; Giustozzi and Mangal, "Building Momentum."
8 Giustozzi, "Thirty Years of Conflict."
9 Waldman, "The Sun in the Sky"; Giustozzi, "The Next Congo"; Rashid, *Descent into Chaos*; Farrall, "Interview"; Majidyar, "Iran's Hard"; Steinberg and Woermer, "Exploring Iran"; Small, "Why is China."
10 Interviews with Taliban leaders and cadres, various locations, 2012–2015.
11 Pillalamarri, "Here's the Most Disturbing"; Giustozzi, "Taliban Beyond the Pashtuns"; Giustozzi and Reuter, "The Insurgents."
12 Of particular note is the formation of the High Council of the Taliban Emirate, the first formal split in the Taliban. In 2015–2017 violent infighting between the splinter group and the mainstream Taliban was common.
13 Interviews with Taliban leaders and cadres, various locations, 2012–2015; Giustozzi et al., "Shadow Justice."
14 Giustozzi, "Military Adaptation."
15 Williams, "Taliban Summer Offensive"; Giustozzi, "Military Adaptation"; Giustozzi and Mangal, "Assessing the Taliban's Capture"; Giustozzi and Mangal, "Building Momentum."
16 Giustozzi and Mangal, "Assessing the Taliban's Capture"; Interviews with Taliban commanders, military cadres and logisticians, July 2014–November 2015.
17 See Giustozzi, *War, Politics*.
18 Giustozzi, *Army of Afghanistan*; Urban, *War in Afghanistan*.
19 Personal communications with former officials of the Najibullah regime, London and Kabul, 2008–2010; Corwin, *Doomed in Afghanistan*.
20 For an overview of the ISAF mission, see Fairweather, *The Good War*.
21 Fairweather, *The Good War*; interviews with 16 Afghan army and police officers, Kabul and the provinces, September–November 2015.
22 Interviews with 16 Afghan army and police officers, Kabul and the provinces, September–November 2015; Giustozzi, "Afghan National Army after ISAF."
23 Al-Smadi, "Deep Differences"; Price, "Afghanistan and its Neighbours"; personal communications with foreign diplomats in Kabul, 2013–2016.
24 Fairweather, *The Good War*; "Sallust," *Operation Herrick*; Giustozzi and Kalinovsky, *Missionaries of Modernity*; Mukhopadhyay, *Warlords, Strongman Governor*; Giustozzi, "Resilient Oligopoly"; Giustozzi and Ishaqzadeh, *Policing Afghanistan*.
25 Personal communications with foreign diplomats in Kabul, March, April, September 2015 and January 2016; personal communication with government officials in Kabul, September 2015 and January 2016.
26 Personal communications with foreign diplomats in Kabul, March, April, September 2015 and January 2016; personal communication with government officials in Kabul, September 2015 and January 2016; interviews with 29 army and police officers, September–November 2016 and November 2017; Giustozzi and Ishaqzadeh, *Policing Afghanistan*; Giustozzi, *Army of Afghanistan*.
27 Personal communications with foreign diplomats in Kabul, March, April, September 2015 and January 2016.

28 Personal communications with foreign diplomats in Kabul, March, April, September 2015 and January 2016; personal communication with government officials in Kabul, September 2015 and January 2016; interviews with 29 army and police officers, September–November 2016 and November 2017.
29 Personal communications with foreign diplomats in Kabul, January 2016; personal communication with government officials in Kabul, January 2016.
30 Kubha, "Govt Increases."
31 Personal communications with foreign diplomats in Kabul, January 2016; personal communication with government officials in Kabul, January 2016.
32 Interviews with 13 police and army officers, November 2017.
33 Kate Clark and Borhan Osman, "More Militias? Part 2: The Proposed Afghan Territorial Army in the Fight Against ISKP," Berlin: AAN, September 23, 2017; Sharon Weinberger, Paul Mcleary, "Building the Afghan Air Force Will Take Years," *Foreign Policy*, August 21, 2017; Interviews with 13 police and army officers, November 2017.
34 Giustozzi, "Afghan National Army after ISAF"; www.understandingwar.org/press-media/webcast/surge-afghanistan-command-voices; Farrell and Giustozzi, "Taliban at War."
35 "US General Sees Afghanistan 'Deteriorating'," Voice of America, January 28, 2016; Giustozzi and Mangal, "Building Momentum"; interviews with army and police officers in Helmand, September–October 2015; personal communications with foreign diplomats in Kabul, January 2016; personal communications with government officials, Kabul, January 2016.
36 Interviews with 29 army and police officers, Helmand, Kandahar and other locations, September–November 2016 and November 2017.
37 See for example Aikins, "Our Man in Kandahar."
38 Interviews with 29 army and police officers, Helmand, Kandahar and other locations, September-November 2015 and November 2017.
39 Ibid.
40 Giustozzi, *Empires of Mud*; Mukhopadhyay, *Warlords, Strongman Governors*; Giustozzi, "Resilient Oligopoly"; Personal communications with foreign diplomats in Kabul, January 2016; personal communications with government officials, Kabul, January 2016.
41 Rasmussen, "Afghanistan's Warlord Vice-president"; Adeel, 'Military operations led by Ata'; personal communications with foreign diplomats in Kabul, January 2016; personal communications with government officials, Kabul, January 2016; Mukhopadhyay, *Warlords, Strongman Governors*.

Bibliography

Adeel, Mirwais. "Military Operations Led by Ata Mohammad Noor Sparks Concern," *Khaama News* (June 18, 2015).
Aikins, Matthieu. "Our Man in Kandahar." *The Atlantic* (November 2011).
Al-Smadi, Fatima. "Deep Differences over Reconciliation Process in Afghanistan." Al Jazeera Centre of Studies, January 13, 2014.
Clark, Kate and Borhan Osman. "More Militias? Part 2: The Proposed Afghan Territorial Army in the Fight Against ISKP." Berlin: AAN, September 23, 2017.
Corwin, Philip. *Doomed in Afghanistan: A U.N. Officer's Memoir of the Fall of Kabul and Najibullah's Failed Escape, 1992*. Chapel Hill: Rutgers, 2003.
Fairweather, Jack. *The Good War*. New York: Vintage, 2015.
Farrall, Lea. "Interview with a Taliban Insider: Iran's Game in Afghanistan," *The Atlantic*, (November 14, 2011).

Farrell, Tho and Antonio Giustozzi. "Taliban at war," *International Affairs* (July 2013).

Franco, Claudio and Antonio Giustozzi. "Revolution in the Counter-revolution: On the Efforts to Centralize the Taliban's Military Leadership." *Central Asian Affairs*, (forthcoming).

Giustozzi, Antonio. *War, Politics and Society in Afghanistan, 1978–1992*, London: Hurst, 2000.

Giustozzi, Antonio. *Empires of Mud*. London: Hurst, 2009.

Giustozzi, Antonio. "Taliban Beyond the Pashtuns," The Afghanistan Papers n. 5, Waterloo (Ontario): CIGI (July 2010).

Giustozzi, Antonio. 'The Resilient Oligopoly: A Political Economy of Northern Afghanistan 2001 and Onwards." Kabul: AREU, 2012.

Giustozzi, Antonio. 'Thirty Years of Conflict: Drivers of Anti-government Mobilisation in Afghanistan, 1978–2011." Kabul: AREU, January 2012.

Giustozzi, Antonio. "The Military Adaptation of the Taliban, 2002–2011." In Theo Farrell, Frans Osinga and James A. Russell. (eds.). *Fighting the Afghanistan War: States, Organisations and Military Adaptation*. Redwood City, CA: Stanford University Press, 2013.

Giustozzi, Antonio. "The Next Congo: Regional Competition for Influence in the Wake of NATO Withdrawal." Afghanistan Regional Forum n. 10, Elliott School of International Affairs, George Washington University, September 2013

Giustozzi, Antonio. "The Islamic State in Khorasan: A Nuanced View." London: RUSI, 2016.

Giustozzi, Antonio. *The Army of Afghanistan*. London: Hurst, 2016.

Giustozzi, Antonio. "The Afghan National Army after ISAF," Kabul: AREU, forthcoming.

Giustozzi, Antonio and Mohammed Ishaqzadeh. *Policing Afghanistan*. London: Hurst, 2013.

Giustozzi, Antonio and Artemy Kalinovsky. *Missionaries of Modernity*. London: Hurst, 2016.

Giustozzi. Antonio and Silab Mangal. 'Assessing the Taliban's Capture of Kunduz and Its Ongoing Implications." *IHS Jane's Terrorism and Insurgency Monitor*, October 15, 2015.

Giustozzi, Antonio and Silab Mangal. "Building Momentum – The Taliban's Territorial Manoeuvres in Helmand." *Jane's Terrorism and Insurgency Monitor*, November 5, 2015.

Giustozzi, Antonio and Silab Mangal. "A Gathering Storm? The Islamic State Campaign in Eastern Afghanistan." *Jane's Terrorism and Insurgency Monitor*, November 13, 2015.

Giustozzi, Antonio and Christoph Reuter. "The Insurgents of the Afghan North." Kabul/ Berlin: Afghan Analyst Network, May 2011.

Giustozzi, Antonio, Claudio Franco and Adam Baczko. "Shadow Justice: How the Taliban Run their Judiciary." Kabul: Integrity Watch Afghanistan, 2013.

Kubha, Samim. "Govt Increases Provincial Governors' Authority." *TOLOnews*, January 29, 2016.

Majidyar, A. K. "Iran's Hard and Soft Power in Afghanistan." *American Enterprise Institute*, August 27, 2012.

Marzban, Omid. "Gulbuddin Hekmatyar: From Holy Warrior to Wanted Terrorist," *Terrorism Monitor*, 4 (18) (September 21, 2006).

Mukhopadhyay, Dipali. *Warlords, Strongman Governors, and the State in Afghanistan*. Cambridge: Cambridge University Press, 2014.

Pillalamarri, Akhilesh. "Here's the Most Disturbing Thing About the Taliban Takeover of Kunduz," *The Diplomat*, October 2, 2015.

Price, Gareth. "Afghanistan and its Neighbours: Forging Regional Engagement." London: Chatham House, Afghanistan: Opportunity in Crisis Series No. 9.

Rashid, Ahmed. *Descent into Chaos*. London: Penguin, 2008

Rasmussen, Sune Engel. "Afghanistan's Warlord Vice-president Spoiling for a Fight with the Taliban," *The Guardian*, August 4, 2015.

Rassler, Don. "Situating the Emergence of the Islamic State of Khorasan." CTC Sentinel, March 2015.

Ruttig, Thomas. "Bomb and Ballot: The Many Strands and Tactics of Hezb-e Islami," Afghanistan Analyst Network, February 19, 2014.

"Sallust". *Operation Herrick*. CreateSpace Independent Publishing Platform, 2015.

Seldin, Jeff. "US General sees Afghanistan 'deteriorating'," *Voice of America*, (January 28, 2018) available at www.voanews.com/a/united-states-general-security-afghanistan-deteriorating/3166991.html.

Small, Andrew. "Why is China Talking to the Taliban?," *Foreign Policy* (June 21, 2013).

Steinberg, Guido and Nils Woermer. "Exploring Iran and Saudi Arabia's Interests in Afghanistan and Pakistan: Stakeholders or Spoilers – A Zero Sum Game? Part 1: Saudi Arabia." Barcelona: CIDOB, April 2013.

Tahir, Muhammad. "Gulbuddin Hekmatyar's Return to the Afghan Insurgency," *Terrorism Monitor*, 6 (11) (May 29, 2008).

Urban, Mark. *War in Afghanistan*. Basingstoke: Macmillan, 1989.

Waldman, Matt. "The Sun in the Sky." London: Crisis States Research Centre (LSE), 2009.

Weinberger, Sharon and Paul Mcleary. "Building the Afghan Air Force Will Take Years." *Foreign Policy* (August 21, 2017).

Williams, Elizabeth. "Taliban Summer Offensive Shows Increasing Capability." Washington: Institute for the Study of War, July 29, 2014.

12 Insurgency and violent extremism in Pakistan

Marvin G. Weinbaum

Introduction

Insurgencies against the state are hardly new for Pakistan. The country's history provides several notable armed challenges to provincial and national authority. From the breakaway of East Pakistan to the long simmering rebellion in Baluchistan, there have been serious, violent confrontations with forces of the state. The endemic lawlessness across the country, most notably in the country's largest city, Karachi, has also in many respects resembled an insurgency. This chapter focuses on the threat to the state that has emerged post-2001 from the militancy in Pakistan's tribal belt. It identifies the agents of this insurgency and traces the various campaigns and strategies that have been employed to contain and defeat militants, as well as the post-conflict counterinsurgency measures undertaken in this turbulent region.

But a full discussion of the threat of anti-state forces has also to take into account the militant extremism posed by Pakistan's jihadi and sectarian organizations. They have in common with insurgents in Pakistan's Federally Administered Tribal Area (FATA) many of the same root causes, including a range of economic, social and political grievances against the state. Insurgent and militant extremist groups share often similar drivers and pathways toward recruitment. High unemployment, underdevelopment and poor governance are contributing factors, as are frustrations resulting from corruption and perceived injustices. Like social and peer pressures, material incentives help to explain individual recruitment. Criminal elements and meddling external powers also fuel challenges to the state. No less important, insurgency and extremism have in common a rejection of the basic tenets of the country's democratic constitutional order and a willingness to use force if necessary to impose their vision of a political system fully compliant with Sharia law.

For the foreseeable future, the Pakistan state is unlikely to break apart under the weight of provincially based insurgencies or yield power to radical Islamists. The resilience of its government institutions, however much they underperform, should not be underestimated. Only a relatively small number of those citizens with grievances are prone to take political action, much less to employ violence. Still, existing challenges to state authority must be taken seriously. They are symptomatic of increasing religious-cultural intolerance and growing popular disaffection with the political system. For some time, Pakistan has been open to radical ideas and has pursued policies that have ceded space to extremist groups.

Until fairly recently the country's military and civilian leadership have shied away from fully confronting Pakistan's insurgency on its Afghan frontiers, and have failed to meet the challenges to the country's security posed by its extremist organizations. This chapter discusses how mounting domestic violence and regional strategic calculations have forced many of Pakistan's military and civilian leaders to reassess state policies dealing with terrorism and domestic extremism more broadly. But it will also identify obstacles to effective implementation of counterinsurgency and counterterrorism policies, and why there remain serious doubts about the state's commitment to meaningful change.

Anti-state groups

Insurgent and foreign groups

The most serious challenge to the Pakistan state, since West Pakistan now Bangaladesh acquired independence, remains the armed insurgency in Balochistan. It has in common with other nationalist movements in Sindh and Khyber Pakhtunkhwa provinces long held demands for greater provincial autonomy and identity recognition. But while the others have mostly been non-violent, the Baloch separatists have fought a guerilla war against the state intermittently over much of the country's history. Among these groups are the Balochistan Liberal Army, Lashkar-e-Balochistan and the Baloch Liberation Front. But none of these groups has sought radical change to Pakistan's form of government. At one point or another, they could probably have been satisfied with generous economic or political concessions from a federal government seen as dominated by Pakistan's Punjab Province.

Those groups taking up arms against the state in recent years differ substantially in the threat they pose to the state. All seek fundamental changes in Pakistan's prevailing system of government, namely creation of a more truly Islamic state. Principal among them are numerous groups that have operated under the banner of the Tehrik-e-Taliban (TTP),[1] a loosely connected set of militias led by commanders whose power has derived from their easy access to weapons and money, and whose legitimacy draws from their tribal connections and professed Islamic piety. Taliban forces have taken up arms against the state and its agencies in a traditionally lightly governed FATA in order to impose locally their ideas of Islamic governance as a step toward eventually affecting change in the political and social fabric of Pakistan. Organized in 2007, the TTP grew out of armed resistance to the state by tribally based militias located mostly in North and South Waziristan. Like so many other extremist organizations in Pakistan, the TTP has been a product of the Afghan wars of the 1980s and 1990s and the influence of Afghan Taliban fighters who escaped to Pakistan in 2001.

Foreign combatants also finding safe haven in Pakistan and neighboring border areas with Afghanistan have made common cause with Pakistan's TTP and enjoyed its protection. Most notable, of course, is al Qaeda but other militant foreign groups, among them the Islamic Movement of Uzbekistan (IMU) and the Eastern Turkistan Islamic Movement (ETIM), have similarly cooperated operationally and worked against the Pakistan state. Al Qaeda has helped to finance insurgent activities and train Pakistan's Taliban. It has also been linked to high profile terrorist attacks inside and outside FATA. Now also known as al Qaeda in the Indian Subcontinent (AQIS) the organization probably numbers no more than a few hundred fighters. However, despite suffering from the loss of key leaders killed in US drone attacks, al Qaeda may be strengthening in both Pakistan and Afghanistan with the return to the region of individuals who joined with Islamic State (IS) fighters in Syria and Iraq.[2]

Also a possible recipient of returnees is Islamic State-Khurasan Province (ISKP) or Daish, which has been able to establish control in several provincial districts in eastern Afghanistan and has taken root in Pakistan. As in Afghanistan, ISKP members in Pakistan began appearing at the end of 2013 mainly from insurgent groups disaffected from their respective Taliban leaderships. By one estimate, there are roughly 8,000 IS militants in Afghanistan and 2,000 to 3,000 in Pakistan.[3] From sanctuaries in eastern Afghanistan, ISKP is reported to have joined forces with at least one TTP faction to launch attacks on army border posts inside Pakistan.[4] ISKP has claimed responsibility for several of the most deadly terrorist acts in Pakistan since 2015, attacking religious shrines and military and urban civilian targets. While those taking the ISKP brand in both Pakistan and Afghanistan are thought to operate largely independent of the IS leadership in Iraq and Syria, their members practice many of the same ruthless tactics and share a vision of creating a transnational Islamic entity.

Three smaller militant groups have also left strong marks in recent years. Ansar ul-Sharia Pakistan (ASP), which splintered off from ISKP after reportedly becoming put off by its brutality, mounted its first terrorist attack in Pakistan in February 2017 and is regularly linked to police killings in Karachi and attempts to assassinate political party leaders. Its fighters, drawn from Pakistan's tribal areas, Karachi and Punjab Province, are mostly made up of recruits from ISKP, TTP and other militant groups. Although ASP is not formally affiliated with al-Qaeda, it is believed to have a close affinity with the group ideologically and to be inspired by its leader Ayman al-Zawahiri.[5]

A second group, Jamaat-ul Ahrar (JuA), emerged in 2014 as a faction of the TTP but appears as well to have links to Islamic State and possibly to al Qaeda. Since 2014 JuA has claimed to be behind widespread attacks, including in 2016 and 2017 suicide attacks in FATA and Lahore. Both JuA and ASP demonstrate so clearly the fluidity among the anti-state groups and blurring of territorial and ideological boundaries between them. The third group, Jandullah, unlike its Iranian Baloch, militant cousin, targets Pakistan's government institutions and civilians and has carried out suicide attacks especially against Shia. Pakistan's Jandullah is known to operate closely with both TTP and Islamic State.

Jihadi and other extremist organizations

Domestic extremist groups offer a related challenge to the Pakistan state. They are generally of three types. The first includes jihadi organizations over which the Pakistani state through its intelligence agencies exercises considerable influence. They are able to fundraise and openly propagate their radical views. Their leaders enjoy considerable freedom of movement. Several are the progeny of Pakistan's Inter-Service Intelligence (ISI). Meant to serve the military as armed proxies in Afghanistan and Pakistan, they continue to be treated as assets. The most notable example of this type of organization is Lashkar-e-Taiba (LeT), whose militant wing hides behind extensive and often popular community social and charitable activities. Combined with strong organizational skills and effective messaging, LeT is able to recruit and motivate its followers.

A second type also normally takes directions from the ISI. But its organizations stand apart in the frequency with which they undertake operations not fully sanctioned by the intelligence services. Jaish-e-Mohammad (JeM), an organization originally created by Pakistan's ISI to consolidate jihadi operations in Kashmir, fits this description. They are also especially likely to forge alliances with groups outside the government's orbit of control. Although murky, a January 2, 2016 attack on an Indian airbase in Pathankot provides an example of a probable unauthorized action undertaken by JeM designed in this instance to disrupt efforts to restart a comprehensive dialogue between Pakistan and India.

Those jihadi and other domestic extremist group organizations operating more often outside the control of the ISI make up a third type. The militantly sectarian Lashkar-e-Jangvi (LeJ) and the volatile Red Mosque followers of Maulana Abdul Aziz are cases in point. The actions of these groups have at times been in line with government policy, and Pakistan's intelligence services, principally the ISI, are not hesitant in using them for their own purposes. But more typically these groups engage in activities that clash directly with state policies and mount attacks from which the Pakistan government is anxious to be disassociated.

Pakistan's first serious encounters with domestic Islamic extremists occurred in the 1990s with a rash of targeted sectarian killings against Shiite citizens. Terrorist bombings, including suicide attacks, began in earnest within the last decade and are marked most notably by the 2008 massive bombing of Islamabad's Marriott Hotel. From 2003 through to mid-2017, nearly 22,000 civilians were killed as a result of terrorist violence, and more than 6,800 security force personnel have died at the hands of militants.[6]

Pakistan has officially banned over 200 militant organizations, including the leading foreign terrorist groups.[7] But most are either currently inactive or are organizations that have re-titled themselves in the hope of avoiding being labeled illegal and face the possibility of having their assets seized. Many groups are in fact factions that have split away from parent groups. At present, there are estimated at least 60 jihadi and other Islamic extremist organizations that together form the hub of terrorism in Pakistan.[8] Many of them form a seamless network of anti-state groups that collaborate in joint operations. Nearly all the groups identify with rigidly conservative Salafi Islamic precepts. Most are also sympathetic to the TTP and several have carried out attacks inside Pakistan on its behalf.

Many of the very forces responsible for domestic violence and that aspire to bring conceivably existential change to Pakistan are ones that the state has itself unleashed. They were launched and nurtured in the state's use of militant groups as strategic pieces in an India-centric foreign policy that employed surrogates to carry out Pakistan's policies in Kashmir and Afghanistan. To foster those policies but also as a means of domestic control, state institutions allowed militant elements to expand domestically and spread their roots in the society.

Counterterrorism and counterinsurgency

Strategies

Pakistan's insurgencies in its border areas have most of their roots in the jihad in Afghanistan during the 1980s when there began a gradual transformation of leadership structure within tribal groups. The traditional authority figures whose influence lay in their social status and relationship with agents of the state gradually gave way to a new breed of leader. This trend in FATA was accelerated in the 1990s, as Pakistan's material support of the Taliban steadily overran forces loyal to the Afghan government. Much as it had a decade earlier, Pakistan supplied military advisors and equipment and served as the conduit for financial backing for the Taliban.

Pakistan's internal security policy after 2001 and until modified following 2014 was built on three related strategies – selectivity, gradualism and containment – for keeping in check its insurgencies and violent extremist organizations. These strategies grew out of successes and failures over time in trying to manage the country's endemic tribal, ethnic and sectarian conflicts. They were shaped as well by having to craft counterinsurgency policies that could avoid alienating blocs of politicians, the country's religious establishment, and other political obstacles. Policies had also to be sensitive to public reactions to military actions.

Against tribally based insurgencies, the state has encouraged internal factionalism. Much as the British had earlier, the Pakistan army has practiced a policy of divide and rule using rewards and punitive actions to play off militant tribal leaders against one another. Where possible, national and local security forces have settled for containment and avoided broad military confrontation. Their caution grows out of the difficulties encountered in subduing well-armed and motivated tribal insurgents fighting on favorable physical and social terrain. It also reflects the military's frequent lack of confidence that it could take and hold territory on insurgents' home turf. Overall, the army-led Frontier Corps has been more effective than regular army units in winning community trust. Little is expected of the poorly equipped, poorly trained and underpaid local forces.[9]

The military has normally followed a phased approach toward insurgencies that is mirrored to some extent in its efforts to curb domestic violent extremist organizations. There have been long periods of low intensity intermittent confrontation punctuated by the launching of well-publicized relatively brief campaigns. The generals have generally refrained from bolder actions until confident that they had broad popular backing. Even then, the military has appeared to embark on major operations only when, in the wake of high-profile domestic terrorist attacks on soft targets, it felt pressure to act. Despite the regular urging by the United States that Pakistan move to aggressively crush the country's extremist elements, this pressuring has apparently never figured strongly in Pakistan army decisions to undertake its campaigns against militants.

Military operations

In late 2001 and early 2002, with the assistance of U.S. intelligence engaged in Operation Enduring Freedom, Pakistan's security forces were expected to cut the escape routes into South and North Waziristan and round up fleeing Al Qaeda fighters. While several high-profile al Qaeda figures were detained and handed over to the Americans, the Afghan Taliban leadership was able to cross into Pakistan with some impunity. The presence of these fighters and their leaders in Pakistan soon contributed to fostering already budding homegrown insurgencies in several of the tribal agencies, most notably in South Waziristan, where Nek Mohammad, a former mujahid fighter in Afghanistan, had formed one of several militias. The ascendance of these militias coincided with the disappearance of much of the traditional tribal leadership in FATA. Those government-approved local leaders who had not already fled were often killed.

Over the period between 2004 and 2008 regular army-led Frontier Corps and Frontier Constabulary troops with some air support, conducted operations against these insurgent forces in South and North Waziristan. The success of the operations could generally be termed as modest or poor. Territory was cleared but not held as insurgent fighters returned. Top leaders nearly always managed to escape the army's sweep. As is also often pointed out, the Pakistan army is ill prepared for the kind of warfare presented by anti-state, especially terrorist, groups.[10] Although the military's performance in the frontier improved over time, its troops had been trained to fight the Indian army in set battles, not an enemy adept in guerrilla tactics. Security forces complained of a lack of transport, attack helicopters, and strong communications; and the Frontier Corps is especially poorly equipped. In FATA, troops from Punjab are often seen as not culturally attuned and thus unable to gain the confidence of Pashtun tribal populations. One visible impact of the military campaigns to clear FATA of insurgents has been the massive displacement of residents.

A June 2004 peace agreement in South Waziristan with insurgents was the first of several and set a pattern for later deals with insurgents. The terms usually called for the insurgents to promise not to attack the state in exchange for the army not interfering in tribal or local affairs. Insurgent fighters were able to keep their arms. Tellingly, each deal by the military was negotiated from a position of relative weakness and every agreement was soon violated and broke down. In the end, the military position of the insurgents was usually strengthened.

In general, these campaigns against militants generated little popular attention or enthusiasm. The army's actions in 2007 in Islamabad and 2009 in the Khybar Pakhtunkhwa's (KP) Swat district were the exceptions. For some time, the government had shown great tolerance for the occupation by Islamist militants of Islamabad's Lal Masjid (Red Mosque). Eventually, the defiance of the government by its leaders and the provocative behavior of their followers became so blatant that in November 2007 under order from President-General Musharraf, the army laid siege to the mosque and then heavy handedly raided it to remove its occupiers.

Similarly, the military had stood by while those aligned with the TTP were spreading their radical Islamic agenda beyond the delineated tribal boundaries of FATA into settled areas of the Northwest. Tehreek-e-Nafaz-e-Shariat-e-Mohammad (TNSM) fighters led by Sufi Mohammad and aligned with the Hek-imullah's TTP were able during 2008 and 2009 to take control of the Swat District, a picturesquely lush valley and forested highlands. Sufi was already a controversial figure. After the 2001 U.S. military intervention in Afghanistan, he recruited some 8,000 Pakistanis to fight with the Taliban, many of whom would never return.

The National Assembly endorsed in April 2009 a political agreement that allowed for the TNSM's administrative takeover and the imposition of strict Islamic rule, including the primacy of Sharia courts. But the occupiers' severe governance practices soon alienated most of the local population. When its harsh rule caught the attention of the wider public, popular sentiment went between March and May from strong approval to equally strong disapproval of the deal on Swat. The country's politicians and military recoiled at Sufi's denunciation of the parliamentary system and his threat to carry his movement to Islamabad. After TNSM-aligned fighters appeared to be following through on their leaders' threat by occupying Swat's neighboring district of Buner, the army, now with federal government backing, was given a free hand to forcibly remove Swat's new rulers. The success of the military in Operation *Rah-e-Rast* that began in May 2009 was a clear demonstration of a willingness to use blunt, often indiscriminate firepower to dislodge Islamic extremists, even if it resulted in creating a massive displacement of people.

Even when the army had been able to rout insurgents as in Swat and manage to hold areas taken, its attempts to employ the principles of a COIN strategy such as used in Iraq in 2007 fell short in the prescribed all-important building phase. With the military engaged largely in scorched-earth warfare resulting in heavy casualties, this key element of the COIN strategy received little emphasis. A population-centric approach is designed mainly through development programs to win over the loyalty, trust and confidence of local populations. To gain this popular support, civil authority is expected to take the lead from the military in providing educational and health facilities, greater economic opportunities, and a quality of governance that can inspire confidence. But Pakistan's governments when given the opportunity have seldom shown either will or capacity.

With a new government in power after parliamentary elections of 2013, there was renewed interest in reaching a political settlement with the insurgency in the FATA. The new Sharif government was anxious to engage with the TTP. When talks broke down after several fruitless negotiating sessions, the army felt free to mount its long-planned air and later ground offensive against the insurgents' stronghold in North Waziristan in operation *Zarb-e-Azb*. The action followed on the heels of an early June 2014 attack on Karachi's Jinnah International Airport that rallied most political elements behind the decision. But the watershed event that brought the political elites and public together behind the military's apparent resolve to crush the TTP was the Peshawar Army Public School massacre in December 2014 in which 132 children and nine others were killed. It was the bloodiest terrorist attack the country had ever suffered. Operation Zarb-e-Azb – named for the sword of the prophet Muhammad – in North Waziristan would turn out to be Pakistan's largest counterinsurgency operation.

Operation Zarb-e-Azb

The objective of Operation Zarb-e-Azb, according to a statement by the Inter-Services Public Relations (ISPR), the Pakistan military's public affairs wing, was to target "foreign and local terrorists" hiding in sanctuaries in North Waziristan.[11] These were identified as TTP fighters and members of al Qaeda, whose organization had been training and facilitating TTP fighters to carry out attacks in Pakistan in exchange for hosting al Qaeda militants in TTP safe areas. Other al Qaeda-allied groups announced to be targeted were Uzbek militants with IMU, the Islamic Jihad Union (IJU), Uighur separatists belonging to ETIM, as well as various Tajiks, Chechens and Arab fighters. The campaign in North Waziristan was complicated by the presence in the agency of ISI-protected Afghans belonging to the Haqqani network. So too was the fact that some of Pakistan's divide and rule policy's "good" militants were bedded down in North Waziristan.

The Pakistani army deployed from 30,000 to 60,000 ground troops to North Waziristan, including special operations forces from its Special Services Group and armored and artillery units, as well as aerial units in close air support, and its own drones to carry out surveillance and strikes. In the first week of operations alone, the army targeted primarily Uzbek militants in clashes and air strikes in areas like Shawal and Datta Khel, the latter being a primary stronghold of fighters loyal to Bahadur. Most ground forces focused on clearing militant bases from the agency's main population centers in Miramshah and Mir Ali before shifting to more mountainous and treacherous areas closer to the border with Afghanistan.

Operation *Zarb-e Azb* ultimately triggered a mass exodus of militants. It was reported that many local and foreign fighters left before operations began for areas near the border with Afghanistan.[12] Some claim that Pakistan's army deliberately over a two-week period abandoned key posts along North Waziristan's border.[13] This action that could have been intended to help groups like the Haqqani Network slip across the frontier to avoid getting caught up in the operations, also apparently allowed many local militants and foreign fighters to escape. In attacking their strongholds the Pakistan army would subsequently inflict heavy casualties on the remaining fighters, but the operations seemed more intent on clearing areas than on killing militants.

The Pakistani military had by January 2016 declared Operation Zarb-e-Azb a "phenomenal success." The military's public relations office claimed that at least 3,400 militants had been killed and 837 hideouts destroyed over the previous 18 months.[14] But the military command's claims of complete victory in the FATA were somewhat overdrawn. North Waziristan had been cleared of Taliban and Haqqani network combatants, and most territory across the tribal areas was free of organized resistance However, in early 2018 pockets of fighters reportedly were holed up in the Khurram agency and other scattered locations, and the main force of TTP fighters and other militants had managed to take refuge close by in Afghanistan.[15]

Pakistan's intense ground campaign, use of heavy artillery and the often-indiscriminate aerial bombardments by high-performance aircraft in Zarb-e-Azb led to extensive losses of civilian life in the tribal areas. The Pakistani military has been accused of human rights abuses during the campaign, including summary executions, and misconduct of military tribunals against civilians. The media blackout imposed on the region has made these claims difficult to verify. Clearly evident is the fact that operations also took a heavy toll on the local infrastructure. In areas near Miranshah and Mir Ali, roads, homes, and what few civil institutions still existed in the region were reduced to hollow shells by aerial attacks and ground operations.

While the violence has taken its heavy toll, there are also promising indicators of reconstruction that could now contribute to a successful COIN strategy. Under the army's direction, new roads and buildings have been constructed in North Waziristan for the returning of more than 1.5 million displaced civilians. Health facilities, playgrounds and even a sports stadium are being built and plans are underway for bringing schools and electricity to remote villages.[16] But the long-term effects of Zarb-e-Azb will be limited unless the space once occupied by militant groups is replaced with accountable and responsive local political institutions. Efforts to amend prior laws for the tribal areas including the Frontier Crimes Regulations have been lagging. Attempts to develop a civil authority responsible to the Pakistan government repeatedly hit setbacks. Pakistan's failure to integrate these regions with the central government both politically and economically will further undermine efforts to curtail extremism in these regions. Fostering a cooperative population in the tribal areas is essential for reducing support for extremism and generally pacifying the region.

Ongoing security concerns: from North Waziristan to Punjab

In operation Zarb-e-Azb, the military has apparently achieved some success in Pakistan's long-term fight against extremism. Despite TTP and foreign militants' claims of responsibility for deadly attacks against Pakistanis, terrorist violence in Pakistan has visibly dropped in recent years. A study by the South Asia Terrorism Portal reported that Pakistan in 2015 suffered the lowest number of suicide attacks and deaths from terrorism since 2006.[17] Civilian deaths due to terrorism fell from a virtual all-time high of 3,001 in 2013 to 1,781 in 2014 and 940 in 2015 – roughly a 70 percent decline from its peak and the lowest number since 2006.[18] Sectarian incidents and casualties across Pakistan had declined 50 percent since 2012 and by approximately 75 percent in Balochistan.[19] While the pattern of fewer incidents has held, particularly a decline in urban violence, a series of high-profile attacks in 2016 and 2017 have again raised fears. Soft target attacks such as those at Bacha Khan University in Charsadda in January 2016 and in markets in Parachinar and Lahore in June and July 2017 remain easily within the reach of the militants.

Further progress depends on the ability to coordinate with Afghanistan on cross-border operations. With a history of mistrust and suspicion, the Afghan government appears to lack the will and its security forces the capacity to interdict or flush out Pakistani TTP and foreign militants who have found safe haven. Moreover, the surge of militants from Pakistan into accessible, mostly ungoverned Afghan provinces has probably also exacerbated the steady deterioration of security in Afghanistan. Foreign militants from North Waziristan are alleged to have assisted the Afghan Taliban in the stunning takeover of Afghanistan's northern city of Kunduz during two weeks in September 2015.[20]

In Pakistan, military operations have failed to destroy the leadership and organizational networks of key militant groups. Although operations severely disrupted their operations, the mass exodus of most TTP militants protected the group from being irreversibly destroyed and allowed them to preserve their militant networks. The senior leaders of militant groups recently based in North Waziristan have been largely untouched by these operations. Possibly aside from Omar Khalid Khorasani and Hafiz Gul Bahadur,[21] no major commanders, including Maulana Fazlullah and Adnan Rashid, have been killed or captured during the campaign. With their leadership largely intact, these groups retain both the capability and will to strike Pakistan, either through reestablished safe havens in Pakistan or from their new sanctuaries in Afghanistan.

Operation Zarb-e-Azb's narrow focus on North Waziristan has also drawn criticism as militant and sectarian movements have thrived in other parts of the country. Even while the army has been carrying out operations in the tribal areas, it has been seen as doing little to thwart sectarian attacks in Punjab and throughout Pakistan. In Punjab, groups often referred to as the Punjabi Taliban that include Jaish-e-Mohammed and Lashkar-e-Jhangvi as well as the more state-friendly Lashkar-e-Taiba retain their ties to the TTP and have been known to carry out attacks in its name. Militants often move freely among the various groups. Large numbers of those that have fought Pakistan's military in the tribal areas are thought to come from Punjab.[22]

Elsewhere in Pakistan, centers of extremist activity are also believed to be flourishing. Settled areas outside FATA provide a steady stream of recruits for insurgent groups. Funding from illicit activities such as kidnapping and extortion are widespread. Massive numbers of Pashtuns from the tribal belt, many affiliated with the TTP, have relocated to Karachi, where they provide a strong financial support group for the insurgency. For their safety, key figures in al Qaeda and Afghan Taliban are believed to have gone underground. Until 2016 many of Karachi's neighborhoods were beyond the writ of the state.

Signs of a new realism

Despite years of violence in its cities and attacks against institutions of the state, Pakistan has long remained unwilling to acknowledge that it faces a serious, perhaps existential threat from the country's armed groups. Nor has it been ready to own up to the fact that its own policies of long neglect in the FATA and forbearance toward jihadi and sectarian organizations elsewhere bear much of the responsibility. Yet there appears of late to be a growing understanding within both Pakistan's military and civilian leadership of the need to reevaluate the country's policies. Rising violence countrywide from armed groups has made it increasingly difficult to minimize the domestic threat. Military officials now seem more ready to acknowledge that even while the perils of Indian aggression against Pakistan remain undiminished, the most immediate danger to the country comes from within. Rather than relying so strongly on manipulation and deal making to contain it, the exercise of greater force is required.

There are some hopeful signs of a greater resolve to stand up against intolerance and extremism. It has become a mantra of the military that extremism in all its forms threatens Pakistan in a policy of zero tolerance toward terrorists.[23] Modest indicators of progress are seen in increased enforcement of regulations against hate speech, and restrictions on the use of mosque loudspeakers to prevent public incitement. By official count, over 250 suspected and unregistered madrassas have been shut down countrywide.[24] Greater controls have also been put on weapons sales. The February 2016 decision, despite strong conservative opposition, to hang the assassin of former Punjab governor Salmaan Taseer – he had campaigned for the reform of the country's blasphemy law – took courage because of political costs and potential for militant blowback. The passage by the PML-N controlled Punjab Assembly of a law to protect women against violence and the promise of legislation to discourage honor killings was also highly controversial.[25] Among long-banned but tolerated groups, sectarian LeJ, jihadi JeM and pan-Islamic Hizb-ut-Tahrir are officially now targeted. In an effort to end sectarian violence, many clerics preaching the hatred promoted by LeJ have been arrested. And under intensive U.S. and international pressure, the government in January 2018 confiscated the properties and curtailed the financing of the banned Jamaat ud-Dawa and associated Falah-Insaniant Foundation.[26]

But the broadest campaign to address domestic violence has come with the combined effort of the army and local law enforcement countrywide in Operation *Radd-ul-Fasaad.* Begun in February 2017 its stated aim is to root out terrorist cells. As is so often the case, it has come only after an up-tick in violence in cities. Having felt a period of confidence that domestic attacks by extremist groups had been sharply curtailed by Zarb-e-Azb, a series of suicide attacks against Shias and other soft targets helped precipitate the military's decision to act to dismantle these groups' strategic assets.

Still, the military's long-term commitment and sincerity in confronting the country's militant extremists draws skepticism. Much more needs to be done to destroy their organizational infrastructures. There is little evidence of Pakistan's eagerness to fully disown those organizations like LeT with which the country's military has been so intimately connected. No serious charges have yet been brought against Hafiz Saeed its leader who, claiming to lead only LeT's political and charity offshoot, Falah-e-Insaniat Foundation, also is believed to command its militant wing. Saeed had been allowed to speak in a manner that would likely have led to the trial of almost anyone else. Similarly, the Red Mosque's Mullah Aziz continues to preach intolerance and violence in anti-state sermons. In answer to criticism, spokesmen for the government and military promise that LeT militants will be disbanded along with all other groups engaged in terrorism. Particularly in dealing with LeT, however, security forces seem determined to move against their long-term asset, at most, cautiously and gradually. Doubts about the government's resolve to stand up against the forces of extremism were revived in December 2017 with concessions to a newly formed religious party protesting an alleged softening of the country's blasphemy law. After weeks in which demonstrators were allowed to paralyze Islamabad, the regime, possibly prodded by the military, capitulated to demands that included the firing of the Federal Law Minister and a strengthening of the blasphemy law.[27]

Having for so long minimized the perils of internal extremism, the civilian government was slow in developing a full-fledged national counterterrorism strategy. A basic legal framework for dealing with terrorism, the Counterterrorism Act (ATA), was enacted in 1997. The ATA was intended to overcome basic problems in the country's criminal justice system with its long delays in bringing cases to trial and low conviction rates. Public confidence had badly eroded.[28] ATA was essentially revised by the parliament's January 2015 approval of Pakistan's 21st Amendment. It provided that special military courts would for two years handle terrorism cases. It allowed these courts to work in secret and for the subsequent execution of hundreds of those accused. Meanwhile, extrajudicial killing by the military is also said to be still ongoing.[29]

A National Internal Security Policy (NISP) announced in February 2014 was the first ever to address the need for an integrated security approach. It proposed dealing with terrorism, including violent extremism and sectarianism, through involvement of all affected sectors of society, better methods of monitoring, and the separation of terrorists from their support systems.[30] But it took the nationally experienced shock from the Peshawar Army Public School killings, for the government to propose a comprehensive blueprint for curbing terrorism and eradicating violent extremism. A 20-point National Action Plan (NAP) was quickly enacted that includes the call for establishing special military courts and lifting the moratorium on executions. NAP specifically mentions disallowing armed militias, stopping religious extremism, regulating madrassas, and cracking down on hate speech in print and on electronic media, including the Internet. The Plan has also promised to breathe life into a moribund National Counter Terrorism Authority (NACTA) that was created in 2009 to act as Pakistan's internal counterterrorism intelligence agency as well as provide a force to carry out special high-risk operations.

However, most of the objectives of NAP have been piecemeal and not pursued with any great determination. The government has been slow in budgeting NACTA and establishing a NAP-promised dedicated counterterrorism force. Aside from well-publicized efforts against banned armed groups, little has been done to close down these groups' financial fronts and communications networks. The prohibition on media coverage of terrorist outfits and crackdowns on media that promote sectarianism have proven to be difficult to enforce. Regulating madrassas, some of which have close ties to the TTP and jihadi groups, has been one of the most difficult challenges because of the resistance of Pakistan's powerful religious establishment. While supposedly a countrywide crackdown has taken place, reportedly as of February 2017 only two of Punjab's 13,000 declared madrassas, had actually been closed.[31] Without closer coordination between the federal government and provincial apex committees critical to implementation, many of the most important objectives of NAP cannot be realized. Only with the coordination and guidance that an activated NACTA is designed to provide can provincial level army-trained counterterrorism forces be expected to prevent attacks such as those in an amusement park in Lahore and at Bacha Khan University.

Looking ahead

Future developments in the tribal areas hinge strongly on the military's maintaining pressure on armed extremists. But the use of force alone, while a necessary component of counterinsurgency and counterterrorism strategies, is insufficient. Heavy-handed and indiscriminate operations create resentments against the military and state, and foster extremist recruitment. Better training of security forces and more discriminating weapons may help reduce collateral damage. Achieving success in FATA, however, requires also dealing with the deeper causes of insurgency. Much of it depends on gaining the confidence of local populations. A needed follow-up to Zarb-e-Azb includes introducing a job-creating economy, building infrastructure, and delivering political reforms to provide for a greater degree of transparency and accountability. Admittedly, all this may be a tall order for civilian authority. It is being asked to deliver in FATA a level of basic services and good governance that it has largely failed to provide for the rest of the country.

For all of the immediate threats emanating from Pakistan's borderlands, going forward, the greatest danger to its constitutional democratic system may lie elsewhere. While the frontier-based groups have taken credit for most of the terrorist attacks on the armed forces and civilians and have received much of the attention, the TTP are seen by most in Pakistan as an ethnic minority force. As such, the likelihood of its becoming the vanguard of a militantly Islamic insurgency countrywide is doubtful. A more serious challenge comes from those extremist groups embedded within the country's heartland, Punjab province. South Punjab is the organizational home of several of the country's leading jihadi organizations and most powerful Islamic movements, all of which aim to replace the current order with an idealized state governed by Sharia law. The province's mosques and madrassas, among Pakistan's largest and most influential, frequently serve up recruits for violent extremist groups, as do ostensibly nonpolitical organizations like Punjab-based Tablighi Jamaat, a proselytizing movement preaching moral rearmament.

In the campaign against the TTP and its allies, Afghanistan critically matters. This is especially true now that the Pakistan army's actions in FATA have pushed the Taliban across the border into Afghanistan. Without Pakistan receiving cooperation from Afghanistan, namely in the form of intelligence sharing and coordinated if not joint border operations, neither country can hope to staunch cross-border infiltration and raids. The necessary policy coordination is impossible without a lowering of suspicion and distrust in both countries that will be difficult to achieve without concrete confidence-building measures. The Pakistan government asks that the Afghan army dislodge the TTP from the sanctuaries it has established along the border. For its part, the Afghan government wants the eviction if not the arrest of Afghan Taliban on Pakistan's soil. But even with good intentions, the degree to which either country has the means to prevent infiltration along their porous border is open to debate.

More broadly, many among Pakistan military's senior leadership have come to recognize the converging interests of Pakistan with Afghanistan in combating insurgency and terrorism. Unlike the 1990s when there was no Taliban insurgency in Pakistan, strategists are concerned about the impact of an outright victory by the Afghan Taliban on Pakistan's fight against its own militants. While the immediate military goals of the two Taliban may differ, they have a history of cooperation and share similar prescriptive Islamic agendas. At the same time, the Pakistan military is reluctant to break its close links with the Afghan Taliban. An armed force of trusted Pashtuns remains as the centerpiece of a hedging strategy should an embattled Afghanistan break apart and Pakistan need the Taliban to help it secure a sphere of influence across its border. Paradoxically, the continued patronizing of Afghan Taliban insurgents undermines chances for the kind of stable, unified Afghanistan that can also be to Pakistan's advantage.

For the time being at least, the future of Islamic State in the region is problematic. As in Afghanistan, Pakistan's Islamic State has had to contend for territory and legitimacy with more established anti-state entities such as the Taliban and al-Qaeda. Since Islamic State's recruits have largely come from the ranks of defectors, its growth potential may well be set by the internal cohesion of the Taliban organizations in the two countries.

Whether future insurgencies can be prevented and the growth of radicalism reversed depends strongly on the emergence of civilian leadership willing to stand up and assume political and even physical risks. The task is to begin to push back against radical narratives, many of them based on conspiracy thinking encouraged by a market-driven media, that for so many Pakistanis have come to define the political world. Effective counter-narratives will require greater attention by the state to the use of strategic communications and the enlisting of credible religious figures and secular civil society organizations to reinforce messages aiming at de-radicalization. Additional legislation may be needed, but more important is the full enforcement of existing laws.

The military and elected government are not infrequently in disagreement over how insurgency and terrorism are to be handled. On matters of national security the well-recognized entrenched civil-military imbalance between elected officials and senior military is often clearly in evidence.[32] A major token of civilians yielding of political space to the military came with the government's acceding to the military's demand for judicial authority in terrorism-related cases. Meanwhile, the military also keeps tight reins on the media, though most of this is accomplished through self-censorship. The appointment in October 2015 of a general as Prime Minister Nawaz Sharif's security advisor and his retention by interim Prime Minister Shahid Khaqan Abbasi reflects the military's ability to occupy a seat within the highest government councils.

As justification for its encroachment in counterterrorism policy areas the military contends, not entirely without merit, that it has little choice but to assume responsibilities when civilians have failed to perform adequately or refuse to assume responsibility. Clear differences exist between the military and civil authorities over the pace and manner of implementation of the NAP. These disagreements became highly transparent in the standoff in early April 2016 between the senior military and Sharif government over whether army-led Rangers should be allowed to launch a counter-terrorism campaign in South Punjab to purge some of the country's deadliest groups. The government argued that civilian law enforcement forces alone should undertake the operations but finally agreed to what the military's public relations called a "coordinated operation."[33]

The troubled relationship between civilian authority and the armed forces was publically exposed in an English language newspaper columnist's report of an mid-October 2016 meeting in which Shabaz Sherif, Punjab Chief Minister, charged the military with obstructing police actions against groups like LeT and JeM. This strong accusation appeared to confirm claims by Pakistan's critics that the country's army was abetting terrorists. Anxious to smooth over the sensationalized rift, a clearly embarrassed government quickly denied that the confrontation had ever occurred.[34]

Government policies can contribute greatly in stabilizing Pakistan through instituting inclusive and more equitable governance, and by enhancing people's overall wellbeing. But when all else fails, it is the military to which most Pakistanis look to protect the country from the destructiveness of violent extremism. Despite all the military has done to distort the country's domestic priorities, undermine its democratic processes, and manipulate its foreign policies, in the end the military alone is able to keep the Pakistani state whole.

Conclusion

There is little doubt that in the collective consciousness in Pakistan, the threats of domestic extremism and dangers of terrorism have risen substantially in recent years. A public that has always recoiled from domestic violence seems more willing to hold some groups directly responsible. The rhetoric of the country's leaders, military and civilian, has changed as has their readiness to intervene more forcefully in tribal regions and to carry the fight against militants to Karachi and Punjab Province. Yet there is still much remaining to be done in formulating comprehensive counterterrorism policies to diminish the threat and address the causes of what could become an existential conflict over the future of Pakistan. Despite a seeming newfound willingness to clip the wings if not cripple several jihadi groups, the policy of distinguishing terrorist groups as either good or bad remains.

Insurgency in Pakistan's border regions is likely to be present for some time. The TTP and other anti-state forces cannot be entirely defeated so long as they are still able to find sanctuary across the border in Afghanistan. Even with a more cooperative Kabul government, its weakening military posture makes the dislodging of Pakistan's enemies on Afghan soil improbable any time soon. Earlier policies of containment of militants in the tribal areas through a combination of limited political deals and punitive military actions are unlikely to be repeated. Nor is a grand bargain with the Pakistani Taliban in the cards, or any other shortcuts to ensuring a pacified FATA.

That leaves as the best prospect for Pakistan the unglamorously slow process of implementing political, legal and social reforms along with economic improvements in its border areas. It amounts to a strategy of institution building that is intended both to reduce the incentives to recruitment to anti-state groups and the reintegrating of fighters back into their communities. It assumes that there will probably always be an irreconcilable core insurgent leadership but one that can eventually be isolated and so weakened as to be more irritant than threat. Importantly, much of this approach to achieving success against Pakistan's Taliban and other insurgents also applies to defanging domestic extremist groups, both those currently threatening the state and those that may do so in the future.

Notes

1 Rumi, "Charting Pakistan's Internal Security Policy."
2 Rana, "The Terrorist Bluff."

3 Giustozzi "The Islamic State in 'Khorasan' A Nuanced View."
4 Zulfiqar, "Alarm in Kurram over IS lairs in Paktia," April 3, 2016.
5 Hashmi, "Terror Outfit Ansar-ul-Sharia Behind Recent Attacks in Karachi: Sources."
6 South Asia Terrorism Portal. "Fatalities in Terrorism Violence in Pakistan, 2003–18.
7 The *Express Tribune*, "212 Organisations Formally Banned by Pakistan," June 28, 2015.
8 Ahmed Rashid, "The Army Steps In," *New York Review of Books*, March 31, 2016.
9 Jones and Fair, "Counterinsurgency in Pakistan," RAND Corporation, Seth G. Jones and Christine Fair, 2010, XIV.
10 See Jones and Fair, pp. 34ff. for list of counterinsurgency campaigns and p. 76 for a descriptive summary of operations. Also see Daud Khattak, "Evaluating Pakistan's Offensives in Swat and FATA," Combating Terrorism Center Sentinel at West Point, October 31, 2011.
11 Inter Services Public Relations (ISPR), Press Release No. PR123/2014, June 15, 2014.
12 Shah et al., "Militants Slip Away Before Pakistan Offensive."
13 Ilyas, "Will Pakistan go all out against militants?"
14 Saleem, "Why Pakistani Army's Anti-terror Campaign Falls Short."
15 *The Economist*, "Mopping up."
16 Ibid.
17 *The Nation*, "Charsadda attack was zarb-e-azb blowback: PM."
18 Rafiq, *The Diplomat*, "What Happened to ISIS's Afghanistan–Pakistan Province?" February 2, 2016.
19 Lalwani, "Actually, Pakistan is Winning Its War on Terror."
20 Obaid Ali, Afghanistan Analyst Network, Kabul, January 30, 2016.
21 Sherazi, "Key Commanders of the Gul Bahadur Group Killed in Datakhel Strikes: Reports."
22 Rodriguez, "Taliban Taps the Punjab Heartland," November 16, 2009.
23 Gannon, "In Pakistan, Tackling Extremism is a Minefield," April 6, 2016.
24 Haq, "Over 250 Madrassas Shut Down Countrywide."
25 Gabol, "Punjab Assembly passes Protection of Women Against Violence Bill."
26 Bhatti and Khan, "Pakistan Bans Donations to Jamaat-ud-Dawa, Organisations on UNSC Sanctions List."
27 Siddiqui, "What's Behind the Islamists Protests in Pakistan."
28 Parvez and Rani, "An Appraisal of Pakistan's Counterterrorism Act."
29 Rashid, "The Cost of Pakistan's War on Terrorism."
30 *The Nation*, Text of Pakistan's National Security Policy 2014–2017, February 27, 2014. See also Rumi, Special Report 368.
31 Syed "2,327 'Suspected' Madrassas Shut Down Countrywide, NAP Implementation Documents Reveal."
32 See Michael Kugelman, "Since the Peshawar school massacre Intolerance has in Pakistan," *Wall Street Journal*, December 16, 2015.
33 Yousaf, "Army Assisting in South Punjab Terror Purge."
34 Siddiqa, "Cyril Almeida's Story Points to Old Fault Lines and New Strains in Pakistan Army-govt Relationship."

References

Ali, Idrees. "Pentagon Not to Pay Pakistan $300 Million in Military Reimbursements," *Reuters*, August 4, 2016.
Ali, Obaid. "The 2016 Insurgency in the North." Afghanistan Analysts Network, January 30, 2016.
Ali, Zulfiqar. "Alarm in Kurram over IS Lairs in Paktia." *Dawn*, April 3, 2016.

Bhatti, Asif and Azam Khan. "Pakistan Bans Donations to Jamaat-ud-Dawa, Organisations on UNSC Sanctions List." *Geo News*, (January 1, 2018) available at www.geo.tv/latest/174798-pakistan-plans-to-seize-control-of-jamaatud-dawa-charities-assets-sources.

Bodirsky, Daniel. "Beyond Operation Zarb-e-Azb in Northwest Pakistan." *IISS Voices*, December 4, 2015.

Gabol, Imran. "Punjab Assembly Passes Protection of Women Against Violence Bill," *Dawn*, (February 24, 2016) available at www.dawn.com/news/1241626.

Gannon, Kathy. "In Pakistan, Tackling Extremism is a Minefield," *Washington Post*, April 6, 2016.

Gishkori, Zahid. "212 Organisations Formally Banned by Pakistan," *The Express Tribune*, (June 28, 2015) available at https://tribune.com.pk/story/911295/212-organisations-formally-banned-by-pakistan/.

Giustozzi, Antonio. "The Islamic State in 'Khorasan' A Nuanced View." *RUSI Commentary*, February 5, 2016.

Gul, Pazir. "Gul Bahadur Group Asks People to Leave N. Waziristan," *Dawn*, May 31, 2014.

Hashmi, Talha. "Terror Outfit Ansar-ul-Sharia Behind Recent Attacks in Karachi: Sources," *Geo News*, (July 22, 2017) available at www.geo.tv/latest/150528-terror-outfit-ansar-ul-sharia-behind-recent-attacks-in-karachi-sources.

Haq, Riazul. "Over 250 madrassas Shut Down Countrywide," *The Express Tribune*, (February 25, 2016) available at https://tribune.com.pk/story/1053991/national-action-plan-over-250-madrassas-shut-down-countrywide/.

Humayun, Fahd. "NAP in Southern Punjab," *The News International*, February 6, 2016.

Internal Displacement Monitoring Center (IMDC). "Pakistan IDP Figures Analysis." December 2015.

Institute for Conflict Management. "FATA Assessment 2015." South Asia Terrorism Portal.

Inter Service Public Relations (ISPR). "Press Release No. PR123/2014." June 15, 2014.

Jones, Seth G. and C. Christine Fair. "Counterinsurgency in Pakistan." Santa Monica, CA: RAND Corporation, 2010. www.rand.org/pubs/monographs/MG982.html.

Khan, M. Ilyas. "Will Pakistan Go All Out Against Militants?" *BBC News*, June 16, 2014.

Khattak, Daud. "Evaluating Pakistan's Offensives in Swat and FATA." *Combating Terrorism Center Sentinel at West Point*, October 31, 2011.

Kugelman, Michael. "Since the Peshawar School Massacre Intolerance has Increased in Pakistan," *The Wall Street Journal*, December 16, 2015.

Lalwani, Sameer. "Actually, Pakistan Is Winning Its War on Terror." *Foreign Policy*, December 10, 2015.

Parvez, Tariq and Mehwish Rani. "An Appraisal of Pakistan's Counterterrorism Act." *United States Institute of Peace*, Special Report 377, August 2015.

Rafiq, Arif. "What Happened to ISIS's Afghanistan–Pakistan Province?" *The Diplomat*, February 2, 2016.

Rana, Muhammad Amir. "The Terrorists' Bluff," *Dawn*, July 3, 2017.

Rana, Shahbaz. "Seeking assistance: Zarb-e-Azb to cost $1.3b, Dar tells US legislator." *The Express Tribune*, February 19, 2015.

Rashid, Ahmed. "The Cost of Pakistan's War on Terrorism," *Financial Times*, November 15, 2015.

Rashid, Ahmed. "The Army Steps In." *New York Review of Books*, March 31, 2016.

"Return of North Waziristan IDPs Continues at Snail's Pace," *Dawn*, May 4, 2015.

Rodriguez, Alex. "Taliban Taps the Punjab Heartland," *Los Angeles Times*, November 16, 2009.

Rumi, Reza. "Charting Pakistan's Internal Security Policy." *United States Institute of Peace. Special Report* 368, May (2015.) www.usip.org/sites/default/files/SR368-Charting-Pakistans-Internal-Security-Policy.pdf.

Saleem, Aasim. "Why Pakistani Army's Anti-terror Campaign Falls Short." *Deutsche Welle*, January 28, 2016.

Shah, Saeed, Safdar Dawar and Adam Entous. "Militants Slip Away Before Pakistan Offensive," *Wall Street Journal*, July 17, 2014.

Sherazi, Zahir Shah. "Key Commanders of Gul Bahadur Group Killed in Datakhel Strikes," *Dawn*, December 7, 2014.

Siddiqa, Ayesha. "Cyril Almeida's Story Points to Old Fault Lines and New Strains in Pakistan Army-govt Relationship," *The Indian Express*, (October 14, 2016) available at http://indianexpress.com/article/opinion/columns/dawn-journalist-cyril-almeida-banned-pakistan-civilian-govt-military-meet-3081347/.

Siddiqui, Niloufer. "What's Behind the Islamist Protests in Pakistan?" *Washington Post*, December 8, 2017.

South Asia Terrorism Portal. "Fatalities in Terrorism Violence in Pakistan, 2003–18," available at www.satp.org/satporgtp/countries/pakistan/database/casualties.htm#.

Sulaiman, Sadia. "Hafiz Bul Bahadur: A Profile." *The James Foundation*, Terrorism Monitor, 7 (9), December 8, 2015.

Syed, Azaz. "2,327 'Suspected' Madrassas Shut Down Countrywide, NAP Implementation Documents Reveal," *Geo News*, (February 18, 2017) available at www.geo.tv/latest/131542.

Syed, Baqir Sajjad. "Operations in Fata Nearing Completion," *Dawn*, March 15, 2016.

The Express Tribune "Zarb-e Azb operation: IDPs protest food shortage," June 24, 2014.

The Economist, "Mopping up," March 3, 2018.

The Nation. "Text of National Security Policy 2014–2018," February 27, 2014.

The Nation. "Charsadda Attack was zarb-e-azb Blowback: PM," January 22, 2016.

"U.S. Congressional Research Service." February 24, 2016, https://fas.org/sgp/crs/row/pakaid.pdf.

Yousafzai, Sami. "A Taliban-Russia Team-Up Against ISIS?" *The Daily Beast*, October 26, 2015.

Yousaf, Kamran. "Army Assisting in South Punjab Terror Purge," *The Express Tribune*, April 7, 2016.

13 Counterinsurgency in Pakistan

The role of legitimacy

Anatol Lieven

Contrary to the predictions of many experts, the campaign by the Pakistani Army and state against the Pakistani Taliban and their allies over the past decade has proved successful in its main objectives. This chapter deals with this success, and not with the insurgency in Pakistani Balochistan, which while at a much lower level than that of the Pakistani Taliban, is still ongoing and seems likely to last for a very long time.

The Pakistani achievement can therefore be placed among other recent examples of successful counterinsurgency. On the other hand, the fact that extremely serious terrorist attacks have continued in Pakistan despite the success of counterinsurgency is both a warning to maintain an intellectual distance between the concepts of insurgency and terrorism, and a grim indication that large-scale and morally unrestrained terrorism is now most probably a permanent threat in the world as a whole, irrespective of success against particular insurgencies and rebellions.

The civil wars in Pakistan also raise a number of interconnected questions for the study of insurgency and counterinsurgency in general. Of these the most important is the issue of legitimacy. A number of different factors contributed to the hesitations in developing a concentrated effort against the rebels before 2008 and again in the period 2011–2014.[1] The most important factor of these was the profound unpopularity of the campaign in the eyes of the mass of the Pakistani population, most elected politicians and political parties, and a large part of the armed forces themselves.

These feelings were rooted in the perception that this was a war that had been forced on Pakistan by the United States, and that the insurgents were not really rebelling against Pakistan but only trying to help the legitimate war of resistance in neighboring Afghanistan. Only when both the ambition and the atrocities of the militants became too great to ignore in 2008–2009 did public opinion begin to swing against them; and it was this shift that more than anything else underpinned the subsequent success of the counterinsurgency campaigns.

Background to the insurgency

Most Pakistanis were brought up to see a direct continuity of religiously inspired Muslim resistance to infidel invaders, from Syed Ahmed Barelvi in the 1830s through to the resistance of the Afghan Mujahedin against the Soviet occupation of Afghanistan in the 1980s. This legacy became of crucial importance after 9/11 and the US invasion of Afghanistan.[2] A majority of Pakistanis, and an overwhelming majority of Pakistani Pashtuns, saw a direct parallel between the Soviet occupation and the US presence after 2001.[3] By some accounts, up to 20 percent of the rank and file of the Pakistan Army are Pashtuns, if the exclusively Pashtun Frontier Corps is included. The Pakistan Army at the start of the insurgency was thus in the position of a post-colonial army recruited from among people who identified strongly with their own anti-colonial tradition.

Brigadier Shaukat Qadir quotes Nek Mohammed Wazir, leader of the militants in South Waziristan, addressing a Jirga (tribal council) including senior Pakistani officers in early 2004:

> Fifteen years ago, it was jihad, sanctioned by the Pakistani and Saudi governments and supported by the USA, to kill the pale-faced infidel who had occupied Afghanistan by force. Why is it that today, when we have another pale-faced infidel occupying Afghanistan, we are being asked to kiss his boot when he kicks us, instead of killing him for it?[4]

Resistance to the U.S. invasion by both the Afghan Taliban and their Pakistani allies was widely seen as a "defensive jihad," and as such, support for it was incumbent on *every* Muslim, and not only on the Afghans.[5] This belief grew even stronger after the US invasion of Iraq in 2003, which seemed to Pakistanis to confirm a pattern of US anti-Muslim aggression. Combined with other factors, this led a level of anti-American feeling in Pakistan which was among the highest in the entire world.[6]

As a leading member of the Awami National Party told the author in Peshawar in August 2008, "a key problem for us in fighting the militants is that every Pashtun has been brought up from his cradle to believe that to resist foreign occupation is part of what it is to do Pashto" (in other words, to follow the Pashtun code of behavior).[7] The greatest advances of the Pakistani Taliban in 2007–2009 also coincided with the collapse of the Musharraf regime during those years and the highly complicated and fraught transition back to a Pakistani form of democracy. Inevitably, the counterinsurgency effort suffered very badly as a result.[8]

Background to the counterinsurgency

At the start of the insurgency, a plurality of Pakistanis (41 percent) polled by the Pew Research Center said that suicide terrorism was often or sometimes justified, compared to 35 percent who said that it was never justified. By the end of 2017, the proportion of those saying that suicide terrorism could often or sometimes be justified had dropped to 9 percent, with 68 percent opposed.[9] This shift is however less impressive than it seems, since my own (admittedly small-scale and informal) opinion surveys in the years after 2007 revealed that while a majority of respondents did indeed denounce terrorism against fellow Pakistanis, a great many of these added that it was not the "real" Pakistani Taliban who were responsible for these attacks, but the "Indian-backed" or "American-backed" Taliban, and that military offensives against the Pakistani Taliban were therefore unjustified. Only in 2014–2015, with intense military pressure on politicians and the media to abandon any sympathy for the Pakistani Taliban, and especially after the massacre at the Army Public School in Peshawar in December 2014, did an overwhelming majority of Pakistanis begin to denounce the Pakistani Taliban as a whole.

The original Pakistani military intervention in South Waziristan in March 2004 was also damaged by the fact that it was the product of a botched-up compromise between incompatible strategies. On the one hand, the Musharraf government was under intense pressure from Washington to take military action.[10] On the other, the Pakistani Army and especially the Inter-Services Intelligence (ISI) had close links to the Afghan Taliban, and in the case of the ISI, to some of the Pakistani militant groups in FATA as well, who had previously been ISI proteges in the Afghan insurgency of the 1980s and the Kashmir insurgency of the 1990s. Senior military and police officers have told me that throughout the period of military operations against the militants in Swat and FATA they faced a problem of ISI interference to protest particular allies.[11]

Musharraf, the high command, and ISI were well aware of the unpopularity of a move against the Afghan Taliban. Thus one peace deal in 2004 explicitly stipulated that while the houses of tribesmen giving shelter to foreign militants were to be destroyed, the Afghan Taliban were not to be considered "foreigners."[12] The initial operations in 2003–2004 therefore involved very limited numbers of troops, directed only against militants from the former USSR plus any Al Qaeda groups in the area.[13]

Even a very limited operation seen as ordered by the USA was however bitterly unpopular. Moreover, the use of Pakistani regular forces in FATA was unprecedented. After independence, the Pakistani state ordered the withdrawal of all Pakistani regular forces from FATA as a way of appealing to tribal loyalty. Henceforth, internal and border security would be maintained by the Frontier Corps, drawn from the Pashtuns themselves.[14] The high command also seem to have been unaware of the extent to which the Pakistani system of governance in FATA had become enfeebled over the years. This was (and on paper largely remains) a system of "indirect rule" created by the British Empire to manage the Frontier tribes. Administrative authority was held not by regular Pakistani officials but by Political Agents, working in turn through *maliks* or local tribal elders supposedly exercising authority over their clans.[15]

Over the years however the authority of the maliks had been greatly weakened by a vast increase in their numbers due to political patronage, and by a growing sense in the populations that they had become corrupt, self-serving appointees of the Pakistani state, rather than representatives of tribal interests and feelings – much the same mixture that helped destroy the position of such intermediary figures in the former colonial world (the *cadis* of French Algeria for example). The weakness of the maliks contributed to the ease with which the militants in FATA were able to kill them or drive them out, without their fellow tribesmen rallying to their defense.

As a result of these combined factors, while the original operation in South Waziristan was a success in its own terms, leading to the capture of several hundred Uzbek, Tajik and Chechen militants, it also sparked an uprising by local Pashtun militants. An additional reason was the accidental killing of women and children by a military lacking in modern equipment and command and control. This remains a problem up to the present, but from 2009 the Army has reduced civilian casualties by the simple expedient of urging or compelling most of the population to leave areas where offensives are about to be launched.

Chief errors

The Musharraf administration's decision to conduct a small operation in South Waziristan against a highly limited group of targets contributed to the first and classic mistake of Pakistan's counterinsurgency campaign (like those of the USA in Afghanistan and Iraq): the failure to deploy enough troops. This was also due to the Pakistani high command's continued prioritizing of India as the greatest security threat – a stance which was reinforced by the furious Indian response to the attacks by Pakistani-based terrorists on Mumbai in October 2008.[16]

Only in 2009 did the military come to acknowledge the Pakistani Taliban as the most urgent threat. Thereafter, they deployed overwhelming numbers of troops for every operation. In 2014 there were 160,000 regulars and almost 60,000 men of the Frontier Corps deployed in Federally Administered Tribal Area (FATA) (which has a population of around four million people, one-third of whom were at that stage displaced), and by the time of my visit to South Waziristan in April of that year, the force there numbered 28,000 men, comprised of a reinforced army division and three wings of the Frontier Corps.[17] This was in a population which, before the fighting, had numbered around 530,000, but by 2014 was estimated to be around 300,000, with the rest of the internally displaced persons (IDPs).

A second classic early failure was the omission to strengthen the paramilitary and police forces. Here however the failure was due mostly to the very nature of the Frontier Corps, the Frontier Constabulary (the police force of FATA) and the Pakistani police in general. The Frontier Corps and Frontier Constabulary suffered from great demoralization after 2003 as a result of divided Pashtun loyalties and the counterinsurgency's lack of popular legitimacy. For several years, from 2004 to around 2009, they became largely ineffective as fighting forces. Demoralization in the case of the Frontier Corps was increased by an (accurate) feeling of being the poor relations of the Pakistani Army, with poorer equipment, pay and conditions.[18]

The Frontier Corps was restored from 2008 on, due to the general transformation of the state and military's approach to the conflict, the belated provision of improved equipment (including most importantly body armor, the lack of which had cost so many lives) and in the crucial years of 2008–2010 to the able, deter-mined and idiosyncratic leadership of its new commander, Major General Tariq Khan, whose ability to inspire his men was increased by his own family back-ground as a tribal aristocrat.

In the case of the Pakistani police another factor was also present. Rather remarkably for a country which has experienced three lengthy periods of military rule, the police in Pakistan have always come under the authority of their respec-tive provincial governments, not the center. In 2002, as part of his efforts to legitimize his military regime, President Musharraf held federal and provincial parliamentary elections. The result in the North West Frontier Province (NWFP) was the formation of the provincial government by a coalition of Islamist parties, the Muttahida Majlis-e-Amal (United Action Council or MMA), which had cam-paigned on a platform of bitter opposition to Pakistan's alliance with the USA. Not surprisingly under such a government, the police of the NWFP were less than wholehearted in their opposition to Islamist militancy.[19]

The election of the MMA also reflected a much wider feature of Pakistani public opinion. It is a striking fact that with the exception of the Muttahida Qaumi Movement (MQM), representing the Mohajir population in Sindh, every major party which stood in the elections of 2008 did so on a platform of making peace with the Pakistani Taliban. In the elections of 2013, this was also true of the Muslim League (PMLN) under Nawaz Sharif, which won at a national level and in Punjab, and of the Pakistan Tehrik-e-Insaf, or Justice Party (PTI) of Imran Khan, which won in Khyber Pakhtunkhwa.[20]

The belief that the militant groups in Swat were only trying to help the Afghan Taliban may very well have been correct in 2002–2006.[21] Thereafter however insurgency spread from FATA into certain districts of the NWFP, and Islamist terrorism began to affect the whole of Pakistan. The trigger for an immense growth of the rebel groups (which in September 2007 came together to form a loose alliance called the Tehrik-e-Taliban Pakistan or TTP) was the deci-sion by the Musharraf administration to storm the Red Mosque (Lal Masjid) complex in Islamabad, an Islamist radical center whose students in previous months had engaged in a campaign to impose Sharia law in the capital.[22] The bloody military operation caused a revulsion of feeling against Musharraf, and a surge in radicalism and ruthlessness among the rebels.[23] Their leaders now declared that their objective was no longer simply to protect the Afghan Taliban and the Afghan jihad, but to carry out an Islamic revolution.[24]

After the elections of 2008, the new national government of the Pakistan Peo-ple's Party (PPP), and the NWFP provincial government of the Awami National Party (ANP) negotiated the Nizam-e-Adl (System of Justice) agreement with local militants in the NWFP district of Swat, granting what was supposed to be their most important condition for peace, the introduction of Sharia law in Swat. This agreement had the support of a large majority of the population in Swat and in Pakistan as a whole.

Clearly believing that they had the Pakistani state and army on the run, the militants then made a crucial error by attacking the neighboring district of Buner

– a mere 60 kilometers from Islamabad – and driving out the police and local authorities. This move made clear to the government and high command, and significant sections of public opinion, that the militants' objective was indeed revolution. This incursion came on top of widely publicized incidents in Swat which did much to transform at least educated public opinion, including a declaration by a supposedly more moderate Swati militant, Sufi Mohamed, that democracy is contrary to Islam.[25] As a result, for the first time the counterinsurgency began to enjoy a measure of national and nationalist legitimacy as a struggle to defend Pakistan.

The offensives against the Pakistani Taliban

The stage was therefore set for a series of military offensives which between 2009 and 2015 reconquered almost all the insurgent-controlled territory.[26] There was however something of a pause in the military offensives between 2011 and 2014. Two factors seem to have been responsible for this. The first was the unwillingness of the military to occupy North Waziristan, since this was the base of some of the Pakistani military's closest Afghan allies, the Haqqani network. According to journalists and IDPs from FATA with whom I spoke in July 2016, when the military finally moved into North Waziristan in 2014, the Haqqanis were allowed – or helped – to move by a complicated route to the Tirah valley, from which they have continued their campaign against the US-backed Afghan state. This could not have occurred without at the very least the acquiescence of the Pakistani military.[27]

The second reason was the acute deterioration in relations between Pakistan and the USA which took place as a result of successive incidents in 2011 including the location and killing of Osama bin Laden on Pakistani soil by the USA. The resulting Pakistani public anger distracted attention from the fight against the Pakistani Taliban and also – in the confused Pakistani public mindset – gave them a certain new legitimacy as enemies of America. As a result, the PMLN and PTI stood for election in 2013 on a platform of negotiating peace with the militants. In the case of the PTI, to judge by my own researches, a great many Pashtuns who joined this new party did so at least in part because of this stance. Every new PTI voter with whom I spoke before the elections gave it as one of the key reasons for abandoning the PPP and ANP – who in 2009 gave their support to the anti-Taliban campaign – and joining Imran Khan.

Shortly before his retirement in 2014, then Chief of Army Staff (COAS) General Ashfaq Kayani attacked these sentiments in a speech on Martyr's Day. In an implicit offer of a bargain with the next governments (due to be elected the next month), he declared the military's commitment to democracy, but demanded an end to political and media equivocation concerning the need for tough counterinsurgency:

> Our external enemies are busy in igniting the flames of this fire. However, despite all this bloodshed, certain quarters still want to remain embroiled in

the debate concerning the causes of this war and who imposed it on us. While this may be important in itself but the fact of the matter is that today it is Pakistan and its valiant people who are a target of this war and are suffering tremendously. I would like to ask all those who raise such questions that if a small faction wants to enforce its distorted ideology over the entire Nation by taking up arms and for this purpose defies the Constitution of Pakistan and the democratic process and considers all forms of bloodshed justified, then, does the fight against this enemy of the state constitute someone else's war? Even in the history of the best evolved democratic states, treason or seditious uprisings against the state have never been tolerated and in such struggles their armed forces have had unflinching support of the masses; questions about the ownership of such wars have never been raised. We cannot afford to confuse our soldiers and weaken their resolve with such misgivings.[28]

The negotiations initiated by the new governments were therefore opposed by the military. They were brought to an end by pressure from General Kayani's successor as COAS, General Raheel Sharif, after February 2014, when the Pakistani Taliban murdered 23 prisoners from the Frontier Corps whom they had been holding hostage. As in 2009, the military also exerted pressure on the media to take an unequivocal stance against the militants. As Lt General Khaled Rabbani, commander of the XI Corps and responsible for the operations in FATA, told me in April 2014, while talks with the Pakistani Taliban were still continuing:

> As far as I am concerned there is no ceasefire. My orders from the Chief of Army Staff are to continue the battle without having to ask additional permission. The only negotiations can be for surrender.... There is no question of giving the militants back control of any territory. This would mean abandoning our civilian allies to be killed by the terrorists. That happened before, but will not be allowed to happen again.... An order to make a deal with the terrorists would be illegal, and as commander of XI Corps I only accept legal orders.[29]

As in 2009, however, but to a much greater extent, the change in public attitudes was produced by a subsequent atrocity by the militants: the massacre of 141 children and teachers at an Army Public School in Peshawar on December 16, 2014. After this, even the Pakistani Islamist parties ceased to advocate peace. I visited Peshawar in April 2014 and June 2015, and was struck by the radical change in public opinion between the two dates. The school massacre led to the adoption by the government and military of a joint National Action Plan against terrorism, pledging action against all terrorists in every province, and co-ordinated action with other countries to cut off terrorist financing.

An additional factor in strengthening the counterinsurgency narrative was that since 2009, more and more of the Pakistani Taliban had taken refuge in

Afghanistan – including the Swati leader Fazlullah, who on November 7, 2013 was chosen as the (largely titular) leader of the Pakistani Taliban in succession to Hakimullah Masud, killed by a US drone strike. The Pakistani government and military alleged – not without some evidence – that the Pakistani Taliban was enjoying shelter from the Afghan intelligence service (NDS) in retaliation for Pakistani shelter for the Afghan Taliban. The NDS in turn is known to enjoy close relations with the Indian intelligence service (the Research and Analysis Wing, or RAW). This allowed the Pakistani military to fit the struggle against the Pakistani Taliban conveniently into the old, well-worn, and (in general) publicly accepted Pakistani narrative of the need to resist threats and pressure from India.

The centers set up in Swat and FATA for the rehabilitation of low-level Taliban prisoners (which I visited in 2010 and 2014) and the FM 96 military-run radio station preached a message of ardent Pakistani nationalism, and argued that by dividing Pakistan and attacking the Army, the militants were (unwittingly, in the case of the rank and file) playing India's game. To judge by my interviews, this message had a considerable effect.[30]

Pakistani military victory over the insurgents has been accompanied by a considerable amount of construction and development work, especially on communications (according to the old local principle that "government follows the roads") and schools. This work, conducted above all by the military's Frontier Works Organisation (FWO) has won high praise from United States Agency for International Development (USAID), which provided much of the financial support.[31]

Lessons of the Pakistani campaigns

The first lesson, from 2003–2008, is that to wage counterinsurgency successfully a state must be willing and able to deploy sufficient numbers and sufficient firepower. It may well be necessary to use these with restraint, but without them the insurgents will simply pick off outnumbered and outgunned units. Since 2009, the military has been careful not only to deploy absolutely overwhelming forces for every operation, but to keep them in the territory concerned for years if necessary until the local situation had fully stabilized.

The Army has made fairly extensive use of extra-judicial executions. Human Rights Watch alleges that in Swat for example there were 238 suspected cases of these between the spring of 2009 and the middle of 2010.[32] This seems plausible, since in June 2009 local journalists and lawyers gave me figures in the range of 110–130 deaths. I was also told by local elders in Matta who were negotiating the surrender of some Taliban militants that one of their conditions was a promise that they would not simply be shot "as often happens to prisoners." Apart from simply conforming to South Asian tradition in such cases, the Army appears to have been motivated by the same reasons cited in private by the police: namely, that a combination of intimidation, sympathy and incompetence often makes it impossible to gain convictions before the courts even when guilt is obvious.

This unfortunately is actually the case – even in the specially formed Anti-Terrorism Courts. Thus in 2012, 269 out of 365 of those prosecuted were acquitted, with the most frequent reason being that witnesses went back on their initial testimony.[33] In 2010 those accused of responsibility for an attack on the ISI headquarters in Lahore were acquitted, and this was upheld four years later by the Lahore High Court.[34] Human Rights Watch subtitled its reports on the Swat killings "Military Abuses Undermine Fight Against Taliban," but this does not in fact seem to have been the case (unlike in Balochistan). On the whole, South Asian experience tends to suggest that people there are rather tolerant of such actions as long as they believe in the overall legitimacy of the state's strategy and as long as they believe that reasonable care is being taken to make sure that those killed are actually guilty. My interlocuteurs in Swat (admittedly mostly from the local educated middle classes) seemed to think these killings a reasonable price to pay to be delivered from the rule of the Taliban.

The Pakistani experience also indicates that in certain specific circumstances, removing the population from areas affected by insurgency and counterinsurgency can work. By 2012, one in five of the population of FATA had been forced at some stage to leave their homes.[35] By July 2015, an estimated 1.8 million people had been displaced from Fata (most of those displaced in Swat in 2009 had returned home within a year or so).[36] This included a large majority of the populations of North and South Waziristan, many of whom had not returned home by mid-2016. To judge by my interviews with IDPs in July 2016 however, the failure of those from South Waziristan to return was less to do with continued fighting, and more to the fact that in their absence their mud houses had collapsed from lack of maintenance. It also seemed that in many cases, the people concerned found that they had got better jobs and a better lifestyle from migrating out of FATA – as generations of tribesmen had done before them.

The tactic of clearing the population from areas of insurgency got a bad name due to the way in which the US "Strategic Hamlet" program in Vietnam and the French *regroupement* program in Algeria failed to "drain the swamp" of insurgent support while increasing the misery and resentment of the rural population.[37] On the other hand, the Vietnamese Communist Party did admit that Relocation for a time "reduced significantly regions of importance for us, and … caused many difficulties."[38] In the case of Swat and FATA, the Pakistani military's efforts seem to have both freed the military's firepower, and saved the lives of numerous civilians.

Above all, the Pakistan Army's campaigns were not accompanied by the looting, random brutalization and most especially extensive rape carried out by US and South Vietnamese forces, by the Afghan police, and by the Pakistani Army in East Pakistan in 1971. Rape has never been credibly alleged in the recent Pakistani campaigns – a tribute to the discipline of the Army when given clear orders. In this case, the orders were given, because, as a Pakistani Lt Colonel told me in Swat in 2009, the entire Army was aware of the catastrophic impact that this crime above all others would have on Pashtun attitudes to the military and to Pakistan, because of the intense Pashtun concern with female "honour." The lesson here would seem to be that like many crimes, the effects of war crimes are to some extent culturally specific. A counterinsurgency force may be able to get away with a considerable number of reasonably targeted extra-judicial executions, where even a very small number of rapes would ruin its whole position.

Another key difference from the past French and American programs is that the tribal populations were encouraged and warned to move rather than being deported by force; that they were better looked after in the places to which they moved; and that no attempt was made to keep them in fortified camps. Instead, they joined longstanding patterns of Pashtun migration, with many ending up joining relatives already living and working in Karachi and elsewhere.[39]

Another lesson is the need for unity of leadership. In the period 2003–2007, the most important division was between on the one hand the Pakistani military government, who were trying to carry out a limited campaign against particular targets, combined with negotiations, and on the other the US government and military, which were using drones against *all* militant targets in FATA. During this period – before the militants developed their goal of the overthrow of the Pakistani state – the Pakistani strategy might possibly have worked had it not been for the determination of the USA to pursue its own campaign against the Afghan Taliban and their allies, including through missile strikes on Pakistani soil.

Since the fall of the Musharraf regime in 2008, the most important division has been between the Army command and successive elected governments. Apart from other reasons for hesitation in attacking the Taliban, the high command was always concerned that without the full support of the national and provincial government, public support for the operation would not be forth-coming and in addition, the generals would have to bear the entire blame if any-thing went wrong.

Nonetheless, from 2009–2011 and – more strongly – from 2014 unity of leadership was achieved in the battle against the Pakistani Taliban, essentially through the governments bowing to the wishes of the Army. The history of Paki-stan's insurgency and counterinsurgency therefore emphasizes the need for clarity and unity of purpose, whether in pursuit of outright victory or a negoti-ated compromise. The continued lack of clarity and unity of purpose when it comes to Pakistani sectarian terrorist groups like Lashkar-e-Janghvi – and still more anti-Indian groups like Laskar-e-Taiba – largely explains their ability to go on operating within Pakistan.

In the end, what the Pakistani experience emphasizes above all is the import-ance of legitimacy: legitimacy of the state concerned and of its counterinsur-gency war. Here outside forces are generally at a severe disadvantage. Legitimacy is frequently connected to longevity – something which is also generally denied to outside forces. A disastrous weakness of any counterinsur-gency campaign is a belief that the presence of state forces is temporary, and that that they are liable to disappear and leave their local allies to be slaughtered, as happened to France's Muslim auxiliaries in Algeria and America's allies in South Vietnam.

The most damaging feature of the truces raised by the Pakistani state and army with the militants in 2004–2009 was the way in which they allowed the militants to murder and expel local people who had sided with the Army. During my visits to South Waziristan in April 2014 and my interviews with elders and maliks from North and South Waziristan in Dera Ismail Khan in July 2016, several of my interlocuteurs emphasized the importance of a new conviction that "this time, the Army is here to stay" – and not just for the immediate future but "for as long as Pakistan exists," as a brigadier promised me.

The final lesson of Pakistan's experience is therefore that a reasonably numerous, well-armed and well-led military defending a legitimate state and enjoying sufficient support from enough of the population is likely to be able to defeat an insurgency, or contain it within reasonable limits; but to ensure against the possibility of recurrence, deep reforms are necessary. These are lessons that states often have to learn over and over again. The Pakistani military and political elites seem to have accepted these lessons in principle; but whether Pakistan's fractured, corrupt and incompetent state system is capable of delivering the necessary results is another matter. The next great test of this will be whether Pakistan is capable both of implementing the $46 billion infrastructure package offered by China, and of ensuring that enough of its benefits flow to the populations of FATA and Balochistan.[40] If social, economic and political development can be added to military victory, then the Pakistani experience will count as one of the great counterinsurgency successes of modern history.

Notes

1 See for example Jones and Fair, *Counterinsurgency in Pakistan*; Schmidle, *To Live or to Perish Forever: Two Tumultuous Years in Pakistan.*

2 See Rana et al., *Dynamics of the Taliban Insurgency in FATA*; Fishman, "The Battle for Pakistan: Militancy and Conflict across the FATA and the NWFP."

3 See Franco, "The Tehrik-e-Taliban Pakistan," 269–292; White, "Pakistan's Islamist Frontier: Islamic Politics and US Policy in Pakistan's North West Frontier."

4 Shaukat Qadir, "The State's Response to the Pakistani Taliban Onslaught," 134. See also Qazi, "Rebels of the Frontier: Origins, Organisation and Recruitment of the Pakistani Taliban," 574–602.

5 For the religious roots of Pakistani Taliban motivation, see Sheikh, *Guardians of God: Inside the Religious Mind of the Pakistani Taliban.*

6 Thus according to a Pew Research Center opinion survey, in 2007 only 15 percent of Pakistanis polled viewed the USA favorably compared to 68 percent who had an unfavorable view (www.pewglobal.org/2007/12/28/the-view-from-pakistan/). In 2013, according to the Pew, following a series of clashes between the USA and Pakistan in 2011, only 11 percent of Pakistanis had a favorable view of the USA (compared for example to 40 percent of Chinese who viewed the USA favorably), while 59 percent of those Pakistanis polled regarded the USA as an outright enemy of Pakistan (www.pewglobal.org/2013/05/10/what-pakistan-thinks/).

7 For the reasons for public sympathy for the Pakistani Taliban, see Lieven, *Pakistan: A Hard Country*, 371–441.

8 Ahmed, *Pakistan, The Garrison State: Origins, Evolution, Consequences 1947–2011* (Oxford University Press Karachi 2013), p. 329; Cohen, *The Idea of Pakistan*; Lieven, *A Hard Country*, 161–203.

9 Wilke, Pew Global Attitudes Project, "View from Pakistan," December 28, 2007, at www.pewglobal.org/2007/12/28/the-view-from-pakistan/.

10 Qadir, op. cit., 132.

11 For ISI strategy, see Kiessling, *Faith, Unity and Discipline: The ISI of Pakistan.*

12 Rana et al., op. cit., 93.

13 Ahmed, *Garrison State*, 332–334.

14 Nawaz, *Crossed Swords: Pakistan, Its Army, and the Wars Within*, 544ff.

15 For a description of the constitutional status and administrative system of FATA, see the website of the FATA Secretariat at https://fata.gov.pk/; Zaidi, "Understanding FATA"; Ahmed, *Pukhtun Economy and Society: Traditional Structure and Economic Development in a Tribal Society*, 305–355. For the decay of this system before and during the insurgency, see Nawaz, *Crossed Swords*, 568.

16 Ahmed, *Garrison State*, 359–365.

17 Information to the author from the Army HQ in Wana, South Waziristan, April 2014.

18 Nawaz, *Crossed Swords*, 545. During my visit to South Waziristan in April 2014, military officers told me of two occasions on which FC units had failed to come to the help of Army units in South Waziristan the years 2004–2007.

19 Abbas, "Reforming Pakistan's Police and Law Enforcement Infrastructure," US Institute of Peace, Washington, DC, 2011, at www.usip.org. From 2008 however the police in the NWFP recovered their morale, also through a mixture of better pay and equipment and inspiring leadership, especially on the part of Assistant Inspector General Safwat Ghayyur, who was later killed by the Pakistani Taliban.

20 See Khan, *Pakistan: A Personal History*, 227–228, 235–260; Qadir, op. cit., 154.

21 This at least is the view of Muhammad Amir Rana and his colleagues at the Pakistan Institute of Peace Studies, one of the best informed centers for the study of militancy in Pakistan. See their report of 2009 cited above.

22 Dolnik and Iqbal, *Negotiating the Siege of the Lal Masjid.*

23 Ahmed, *Garrison State*, 339–341.

24 See Raza, "Revenge Attacks in Swat for Lal Masjid: Cleric."

25 See Reggio, "Sufi Mohamed 'Hates Democracy' and Calls for Global Islamic Rule."

26 See Barcelona Centre for International Affairs, Safiya Aftab, "Pakistan: Overview of Sources of Tension."

27 For links between the Army, the ISI and the Haqqanis, see Brown and Rassler, *Fountainhead of Jihad: The Haqqani Nexus, 1973–2012*, 151–182.

28 General Ashfaq Pervez Kayani, speech on the eve of Qaum-e-Shuhadda, March 25, 2013, at www.thenewstribe.com/2013/04/30/pakistan-army-chief-general-ashfaq-pervez-kayani-speech-on-the-eve-of-yaum-e-shuhadda-2013/.

29 Interview with the author in Peshawar, April 18, 2014.

30 For the content of Pakistani nationalism, see Cohen, *The Idea of Pakistan.*

31 Interviews with US officials in Islamabad, June 2015.

32 See Human Rights Watch, "Pakistan: Extrajudicial Executions by the Army in Swat."

33 See Kharal, "ATC Performance: Most Accused in Terror Cases Acquitted."

34 See Asad, "LHC rejects appeals against acquittals of accused in attacks on Army installations."

35 Shinwari, "Understanding FATA: 2011."

36 According to the estimate in 205 of the Internal Displacement Monitoring Centre at www.internal-displacement.org/south-and-south-east-asia/pakistan/figures-analysis. See also International Crisis Group (2010), "Pakistan: The Worsening IDP Crisis."

37 Horne, *A Savage War of Peace: Algeria 1954–62*, 220–221; Kolko, *Anatomy of a War: Vietnam, The United States and the Modern Historical Experience*, 132–137, 236–241; Krepinevich, *The Army and Vietnam*, 67–68.

38 Kolko, op. cit., 133.

39 Qadir, op. cit., 152.

40 "Xi Jinping Agrees $46 billion superhighway to Pakistan," BBC News April 20, 2015.

Bibliography

Abbas, Hassan. *The Taliban Revival: Violence and Extremism on the Pakistan-Afghanistan Frontier*. New Haven: Yale University Press, 2014.

Aftab, Safiya. "Pakistan: Overview of Sources of Tension," (January 2016) available at www.cidob.org/en/publications/publication_series/stap_rp/policy_research_papers/pakistan_overview_of_sources_of_tension_with_regional_implications_2015.

Ahmed, Akbar S. *Pukhtun Economy and Society: Traditional Structure and Economic Development in a Tribal Society*. London: Routledge, 2011.

Ahmed, Akbar. *The Thistle and the Drone: How America's War on Terror Became a War on Tribal Islam*. Washington, DC: Brookings Institution Press, 2013.

Ahmed, Ishtiaq. *Pakistan, The Garrison State: Origins, Evolution, Consequences 1947–2011*. Karachi: Oxford University Press, 2013.

Asad, Malik. "LHC rejects appeals against acquittals of accused in attacks on Army installations," *Dawn*, (January 3, 2015) available at www.dawn.com/news/1154774/lhc-rejects-appeals-against-acquittal-of-accused-in-attacks-on-army-installations.

Barcelona Centre for International Affairs, January, (2016) available at www.cidob.org/en/publications/publication_series/stap_rp/policy_research_papers/pakistan_overview_of_sources_of_tension_with_regional_implications_2015.

Bangash, Yaqoob. *A Princely Affair: The Accession and Integration of the Princely States into Pakistan, 1947–1955*. Karachi: Oxford University Press, 2015.

Bashir, Shazad and Robert D. Crews. *Under the Drones: Modern Lives in the Afghanistan–Pakistan Borderlands*. Cambridge, MA: Harvard University Press, 2012.

Breseeg, Taj Mohammad. *Baloch Nationalism: Its Origin and Development*. Karachi: Royal Book Co., 2004.

Brown, Vahid and Don Rassler. *Fountainhead of Jihad: The Haqqani Nexus, 1973–2012*. London: Hurst and Co., 2013.

Caroe, Sir Olaf. *The Pathans*. Oxford: Oxford University Press, 1958.

Cohen, Stephen P. *The Pakistan Army*. New York: Oxford University Press, 1998.

Cohen, Stephen P. *The Idea of Pakistan*. Washington, DC: Brookings Institution Press, 2008.

Dolnik, Adam and Khuram Iqbal. *Negotiating the Siege of the Lal Masjid*. Karachi: Oxford University Press, 2016.

Fishman, Brian. "The Battle for Pakistan: Militancy and Conflict across the FATA and the NWFP," New America Foundation Counterterrorism Paper (2010) available at http://counterterrorism.newamerica.net.

Franco, James. "The Tehrik-e-Taliban Pakistan," in Antonio Giustozzi (ed.) *Decoding the New Taliban*. New York: Columbia University Press, 2009.

Giustozzi, Antonio (ed.). *Decoding the New Taliban*. New York: Columbia University Press, 2009.

Grare, Frederique. "Pakistan: The Resurgence of Baloch Nationalism," Carnegie Endowment for International Peace, 2006; International Crisis Group (2010)

Grare, Frederique. "Pakistan: The Worsening Crisis in Balochistan," available at www.crisisgroup.org: 132–137.

Haroon, Sana. *Frontier of Faith: Islam in the Indo-Afghan Borderland*. London: Hurst and co., 2007.

Heathcote, T. A. *Balochistan, the British and the Great Game*. London: Hurst and co., 2015.

Horne, Alistair. *A Savage War of Peace: Algeria 1954–62*. London: Macmillan, 1977.

Human Rights Watch. "Pakistan: Extrajudicial Executions by the Army in Swat," (July 16, 2010) available at www.hrw.org/news/2010/07/16/pakistan-extrajudicial-executions-army-swat.

International Crisis Group. "Pakistan: The Worsening IDP Crisis," (2010) available at www.crisisgroup.org.

Jaffrelot, Christophe. *The Pakistan Paradox: Instability and Resilience*. London: Hurst and co., 2015.

Jalal, Ayesha. *Partisans of Allah: Jihad in South Asia*. Cambridge, MA and London: Harvard University Press.

Jones, Seth G. and Christine C. Fair. *Counterinsurgency in Pakistan*. Rand Corporation 2010.

Khan, Imran. *Pakistan: A Personal History*. London: Bantam Press, 2011.

Kharal, Asad. "ATC Performance: Most Accused in Terror Cases Acquitted," in *The Express Tribune*, Lahore, (September 24, 2012) available at http://tribune.com.pk/story/441412/atc-performance-most-accused-in-terror-cases-acquitted/.

Kiessling, Hein G. *Faith, Unity and Discipline: The ISI of Pakistan*. London: Hurst and Co., 2016.

Kolko, Gabriel. *Anatomy of a War: Vietnam, The United States and the Modern Historical Experience*. New York: Pantheon, 1985.

Krepinevich, Andrew. *The Army and Vietnam*. Baltimore, MD: Johns Hopkins University Press, 1986.

Lieven, Anatol. *Chechnya: Tombstone of Russian Power?* New Haven, CT: Yale University Press, 1999.

Lieven, Anatol. *Pakistan: A Hard Country*. London: Penguin, 2011.

Musharraf, Pervez. *In the Line of Fire: A Memoir*. New York, NY: Free Press, 2008.

Nawaz, Shuja. *Crossed Swords: Pakistan, Its Army, and the Wars Within*. New York, NY: Oxford University Press, 2008.

Qadir, Shaukat. "The State's Response to the Pakistani Taliban Onslaught." In Moeed Yusuf (ed.), *Insurgency and Counterinsurgency in South Asia: Through a Peacebuilding Lens*. Washington, DC: US Institute of Peace, 2014.

Qazi, Shehzad. "Rebels of the Frontier: Origins, Organisation and Recruitment of the Pakistani Taliban," *Small Wars and Insurgencies* 22 (4) (2011), 574–602.

Rana, Muhammad Amir, Abdul Basit and Saftar Sial. *Dynamics of the Taliban Insurgency in FATA*. Islamabad: Pakistan Institute of Peace Studies, 2009.

Raza, Syed Irfan. "Revenge Attacks in Swat for Lal Masjid: Cleric," in *Dawn*, (May 4, 2009).

Reggio, Bill. "Sufi Mohamed "Hates Democracy" and Calls for Global Islamic Rule," *Long War Journal* (February 18, 2009).

Roe, Andrew S. *Waging War in Waziristan: The British Struggle in the Land of Bin Laden*. Kansas University Press, 2010.

Schmidle, Nicholas. *To Live or to Perish Forever: Two Tumultuous Years in Pakistan* New York, NY: Henry Holt and Co., 2009.

Sheikh, Mona Kanwal. *Guardians of God: Inside the Religious Mind of the Pakistani Taliban*. Oxford: Oxford University Press, 2016.

Shinwari, Naveed Ahmad. "Understanding FATA: 2011," Community Appraisal and Motivation Programme (2012), 79–82 available at www.camp.org.pk.

Siddique, Abubakr. *The Pashtun Question: The Unresolved Key to the Future of Pakistan and Afghanistan*. London: Hurst and co., 2014.

Thornton, Thomas Henry. *Colonel Sir Robert Sandeman: His Life and Work on the Indian Frontier*. Hesperides Press, 2008.

Tucker, A. L. P. *Sir Robert Sandeman, Peaceful Conqueror of Balochistan.* London: Society for the Promotion of Christian Knowledge, 1921.

Warren, Alan. *Waziristan: The Faqir of Ipi and the Indian Army, 1936–37.* Oxford: Oxford University Press, 2000.

White, Joshua T. "Pakistan's Islamist Frontier: Islamic Politics and US Policy in Pakistan's North West Frontier." Center for Faith and International Affairs, Washington, DC, 2008.

Wilke, Richard. Pew Global Attitudes Project, "View from Pakistan," (December 28, 2007) available at www.pewglobal.org/2007/12/28/the-view-from-pakistan/.

Wirsing, Robert. "Baloch Nationalism and the Geopolitics of Energy Resources: The Changing Context of Separatism in Pakistan." Strategic Studies Institute, April 2008.

Yusuf, Moeed (ed.). *Insurgency and Counterinsurgency in South Asia: Through a Peacebuilding Lens.* Washington, DC: US Institute of Peace, 2014.

Zaidi, Syed Manzar Abbas. "Understanding FATA," Pakistan Institute of Peace Studies (PIPS), October 2010.

14 Thailand's south

Roots of conflict

Thomas A. Marks

There are times when it seems the Thai southern insurgency will just fade away. Then, as in April (2017), an explosion of violence reminds the world that the challenge remains.

In the case at hand, coordinated gun and grenade attacks across the region left at least two dead and eight wounded, even as they sparked considerable discussion concerning motive. Some felt the violence was designed to prod plodding efforts at negotiations; others simply saw it as more of the same in a conflict which since 2004 has left more than 6,500 dead, most civilians.[1] Regardless, after more than a decade of counterinsurgency – often characterized as counterterrorism – the question looms: just what is Bangkok doing or, phrased in a different manner, what should it be doing?

Discussion within Thailand, once robust, has been decidedly muted since the military takeover in May 2014. This has left external views to dominate assessment. This is not altogether unusual, as will be clear in the discussion below, but it bears highlighting that Thai commentators, both official and civil, have often been far more represented, particularly at conferences, than they are now.

Academic views (those that will concern us here) of the conflict in the SBP (Southern Border Provinces) of Thailand are an accurate reflection of policy stances. These views focus overwhelmingly upon the legitimacy of the structures and mechanisms whereby the Malay-speaking Muslim minority (a minority even within Thailand's Muslim population) is incorporated into the larger Thai-speaking, Buddhist kingdom. Even as a strong body of foreign thought supports some form of autonomy as a "solution" for the SBP challenge, Thailand's strategic insistence upon national integrity has avoided serious challenge.

The generally correct conduct of security forces engaged in counterterrorism and counterinsurgency has resulted in critique remaining focused upon operational shortcomings. Acceptable perhaps in the short-run, this posture creates long-term strategic vulnerability.

First, in the absence of moving toward a solution, the state asks its population to fight *against* but not *for* something. This fails to generate legitimacy and forces the insurgents toward ever more radical options. Second, these radical options are certain to have second and third order consequences beyond immediate damage (especially to the tourist industry). These could include a demand for a turn to harsher security measures, which would inevitably alienate members of the SBP population and support heightened calls for external intervention, most prominently by the likes of the Organisation of Islamic Cooperation (OIC) but possibly also by Muslim states and populations (especially those within Southeast Asia). The way forward is a political solution of incorporation built upon legitimacy gained by popular empowerment.

Search for roots of conflict

The Vietnam War era saw a deep body of knowledge emerge concerning Thailand, but changes in context and focus have led to a decline in both the quantity and quality of scholarship available. This has been further impacted by the passing from the scene of those who knew Thailand at first-hand in an operational sense.

The result, therefore, is that much that is said about the insurgency in the SBP either lacks adequate operational grounding or proceeds as if what it has to offer is *sui generis*. Though I have followed events in SBP and have visited the area four times since the current conflict erupted in 2004, I have not been prepared to offer in-depth analysis of an area where my own fieldwork was conducted primarily in an earlier period.[2] This is principally because of the limitations placed upon fieldwork, without which the most basic questions pertaining to an insurgency cannot be addressed; to wit: who joins, who stays, who leaves?

Be that as it may, there does exist a fairly extensive body of literature on the SBP challenge by foreign scholars. This material is overwhelmingly in English, and, as indicated earlier, I know of no equivalent or even similar body of work in Thai or other foreign languages. The English-language material posits three explanations for the SBP situation. These "threads" are not mutually exclusive, but each has a dominant theme. Neither should the illustrative works chosen be considered the only works available, though certainly they are among the most widely available.

Most common is the view which states issues of geography, history, economics, status (social), and politics have marginalized the Malay-speaking Muslims of the SBP, hence what is at hand is the latest of a more-or-less continuous effort to separate the larger Thai-speaking Buddhist kingdom that is Thailand. The violence is a manifestation of historical separatism rather than terrorism per se or a classic insurgency (using terrorism as one of its weapons). This approach is best illustrated by the work of Duncan McCargo, who is Thai-speaking and well-informed on the non-security aspects of the situation in the South. Illustrative works are: McCargo, ed., *Rethinking Thailand's Southern Violence* (Singapore: NUS Press, 2007); McCargo, *Tearing the Land Apart: Islam and Legitimacy in Southern Thailand* (Ithaca, NY: Cornell University Press, 2008); and McCargo, *Mapping National Anxieties: Thailand's Southern Conflict* (Copenhagen, Denmark: NAIS Press, 2012). Necessarily, these works advance Malay-speaking Muslims as a community that is grappling with Thai oppression, which has taken various forms since the imposition of direct rule over "Patani." The solution is posited as some form of autonomy. See e.g., McCargo, "Autonomy for Southern Thailand: Thinking the Unthinkable?" *Pacific Affairs* 83.2 (June 2010), 261–281; and McCargo and Srisompob, "A Ministry for the South: New Governance Proposals for Thailand's Southern Region," *Contemporary Southeast Asia* 30.3 (2008), 403–428.

This is in contrast to the second body of material, which does not focus upon the contested nature of Thai unity but rather examines the challenge to that unity as posed by insurgency using terrorism (or, as often presented, of terrorism which includes aspects of insurgency). The focus is much more upon the threat posed to the state by radical challenge and the approach of the state in meeting that challenge. Mistakes loom prominent in this narrative, because the focus is upon a security battle between contending sides. This necessarily requires examination of Malay Muslim marginalization and alienation. Illustrative works are Rohan Gunaratna, Arabinda Acharya, and Sabina Chua, *Conflict and Terrorism in Southern Thailand* (Singapore: Marshall Cavendish, 2005); and Gunaratna and Acharya, *The Terrorist Threat From Thailand: Jihad or Quest for Justice?* (Washington, DC: Potomac Books, 2013). To this may be added the work of Zachary Abuza, *Conspiracy of Silence: The Insurgency in Southern Thailand* (Washington, DC: US Institute of Peace Press, 2009); and Abuza, *The Ongoing Insurgency in Southern Thailand: Trends in Violence, Counterinsurgency Operations, and the Impact of National Politics* (Washington, DC: INSS/NDU Press, September 2011).

Finally, a third body of material blends the two threads above and distinguishes itself by insisting upon examination of the individual/local level (*micro*) to support the conclusions at the organizational (*meso*) or societal (*macro*) levels. This may confuse, but the scholars concerned are emphatic that meta-narratives, whatever their validity, do not explain just who is doing the rebelling and just what their objectives are. Further, they take issue with the notion that the "Thai South" or "Thai Muslims" are in revolt, highlighting that no published work delimits well geographic areas of alienation or even particular segments of population who are disaffected. Rather, these scholars see a highly complex situation in which numerous factors have come together to result in the present violence. This necessarily makes the solution far more complicated than that advanced by the discussions of either the first or second thread. The approach is perhaps best illustrated by the work of Marc Askew, *Conspiracy, Politics, and a Disorderly Border: The Struggle to Comprehend Insurgency in Thailand's Deep South,*, Policy Studies 29 (Southeast Asia) (Washington, DC: East–West Center, 2007); which was followed by shorter articles as he conducted fieldwork, notably Askew, "Fighting With Ghosts: Querying Thailand's 'Southern Fire'," *Contemporary Southeast Asia* 32.2 (2010), 117–155; Askew, "Review Article – Insurgency Redux: Writings on Thailand's Ongoing Southern War," *Journal of Southeast Asian Studies* 42.1 (February 2011), 161–168; and Askew and Sascha Helbardt's, "Becoming Patani Warriors: Individuals and the Insurgent Collective in Southern Thailand." *Studies in Conflict and Terrorism* 35 (2012), 779–809. Subsequently, Helbardt published a more comprehensive treatment of the issues raised in the "Patani Warriors" article, *Deciphering Southern Thailand's Violence: Organization and Insurgent Practices of BRN-Coordinate* (Singapore: ISEAS, 2015). This appears to be the most complete published treatment of the insurgents available today.

Contending narratives

It should be readily apparent that contained in each of these approaches is the potential to offer support for a particular view of the "Southern Insurgency." The first and second bodies of work are those most often seen as contending, since the first approach questions the legitimacy of the state in its embrace of a restive minority, while the second approach focuses upon separatist outrage. The legitimacy of the state is accepted. If the first approach, then, sees the only alternative to the present situation as some form of autonomy, the second approach focuses upon more astute operational methodologies, whether these lie in better addressing roots of conflict or dealing more effectively with insurgent narrative and strategy. The first approach further is decidedly uncomfortable with the language and approach of counter-terrorism and/or counterinsurgency, while the second approach can be said to have emerged from this language and analytical framework.

Ironically, the third approach, which is the most compelling, has received far less attention than the first and second approaches, possibly because its published works are fewer. This is entirely logical given the difficulty of conducting fieldwork, but it serves to obscure its central tenet: absent the sort of detailed local work reflected in Askew and Helbardt, the claims of the first and second must necessarily be problematic. Indeed, there seems to be no published work that provides a coded SBP map (or even a verbal assessment) that identifies classic "go, no-go, contested" areas. If, for instance, as appears to be the case, Than To *Amphoe* (alternatively, Thanto *Amphur*) is more heavily insurgent-affected than Be Tong *Amphoe* (alternatively, Betong *Amphur*), why is this so? An explanation cannot simply identify disaffection but also loyalty. The one effort of which I am aware that endeavors to do this, by Thomas I. Parks (see bibliography), deals with Satun, which is outside SBP. This said, the third approach does not in any way overlook what it feels is the string of state errors that has made the situation much more difficult.

Rather than go into tactical detail, it suffices to note that all three approaches express a certain frustration with Thai counterinsurgency in the SPB. In this manner, scholars mirror attitudes held at the policy level by external observers, who are overwhelmingly Western or Western-influenced (e.g., Japan). The concern voiced may be found in the growing number of works that touch directly upon the manner in which cultural interaction is based upon mutual benefit and respect.

This is a key matter, because there is no point in even assessing the policy positions of authoritarian states/dictatorships (e.g., Russia and China), since their positions are based upon a frame which holds that if there is a problem in an area, those "causing the trouble" are the problem and must be eliminated (as e.g., in Tibet and Xinjiang or Chechnya). This is diametrically opposed to the Thai CSOC/ISOC approach which seeks the reasons for disaffection. Present operationalization of ISOC strategy in the dual SBPAC-CPM structure is to Thailand's credit. Indeed, there are no works of which I am aware – or policy pronouncements – that label the kingdom as a *systematic* human rights abuser in the SBP.[3]

What is of concern is the perception among some that there is a hardening of the attitude toward the alienated, focusing upon them as the problem as opposed to the challenges of legitimate state-incorporation of a Malay Muslim minority area (in which the minority is the majority). These concerns are reflected in two recent works which touch directly upon the role of Buddhism in the conflict: Michael K. Jerryson, *Buddhist Fury: Religion and Violence in Southern Thailand* and (appropriate chapters in) Vladmir Tikhonov and Torkel Brekke, eds., *Buddhism and Violence: Militarism and Buddhism in Modern Asia*. Such work has been impacted by the role of the Buddhist clergy in recent conflicts, such as that in Sri Lanka, where the nature of intercommunal strife seriously damaged the reputation of the *Sangha* and the faith in general, as is further happening today in Burma/Myanmar. The concern is reflected in charges of legal violations leveled by international human rights organizations (refer to n. 3 of this review).[4] Such charges must be treated seriously and should prompt investigation and action, but they have to date not reached the level seen in the critique of Sri Lanka and Burma.

All serious academic work on the Thai South and on Thai Muslims in general – the sort that informs policymakers when they seek background knowledge – is respectful of both the majority and minority traditions and emphasizes the efforts of local communities to live together. Of particular interest has been exploring the nature, both past and present, of relations between the two communities under discussion here, with the best work disaggregating both to look at local interactions. Strong examples are: Michel Gilquin, *The Muslims of Thailand*, trans. Michael Smithies (Chiangmai: Silkworm Books, 2005); Sugunnasil Wattana, ed., *Dynamic Diversity in Southern Thailand* (Chiangmai: Silkworm Books, 2005); Satha-anand Chaiwat, *The Life of This World: Negotiated Muslim Lives in Thai Society* (Singapore: Marshall Cavendish, 2005); Michael J. Montesano and Patrick Jory, eds., *Thai South and Malay North: Ethnic Interactions on a Plural Peninsula* (Singapore: NUS Press, 2008); Joseph Chinyong Liow, *Islam, Education and Reform in Southern Thailand: Tradition and Transformation*; and Patrick Jory, ed., *Ghosts of the Past in Southern Thailand* (Singapore: ISEAS, 2013).

As the publication dates indicate, there has been a continuous interest in the nature of Thai Muslim society in general and Southern Thai Muslim society in particular as both interface with the larger Thai-speaking, Buddhist population of the kingdom. To the works just listed may be added East–West Center Policy Studies (Southeast Asia) publications (beyond Askew) that deal specifically with the insurgency and are noteworthy for their rigor and empathy: Aphornsuvan

Thanet, *Rebellion in Southern Thailand: Contending Histories*, no. 35 (2007); John T. Sidel, *The Islamist Threat in Southeast Asia: A Reassessment*, no. 37 (2007); and John Funston, *Southern Thailand: The Dynamics of Conflict*, no. 50 (2008). Joseph Chinyong Liow has authored a related assessment in Policy Studies 24, *Muslim Resistance in Southern Thailand and Southern Philippines: Religion, Ideology, and Politics* (2006), as well as Liow and Don Pathan, *Confronting Ghosts: Thailand's Shapeless Southern Insurgency* (Australia: Lowy Institute for International Policy, 2010), which was one of the subjects for review by Askew, "Review Article" (above).

Lest any of this be considered "new," it is useful to highlight much-cited older works still widely available: W. K. Che Man, *Muslim Separatism: The Moros of Southern Philippines and the Malays of Southern Thailand* (Manila: Oxford and Ateneo de Manila University Press, 1990); and David Brown, *The State and Ethnic Politics in South-East Asia* (London: Routledge, 1994). The appropriate fifth chapter (pp. 158–205) in the latter, "Internal Colonialism and Ethnic Rebellion in Thailand," highlights the reality that the focus of the first approach discussed above has long been in existence. While Brown uses the Northeast for his Thai case study, his "internal colonialism" framework addresses not only *Isan* (alternatively, *Isaan*) but also the North and South.

Further, the precise terminology (i.e., "internal colonialism") is now widely used in academic analysis and in spirit informs much policy discussion. It advances the central question: does a diverse polity exist as a consequence of legitimacy or force? To the extent force is used by the center to incorporate peripheral social formations, a claim is advanced that there is a lack of legitimacy. This lack of legitimacy is all the more serious when an "outside" force is judged to be dominating a local population, especially seizing its land and destroying its culture. The consequent lack of legitimacy signals that the structure of the state itself may be challenged, and outside assistance to the peripheral formations can to some extent be considered under the relevant provisions of international law dealing with "liberation struggles."[5]

Conclusion

This leaves us at the point which I have sensed is of greatest concern to Thailand: the possibility of outside intervention, either tangibly (e.g., direct intervention) or intangibly (e.g., information warfare). Trends in international law, as reflected in the existence of the International Criminal Court (ICC), support the tenets of "Responsibility to Protect" (R2P); that is, the sanctity of life (as concerns a national or sub-national population) is paramount, and repression may rise to such level as to justify external intervention (in a variety of forms). This issue surfaced prominently during the conflict in Sri Lanka, especially when the 2009 destruction of the Liberation Tigers of Tamil Eelam (LTTE) insurgency inspired a number of prominent international organizations to demand external intervention to halt the final battle. In the event, rather than entering Sri Lankan national space, the concerned parties launched a campaign (of considerable effectiveness) to isolate Sri Lanka internationally and threaten its public servants with the prosecution inherent to universal jurisdiction.

Nothing even remotely similar to this has surfaced in the SBP case, but a possible strategic move of the insurgents would be to work tactically to create a situation destructive and disruptive enough to internationalize the conflict. This is but a step beyond using destructive methods for the purpose of disrupting the economy, for it simply requires focusing upon provoking a majority backlash such that the minority can claim it is threatened by genocide. Thailand's counterinsurgency doctrine, based as it is on achieving a proper mix of kinetic and nonkinetic approaches, is capable to preventing any such slide into barbarism, but the danger does need to be explicitly recognized.

Similarly, the current kinetic (CPM) and nonkinetic (SBPAC) synthesis has ossified to the extent that the ultimate goal has been forgotten: legitimacy achieved through democratic empowerment. It was this objective that allowed the original CSOC/ISOC operational approach to gain traction and defeat the CPT; its absence keeps alive the possibility of a blow to Thai legitimacy and thus national integrity.

Notes

1 Figures and wording are those of the Reuters report found in "Southern Thailand Attacks Reflect Tension Over Peace Talks, Conflict Monitor Says," *New York Times*, April 20, 2017; available at: www.nytimes.com/reuters/2017/04/20/world/asia/20 reuters-thailand-south-attacks.html?_r=0 (accessed April 21, 2017).
2 My 1977 MA thesis, earned enroute to the PhD at the University of Hawaii, dealt with the reasons southern Thailand historically did not become northern Malaya (now Malaysia); my PhD dissertation, dealing with the insurgency of the Communist Party of Thailand (CPT) necessarily privileged the more central struggles in the Northeast and North at the expense of the South. For published versions of these works see bibliography. I have since the 2004 resurgence of violence returned to the SBP four times to gather information and conduct interviews, assisted by locally based Thai scholars and journalists.
3 This statement can certainly be challenged. A perusal, for example, of the relevant postings on Thailand by Human Rights Watch (see www.hrw.org/asia/thailand) reveals ample challenges to the present actions of the state, to include in SBP, and thus could support a different conclusion. To my reading, though, such postings do not appear to claim that in SBP abuse is systematic or a matter of policy. There may be other cause-oriented positions, though, which require qualification of my assessment.
4 See e.g., Amnesty International, *Thailand: Torture in the Southern Counter-Insurgency* (London: January 2009).
5 The signal case today is the Israeli presence in the Occupied Territories. Most states that condemn Israel as a "settler colonial" society are themselves implicated by the same mechanism in the theory of "internal colonialism" – yet seem quite unaware that in voting against Israel they are laying the foundation for future challenges to their own legitimacy and national boundaries. Historically, for the U.S., the case of Native Americans well illustrates the dynamic. For the legal framework mentioned, see Christopher O. Quaye, *Liberation Struggles in International Law*.

Bibliography

Abuza, Zachary. *Conspiracy of Silence: The Insurgency in Southern Thailand*. Washington, DC: US Institute of Peace Press, 2009.

Abuza, Zachary. *The Ongoing Insurgency in Southern Thailand: Trends in Violence, Counterinsurgency Operations, and the Impact of National Politics*. Washington, DC: INSS/NDU Press, September 2011.

Amnesty International, *Thailand: Torture in the Southern Counter-Insurgency*. London: January 2009.

Askew, Marc. *Conspiracy, Politics, and a Disorderly Border: The Struggle to Comprehend Insurgency in Thailand's Deep South*, Policy Studies 29 (Southeast Asia) (Washington, DC: East–West Center, 2007.

Askew, Marc. "Fighting With Ghosts: Querying Thailand's 'Southern Fire'," *Contemporary Southeast Asia* 32, no. 2 (2010), 117–155.

Askew, Marc. "Review Article – Insurgency Redux: Writings on Thailand's Ongoing Southern War," *Journal of Southeast Asian Studies* 42, no. 1 (February 2011), 161–168.

Askew, Marc and Sascha Helbardt. "Becoming Patani Warriors: Individuals and the Insurgent Collective in Southern Thailand," *Studies in Conflict and Terrorism* 35 (2012), 779–809.

Brown, David. *The State and Ethnic Politics in South-East Asia*. London: Routledge, 1994.

Chaiwat, Satha-anand. *The Life of This World: Negotiated Muslim Lives in Thai Society*. Singapore: Marshall Cavendish, 2005.

Funston, John. *Southern Thailand: The Dynamics of Conflict*. Policy Studies no. 50. Washington, DC: East–West Center, 2008.

Gilquin, Michel. *The Muslims of Thailand*, trans. Michael Smithies. Chiangmai: Silkworm Books, 2005.

Gunaratna, Rohan and Arabinda Acharya. *The Terrorist Threat From Thailand: Jihad or Quest for Justice?* Washington, DC: Potomac Books, 2013.

Gunaratna, Rohan, Arabinda Acharya, and Sabina Chua. *Conflict and Terrorism in Southern Thailand*. Singapore: Marshall Cavendish, 2005.

Helbardt, Sascha. *Deciphering Southern Thailand's Violence: Organization and Insurgent Practices of BRN-Coordinate*. Singapore: ISEAS, 2015.

Human Rights Watch, Thailand; available at: www.hrw.org/asia/thailand (accessed April 21, 2017).

Jerryson, Michael K. *Buddhist Fury: Religion and Violence in Southern Thailand*. NY: Oxford University Press, 2011.

Jory, Patrick, ed. *Ghosts of the Past in Southern Thailand*. Singapore: ISEAS, 2013.

Liow, Joseph Chinyong. *Muslim Resistance in Southern Thailand and Southern Philippines: Religion, Ideology, and Politics*. Policy Studies 24. Washington, DC: East–West Center, 2006.

Liow, Joseph Chinyong. *Islam, Education and Reform in Southern Thailand: Tradition and Transformation*. Singapore: ISEAS, 2009.

Liow, Joseph Chinyong and Don Pathan. *Confronting Ghosts: Thailand's Shapeless Southern Insurgency*. Australia: Lowy Institute for International Policy, 2010.

Marks, Thomas A. *Making Revolution: The Insurgency of the Communist Party of Thailand in Structural Perspective*. Bangkok: White Lotus, 1994.

Marks, Thomas A. *The British Acquisition of Siamese Malaya (1896–1909)*. Bangkok: White Lotus, 1997.

McCargo, Duncan, ed. *Rethinking Thailand's Southern Violence*. Singapore: NUS Press, 2007.

McCargo, Duncan. *Tearing the Land Apart: Islam and Legitimacy in Southern Thailand*. Ithaca, NY: Cornell University Press, 2008.

McCargo, Duncan. "Autonomy for Southern Thailand: Thinking the Unthinkable?" *Pacific Affairs* 83, no. 2 (June 2010), 261–281.

McCargo, Duncan. *Mapping National Anxieties: Thailand's Southern Conflict*. Copenhagen, Denmark: NAIS Press, 2012.

McCargo, Duncan and Srisompob Jitpiromsri. "A Ministry for the South: New Governance Proposals for Thailand's Southern Region," *Contemporary Southeast Asia* 30, no. 3 (2008), 403–428.

Montesano, Michael J. and Patrick Jory, eds. *Thai South and Malay North: Ethnic Interactions on a Plural Peninsula*. Singapore: NUS Press, 2008.

Parks, Thomas I. "Research Note – Maintaining Peace in a Neighbourhood Torn by Separatism: The Case of Satun Province in Southern Thailand," *Small Wars and Insurgencies* 20, no. 1 (March 2009), 185–202.

Quaye, Christopher O. *Liberation Struggles in International Law*. Philadelphia, PA: Temple University Press, 1991.

Sidel, John T. *The Islamist Threat in Southeast Asia: A Reassessment*. Policy Studies no. 37. Washington, DC: East–West Center, 2007.

"Southern Thailand Attacks Reflect Tension Over Peace Talks, Conflict Monitor Says," *New York Times*, April 20, 2017; available at: www.nytimes.com/reuters/2017/04/20/world/asia/20reuters-thailand-south-attacks.html?_r=0 (accessed April 21, 2017).

Amnesty International. *Thailand: Torture in the Southern Counter-Insurgency*. London: Amnesty International, January 2009.

Thanet, Aphornsuvan. *Rebellion in Southern Thailand: Contending Histories*. Policy Studies no. 35. Washington, DC: East–West Center, 2007.

Tikhonov, Vladmir and Torkel Brekke, eds. *Buddhism and Violence: Militarism and Buddhism in Modern Asia*. New York, NY: Routledge, 2013.

Wattana, Sugunnasil, ed. *Dynamic Diversity in Southern Thailand*. Chiangmai: Silkworm Books, 2005.

W. K. Che Man. *Muslim Separatism: The Moros of Southern Philippines and the Malays of Southern Thailand*. Manila: Oxford and Ateneo de Manila University Press, 1990.

Part IV

Victory?

15 Size still matters

Explaining Sri Lanka's counterinsurgency victory over the Tamil Tigers

Sameer P. Lalwani

Introduction

On May 19, 2009, the government of Sri Lanka (GoSL) formally declared victory over the Liberation Tigers of Tamil Eelam (LTTE) – an ethnic separatist and terrorist organization that pushed the frontiers of guerilla tactics and hybrid warfare in its fight against the Sri Lankan state for three decades. The decisive victory and total collapse of the LTTE stunned practitioners and scholars of international security considering how many major powers, with considerably greater resources than Sri Lanka, have struggled to overcome the asymmetric advantages of guerilla insurgencies.

Numerous scholars, policy officials, and even the American Federal Bureau of Investigation (FBI) described the LTTE as "one of the world's most ruthless,"[1] "most dangerous,"[2] "most innovative,"[3] and "most professional and formidable"[4] groups in the world. Led by Velupillai Prabhakaran, the LTTE gained international notoriety by pioneering the use of suicide bombers, female fighters, and even the use of chemical attacks.[5] Thus, scholars of counterinsurgency, terrorism, rebel governance, civil wars, and ethnic politics have all found the Sri Lankan government's decisive triumph over the "A team of terrorist groups" surprising and worthy of further investigation.[6] In fact, Sri Lanka's former Defence Minister Gotabaya Rajapaksa went as far as to label Sri Lanka's success against the LTTE as the "Sri Lankan Model" and encouraged other countries to follow Sri Lanka's direction in their own counterterrorism/counterinsurgency efforts.[7]

Sri Lanka's dramatic military success against the LTTE is so compelling because it defied popular predictions, even just prior to the LTTE's collapse. Strategic assessments prior to or during the final campaign from 2006 to 2009 cast doubt on the prospect of war resumption, battlefield gains, or a military end to conflict.[8] However, against these expectations, the Sri Lankan military decisively crushed the LTTE with brute military force, and since the LTTE's military defeat, violence has not resumed nor conflict recurred. As one scholar points out "the magnitude and decisiveness of the victory is compelling reason enough to study this case."[9]

Several explanations have been advanced to explain Sri Lanka's unexpected success against the LTTE including the government's new employment of force and strategy, the LTTE's organizational weaknesses and missteps, and the contrasting leadership of incumbent and insurgent forces. However, scholars have downplayed a deceptively simple explanation: material preponderance. This article finds that the subtle but significant shift in the material balance of power explains why the final Sri Lankan military campaign succeeded, as opposed to previous efforts.

Even when some scholars have tried to evaluate the explanatory power of different variables, they typically cannot account for variation in military effectiveness across the different campaigns and conflict episodes. Factors generally constant throughout the conflict – like the LTTE's leadership or tactics – fail to account for the GoSL's repeated failure throughout the first five conflict periods before its success in the final phase of the conflict. This article possesses particular utility in that it compares and contrasts several dominant explanations for the LTTE's demise with the ultimate conclusion that material preponderance holds the most explanatory value of all proposed accounts. The analysis poses implications for states hoping to prevail in long-term campaigns against formidable insurgent organizations, such as the Afghan government's ongoing fight against the Taliban.

This chapter proceeds as follows. First, it offers some background on the Sri Lankan civil war. Second, it evaluates the leading explanations of the government's success in the final campaign. Third, it offers an alternative explanation, material preponderance, as the primary reason for incumbent victory. Fourth, the paper marshals some quantitative evidence in support of this explanation and concludes with summary findings and implications.

Conflict background

Emerging from the political oppression in the 1970s, the LTTE was formed in 1976. Within the milieu of various Tamil separatist political sentiment (like Tamil Eelam Liberation Organization (TELO), People's Liberation Organisation of Tamil Eelam (PLOT), Eelam People's Revolutionary Liberation Front (EPLF)) in the 1980s, the LTTE helped to initiate a civil war and fought for more than 30 years against the Sri Lankan state. Its long-term durability stemmed from its high discipline and cult-like ideology that created strong internal cohesion. In addition, external support from the Indian government and Tamil diaspora played an important complementary role in the longevity of the organization.[10]

Early on, the Tigers engaged in guerilla tactics with a strategy of provocation to create a wedge between the Sri Lankan government and moderate Tamils in order to drive them toward more extremist positions.[11] The LTTE ambush of an army patrol in 1983 that killed 13 soldiers served as a suitable trigger for pitting the majority Sinhalese against the Tamil population. Outrage over the ambush ultimately triggered a wave of indiscriminate, anti-Tamil violence, which killed between 400 and 2,000 Tamils. These riots led to burgeoning support, both at home and abroad, for Tamil militant organizations.

Although other rival separatist groups competed for Tamil support, the LTTE eventually weakened them through fratricidal violence and consolidated control over the Tamil separatist movement by 1990.[12] After a brief Indian military intervention from 1987 to 1990, the LTTE garnered a reputation in its fight against the Indian Peace Keeping Force (IPKF) and inherited more serious military hardware after the IPKF withdrawal. As the organization reaped the rewards of sole control over the insurgency, it utilized both guerilla hit-and-run tactics in conjunction with more conventional massing tactics to directly challenge the Sri Lankan Army. It is also during the 1990s that the LTTE's demands escalated to an independent Tamil homeland in the northern and eastern regions of Sri Lanka, which constituted about one-third of the country's landmass (See Figures 15.1 and 15.2).

Figure 15.1 Provincial boundaries of Sri Lanka and key locations.

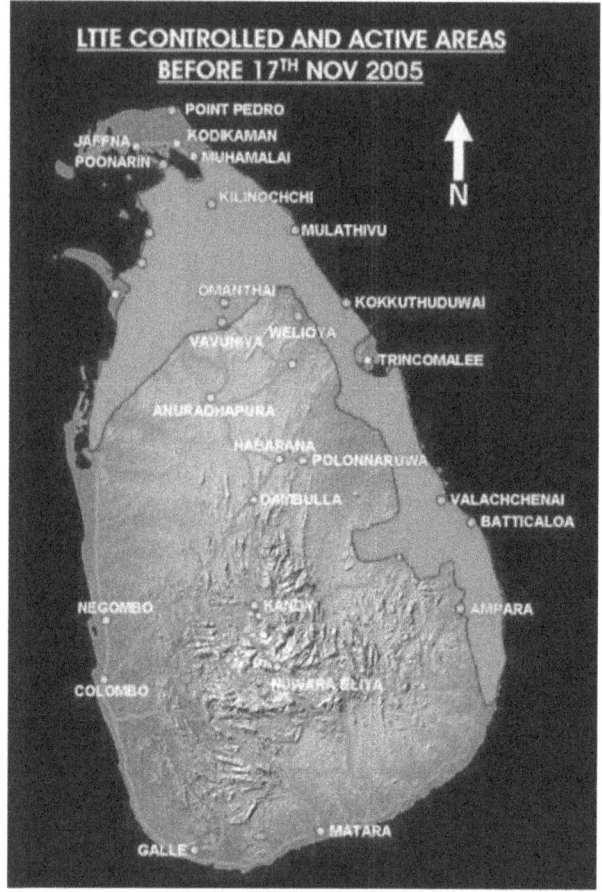

Figure 15.2 Territory controlled by LTTE prior to Eelam IV.

Sources: 'The Battle Progress,' Sri Lanka ministry of Defence, January 18, 2009. Archived February 16, 2009. Accessed January 30, 2012.

Following the IPKF withdrawal, the LTTE inherited a territorial sanctuary and began the process of crude state building. Over the course of the 1990s, the LTTE began setting up "statelike structures and transformed themselves into a more rigidly bureaucratized rational-legal institution."[13] It carried out governance and socialization functions like running schools, a legal system, health care, developing economic infrastructure, military recruitment, and taxation.[14] Because the LTTE was not wholly dependent on a single charismatic authority and in fact had a dense network of cadres with "a rich set of horizontal ties undergirded central organizational processes,"[15] it had proved resilient against attempted or perceived defections by high-level commanders.

In one of the leading studies of the Tamil rebel organizations, Staniland identifies four features to explain the LTTE's resilience and cohesion. These include a small, close-knit social base amongst the lower fishing caste networks, intense ideological discipline, a willingness to use escalatory violence, and strong, savvy leadership from the LTTE's founder and long-time commander Prabhakaran.[16]

In terms of military tactics, the LTTE was a pioneer. It inspired a number of terrorist outfits around the world, including Al-Qaeda in Iraq, primarily for its perfected use of suicide bombings and use of women and children in conflict. Overall, the group conducted over 200 suicide attacks and is credited for at least a dozen high-level assassinations, namely Indian PM Rajiv Gandhi's killing in 1991. Also having assassinated Sri Lankan President Ranasinghe Premadasa in 1993, the LTTE earned the notorious distinction as the only global terror organization to have murdered two world leaders.[17] Despite being such an inventive, militarily effective, and cohesive rebel organization historically, some have blamed the defeat of the LTTE in 2009 on its vulnerability and weakness as an organization. This hypothesis will be explored further below.

Surveying the arc of Sri Lanka's civil war reveals a great deal of consistency throughout the conflict. The Sri Lankan government fought the LTTE for 26 long years before the conflict finally ended in May 2009 with a toll of between 70,000 and 130,000 lives.[18] The conflict can roughly be divided into six phases (Table 15.1): Eelam War I from 1983 to 1987, the IPKF from 1987 to 1989, Eelam War II from 1990 to 1994, Eelam War III from 1995 to 2001, the Ceasefire Agreement period from 2002 to 2005, and Eelam War IV from 2006 to 2009.

Evaluating competing explanations

Eelam IV proved an extraordinary departure from previous GoSL campaigns against the LTTE. Eelam IV involved multiple ground operations, an air and naval campaign, and high intensity pitched battles. What was more striking was that while many doubted military gains from the renewed campaign, few, if any, expected a total GoSL victory and collapse of the LTTE.

Plenty of post-mortems have offered accounts of Sri Lanka's surprising and near-complete victory, and though few offer systematic explanations, one can distill these explanations into one of three concepts of strategic interaction. These include force employment, insurgent attributes, and political leadership.

Force employment

Arguments for why force employment shape military effectiveness offer three different routes. One variant contends that overarching, dichotomous strategic choices shape outcomes, such as between attrition and maneuver warfare.[19] A second specific to asymmetric wars contends the strategic interaction – specifically mismatches between regular and irregular modes of warfare – can undermine effectiveness.[20] A third variant contends professionalization, skill, and

Table 15.1 Phases of conflict*

	1) Eelam I	2) Indian Peace Keeping Force (IPKF)	3) Eelam II	4) Eelam III	5) Ceasefire Agreement (CFA)	6) Eelam IV
Timeframe	July 1983–July 1987	August 1987–March 1990	June 1990–1994	April 1995–December 2001	February 2002–July 2006	July 2006–May 2009
Insurgent approach	Guerilla	Guerilla	Hybrid	Hybrid	Proto-state/guerilla	Hybrid
Incumbent approach	Light conventional	Conventional	Conventional	Conventional	Pause/unconventional	Conventional
Incumbent strategy	Enfeeblement	Attrition	Enfeeblement	Attrition	Pause/cooptation	Attrition
GoSL milt. effectiveness	Low-moderate	Moderate	Low	Low	Moderate	High
Loss-exchange ratio	0.61	2.07	1.23	1.11	1.82	2.95
Civilian fractional loss (est.)	0.78	0.48	0.55	0.42	0.59	0.55

Note

* Data compiled and derived from the following sources: Ferdinando, "A Battering for the Luckless People"; Harihan, "A Tale of Two Interventions"; South Asia Terrorism Portal, "Fatalities in Terrorist Violence"; Hashim, *When Counterinsurgency Wins*, 179; Gunaratna, *Indian Intervention in Sri Lanka*, xv, 315, 465; Bhatt and Mistry, *Cost of Conflict in Sri Lanka*, 18; BBC News, "Sri Lankan Army Deaths Revealed"; Harff and Gurr, "Toward Empirical Theory"; Eckhardt, "Wars and War-Related Deaths, 1900–1995"; BBC News, "Tamil Rebel Leader Ends Isolation"; CNN, "Paper: 20,000 Killed in Sri Lanka Conflict"; Buncombe, "Up to 40,000 Civilians 'Died in Sri Lanka Offensive'."

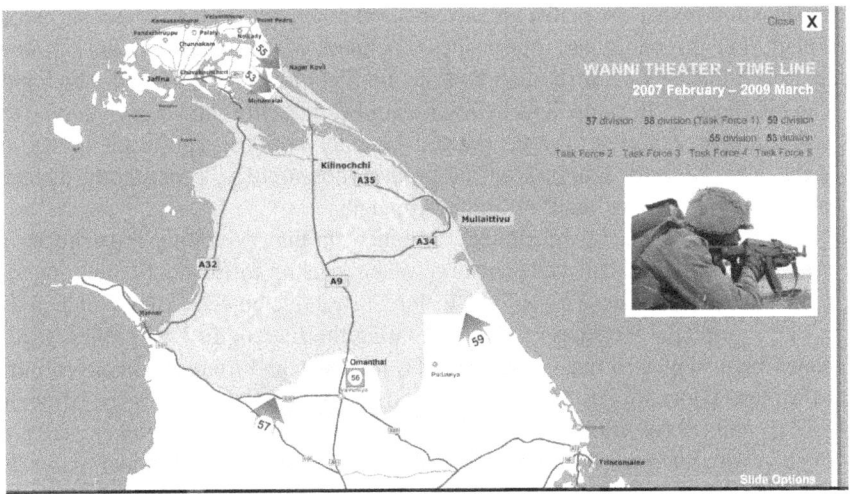

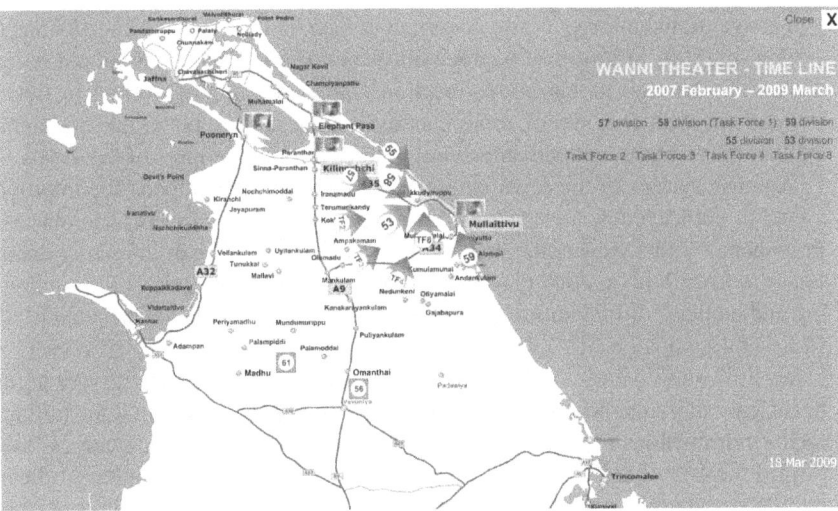

Figure 15.3 Map of Eelam IV campaign – February 2007 and March 2009.

Source: Sri Lanka Ministry of Defence. Accessed January 30, 2012. http://defence.lk/orbat/

tactical and operational proficiency in modern warfare yields military effectiveness.[21]

Following the first route, a number of scholars and analysts attribute military victory success to an overarching strategy of sustained, unapologetic brute force.[22] Some have suggested that the Eelam IV campaign's new strategy of attrition "to cause maximum damage to the LTTE in terms of its human resources capacities and infrastructure" played a decisive role in Sri Lanka's military victory.[23] Although attrition's effectiveness as a counterinsurgency strategy has robust theoretical and empirical support,[24] the problem with this explanation is that Sri Lanka already focused on an attrition approach far before Eelam IV. In particular, both sides operated with a conventional attrition approach during Eelam III such that the conflict "revolve[d] around which side can kill more."[25] In campaigns before Eelam IV, ceasefires and negotiations were not strategic victories but necessitated by interventions, significant losses, or stalemates. Military progress on the ground, not strategy, consistently dictated pauses or the termination of military campaigns.

A second variant of the strategy argument is that exploiting mismatches in strategic interaction can yield victory. When it came to Eelam IV, the government succeeded because it "chose a different strategic objective that matched the LTTE's principal weaknesses while negating their strengths."[26] In other words, some have suggested that after the LTTE had switched from an unconventional to a more conventional posture, the GoSL made a bold choice to go after the LTTE with conventional military power. This argument, however, fails to explain why the government did not dismantle the LTTE much sooner. After the IPKF departure and the onset of Eelam II, the LTTE was fighting at least partly in a conventional manner.[27] The government then had numerous opportunities to exploit this shift in insurgent strategy but for two campaigns, it could not defeat the LTTE. Though its largely conventional approach finally worked in Eelam IV, perhaps with the help of some unconventional tactics, the argument cannot explain the military ineffectiveness and failures throughout most of the 1990s.

Another variant of the force employment explanation contends the improvements in skill, tactics, and operational effectiveness proved critical to success. Proponents of this view argue that the tenure of Commander of the Army Sarath Fonseka from 2006 to 2009 played an important role in improving military training and tactics.[28] While improved tactics appear likely to have contributed to the successful campaign, it's unclear how many of these were really new ideas or better utilization of resources, and how much was independent of a very significant growth in manpower. Some have cast doubt on the notion that the military totally retrained as a skilled force in three to four years. Even as late as 2009 toward the end of the Eelam IV campaign, some former senior military officers intimated that soldiers still lacked the skills necessary to fight.[29] In fact, then Defense Secretary Gota Rajapaksa argued victory was achieved by "the same military that had fought the LTTE in the last 30 years."[30]

Insurgent attributes

The body of research on insurgent attributes generally suggests these can impact conflict dynamics and outcomes via rebel strategic choices,[31] tactical behavior,[32] organization,[33] and political behavior.[34] While there has been less study of the choice of rebel military strategies, some literature suggests insurgents are better poised to win when their strategy is mismatched from the incumbent strategy, that is fighting unconventionally or indirectly against a direct strategy.[35] Another literature contends the use of certain unconventional tactics like terrorism can backfire and compromise legitimacy or lead a state to summon greater resolve.[36] Finally, the type of insurgent organization is expected to have an impact on both cohesion and internal control, combat power, and durability,[37] while retention of a political wing can influence conflict duration and outcomes.[38]

Explanations of the LTTE's demise have implicitly drawn from these theories of insurgent behavior and organization. One camp contends that by fighting conventionally and trying to hold territory, the LTTE abandoned its guerilla fighting roots and became vulnerable to the conventional advantage of the GoSL's superior firepower.[39] In contrast, another explanation contends that the LTTE had pushed asymmetric tactics like terrorism too far, compromising international and local legitimacy while rousing the state to decisive action.[40] A third argues that organizational dependence on diaspora support compromised the roots of the LTTE's legitimacy and power.[41] Finally, a fourth explanation suggests the nature of the LTTE's organization and its underdeveloped political wing hindered its adaptation and incentivized continued military engagements, which led to its destruction.[42]

A consistent problem that arises for each of these insurgent-focused explanations is that they cannot account for the previous 20 years of the LTTE's success. The Tamil Tigers neither lost popular support nor suffered a major defeat from drawing upon horrendous acts of terrorism and employing a full-spectrum of capabilities including conventional warfighting. Furthermore, the organization argument also does not ring true because the LTTE was historically well-organized as a tightly-knit bond network that emerged as the leading insurgent and terrorist organization in the world.

Some scholars charge the LTTE mistakenly departed from its guerilla strategy and instead fought a conventional war in Eelam IV with multiple domains, massed forces, and static defenses.[43] In actuality, these critiques of LTTE strategy seem to lay inordinate blame on the LTTE's conventional capabilities, which were not new and had long proved very effective. First, the LTTE had historically fought across the full spectrum of conflict – including in some conventional modalities. Second, the LTTE fought a hybrid style of warfare, utilizing both conventional and guerilla tactics as far back as Eelam II.[44] The LTTE was a full-spectrum insurgent organization with hybrid capabilities allowing it to move up and down the spectrum between terrorism, guerilla attacks, and conventional warfare. Third, the LTTE consistently harnessed seemingly conventional air and naval forces for unconventional roles both in prior campaigns and during Eelam IV. Only when the GoSL military dramatically expanded at an unprecedented rate did the LTTE's hybrid strategy prove wanting.

If terrorism suddenly spiked before or around Eelam IV, mechanisms of local alienation and state emboldenment might have also intensified. However, the data conflicts with this expectation. As Figures 15.4 and 15.5 indicate, average terrorism incidents in Eelam IV were only roughly 17 percent higher than Eelam II and 17 percent lower than during the IPKF.[45] Incidents peaked at 160 in 2006, but this was only marginally higher than the previous peak of 151 in 1996. Changes in the post-9/11 global order and reduced tolerance for violent non-state

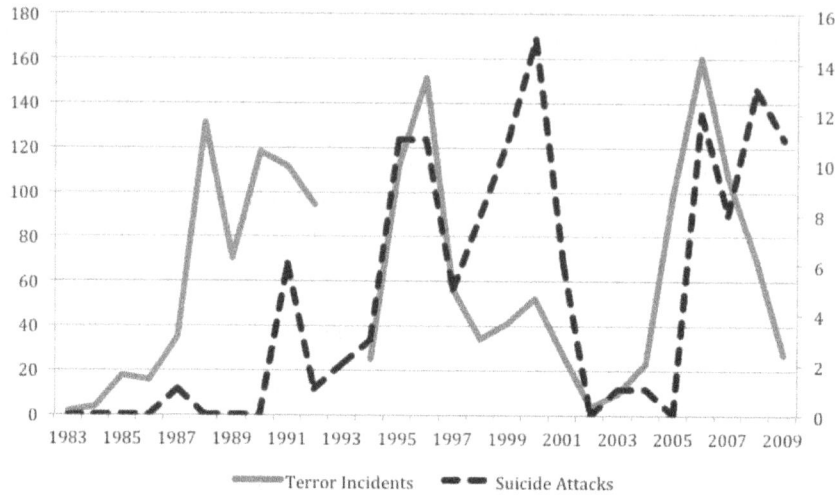

Figure 15.4 Estimated unconventional rebel violence by year.
Sources: Global Terrorism Database; South Asia Terrorism Portal; Pape, 2003.

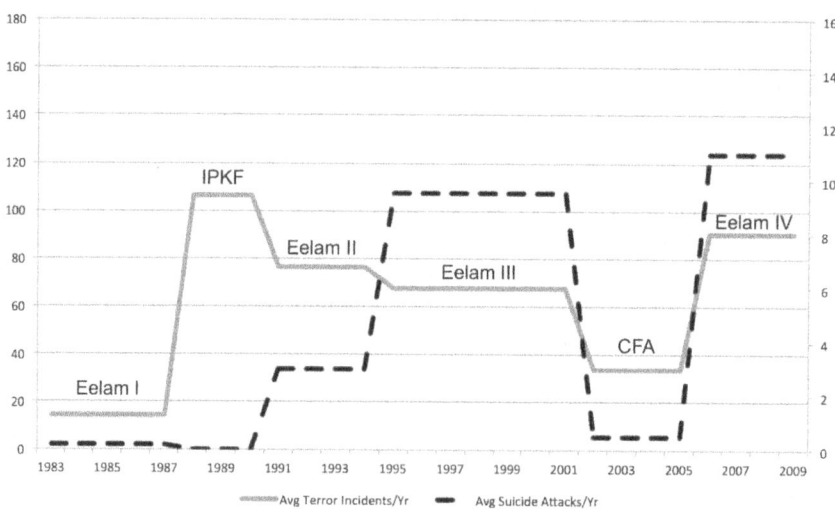

Figure 15.5 Estimated unconventional rebel violence average by phase.
Sources: Global Terrorism Database; South Asia Terrorism Portal; Pape, 2003.

actors did have a discernible effect on the LTTE during Eelam IV, but these effects principally impinged on the LTTE's money and materiel, which constrained its capabilities on the battlefield. Furthermore, the organization argument also does not ring true because the LTTE was historically well-organized as a tightly-knit network that emerged as the leading insurgent and terrorist organization in the world and simultaneously governed territory like a state.

Leadership

A substantial body of work has examined how active leadership can heavily influence state behavior, strategic processes, and war outcomes. The COIN literature in particular tends to focus on "great man" theories.[46] Several analysts implicitly adopt this theoretical framework when assessing Sri Lanka's success against the LTTE. They praise the choices made by GoSL leaders – especially President Mahinda Rajapaksa, Defense Minister Gota Rajapaksa, and Army Chief Sarath Fonseka – in implementing general strategy, designing and resourcing a military for success, promoting officers to successfully execute the campaign, and not buckling to international pressure.[47] However, it is unclear that their political wills were greater than those of prior leaders, that they possessed exceptional leadership qualities, or that they made particularly difficult political choices.

First, it seems unfair to deny the strong political will of prior Sri Lankan leaders and claim that the leadership from 2006 to 2009 was exceptional. President Jayewardene exercised political will by fending off Indian coercion for almost a decade and assuming the massive political risks of effectively collaborating with militant groups by supplying them arms to rid the country of Indian military presence in the late 1980s. President Premadasa also demonstrated resolve when he restarted offensives against the LTTE even though it eventually led to his death from a suicide bomber.

Second, unlike theories of great civilian leaders in war, some evidence suggests that President Rajapaksa did not actively "command" the campaign and instead adopted a hands-off approach that gave the military free reign. Other leaders like President Kumaratunga did actively intervene to sack and promote battlefield commanders during Eelam III, including the promotion of General Fonseka, but this did not achieve the same outcome.[48]

Third, there does not appear to be any evidence that Rajapaksa made hard or costly decisions. Rajapaksa suffered no domestic political costs from a brute force strategy since the affected Tamil civilians were never his political constituency. He suffered few international political costs *during* the campaign, despite some obligatory condemnations. Finally, the economy remained unharmed by his dramatic escalation in warfare.

The counterpart to the leadership explanation is the poor leadership of Velupillai Prabhakaran, the founder and leader of the LTTE, who allegedly become its "chief liability."[49] If Prabhakaran's leadership was the critical variable, why did it take 30 years for his leadership to destroy the LTTE? Accounts that fault Prabhakaran's leadership style simply cannot explain away the prior decades of organizational success. While the LTTE leadership critique does offer some valid points, this is difficult to reconcile with the same leader who propelled the LTTE to success by outbidding rival militant organizations in the 1980s, outlasting the Indian military by 1990, and fighting the Sri Lankan military to a stalemate by 2001. What some identify as the LTTE's strategic missteps actually served as sources of strength and success for much of the conflict. For example, alleged "blunders" of reckless violence like high-profile assassinations, civilian victimization, and fighting Indian forces[50] may have drawn international ire but also allowed the LTTE to overpower rival outfits,[51] attract funding and recruitment,[52] "maintain order and discipline – even overseas,"[53] and retain control over its territory. Although Prabhakaran certainly made errors, misjudgment was an invariant condition throughout the conflict since "at almost every turn, the LTTE made profound strategic miscalculations."[54] Leadership and strategy may offer partial insights into the LTTE's rapid implosion but these variables possess far less explanatory value than an account of material preponderance.

An alternative explanation of material preponderance

While government strategy, insurgent attributes, and leadership are important factors in explaining military effectiveness and outcomes, they all face serious shortfalls when it comes to explaining Sri Lanka's defeat of the LTTE in 2009. This section instead contends that the government's new found material preponderance proved the most critical factor in the outcome of Eelam IV, in contrast to the previous campaigns.

The theoretical literature finds that a number of material factors are associated with military effectiveness and victory in interstate conflict, as well as civil war. Manpower is found to have a significant impact on incumbent victory in counterinsurgency[55] while economic development, capturing something about state capacity, is also found to have a significant effect on military effectiveness across all conflicts.[56] Recent research also finds that external interventions, which improve the relative military capacity of a government irrespective of efforts to bolster government legitimacy, improves the likelihood of government victory.[57]

Manpower and resources

Manpower in particular would prove essential to the GoSL victory. In an interview, Defense Secretary Gota Rajapaksa stated the GoSL analyzed prior campaigns and concluded a key factor was "inadequate numbers" to "hold the ground in the immediate aftermath of battlefield success." General Fonseka also identified the problem as one of numbers.[58] Consequently, they concluded "the solution was to increase force strength."[59] Hashim writes, "one of the most – if not *the* most – important changes was to massively increase the size of the armed forces."[60] With some assistance from external support, the GoSL dramatically expanded the military over five years. The defense budget grew at average annual rate of 30 percent between 2005 and 2009,[61] army manpower tripled to

300,000 troops, and total armed forces personnel increased from 125,000 to 471,000 troops.[62] Meanwhile, the LTTE was hemorrhaging manpower – having lost as much as 40 percent of its forces between 2004 and 2006 – and turned to forcibly conscripting unskilled manpower and even children.[63]

After the Sri Lankan military's massive expansion, it was no surprise it was now able to conduct a final offensive from three directions with 120,000 troops while still leaving plenty of manpower to hold territory it had retaken.[64] For perspective, offensives in earlier campaigns generally involved between 5,000 and 20,000 troops. Before Eelam IV, the military's largest offensive of 40,000 troops successfully retook the northern port of Jaffna but lacked the manpower to prevent the LTTE's retreat into the Wanni interior. Officers and analysts routinely raised the concern of insufficient manpower in earlier campaigns. Gunaratna argues that in Eelam I, "The low strength of the Sri Lankan forces did not permit them to occupy LTTE strong holds, bases and hide-outs, and also to continue operations."[65] After setbacks in 2000, the Sri Lankan Army spokesman faulted personnel shortages and stated, "If we are to make inroads against the Tamil Tigers we need the manpower."[66] In the final stages of the conflict, the force to rebel ratio was estimated around 100 : 1.[67] In Eelam IV, one long time defense analyst argues that holding operations had such saturation of manpower that one "couldn't walk two kilometers without passing five checkpoints."[68] What this suggests then is the dramatic increase in the material balance of power, rather than a shift in force employment or strategy, ultimately made the difference in eliminating the LTTE.

Material preponderance, rather than claims of learning and adaptation, seems to have matched the self-identified non-innovative strategy of "attrition warfare."[69] General Sarath Fonseka stated, "Our main intention is to kill the maximum number of LTTE area leaders and their cadres within LTTE territory."[70] Massive force expansion did enable new tactics such as multiple axes of advance and holding forces, but it also enabled the GoSL to bulldoze adversary forces at a faster rate. In the 2006–2009 campaign, Sri Lankan security forces were being killed at twice the annual rate as previous campaigns, but they were also able to kill LTTE cadres six times the rate of previous campaigns.[71] Toward the end of Eelam IV, the LTTE had to confront the inherent demographic imbalance as it had simply burned through capable manpower and lacked sufficient strength, training, or hardware.[72]

Even seeming changes in strategy, organizational adaptation, and innovation had their roots in basic material conditions. Maritime innovations for instance depended upon sheer material preponderance. For example, the GoSL's maritime interdiction campaign of 'massive retaliation' to deny arms resupply and logistics to the LTTE depended upon an overwhelming material advantage in vessels as high as a 4 : 1 ratio.[73] It is also important to consider the significant build up in naval forces. After 2000, navy personnel jumped by 50 percent, and the patrol and coastal combatant force increased by 233 percent.[74] In addition, the GoSL raised 250 locally built patrol boats to use swarm tactics against LTTE maritime operations. By contrast, the 2004 Indian Ocean tsunami wiped out much of the LTTE's fleet. Toward the end of Eelam IV, the LTTE could hardly muster a maritime challenge. Major engagements with the Sri Lankan Navy (SLN) dropped from 21 in 2006 to 11 in 2007 to only two in 2008.[75]

Unlike other competing explanations, material factors can also account for Sri Lankan successes in the mid-1980s. For example, when the government strategy remained relatively reactive, the balance of forces still favored the government against a still developing but comparatively small LTTE force. Thus the Sri Lankan military "came within a whisker of delivering a knockout punch" in 1987, cut short by Indian intervention.[76] After the IPKF episode, the LTTE had closed some of the gap on the GoSL's significant material advantage and it would not return until the final campaign.

Exogenously driven material changes

In conjunction with the massive buildup of government forces, analysts acknowledge the LTTE had been comparatively weakened and "hollowed" before the fighting of the Eelam IV campaign even began.[77] This is because the GoSL profited greatly from two exogenous shocks and two broader structural changes that materially reshaped the incumbent-insurgent balance of power firmly in favor of the state.

As alluded to previously, the first exogenous shock came in the form of the defection of Eastern province commander Colonel Karuna, one of Prabhakaran's top lieutenants, in March 2004. Upset by his diminishing authority and fearing retribution for his efforts to counter LTTE political centralization, Karuna defected from the LTTE with anywhere from 3,000 to 6,000 cadres to fight a guerilla campaign against the LTTE. The LTTE quickly stamped out Karuna's rebellion, forcing him and his associates to retreat to Colombo, but the loss of manpower, and the intelligence Karuna provided the GoSL severely weakened the LTTE.[78] Though the GoSL did not anticipate or orchestrate the Karuna defection, it did effectively seize upon it,[79] using intelligence gathered from Karuna on LTTE bases and operations to retake the East and launch an effective maritime interdiction campaign. In contrast, though Karuna's defection had a real material impact on LTTE combat capabilities and recruitment in the East, Prabhakaran failed to account for this in his planning.[80] While much-maligned for allowing the insurgents to regroup and rearm, the 2002 ceasefire agreement may have unintentionally intensified the LTTE's internal contradictions between its military and political wings, abetted a power struggle, and enabled this critical defection.

The second exogenous shock followed soon after in the form of the December 2004 Indian Ocean earthquake of historic proportions – the strongest in 700 years – and subsequent tsunami. As one of the worst natural disasters in history, it killed more than 220,000 people in 14 countries, including over 40,000 in Sri Lanka.[81] The tsunami disproportionately devastated the LTTE-occupied coastal regions of eastern Sri Lanka, which wiped out 3,000 Tiger cadres and a quarter of the LTTE's naval forces.[82]

Another important exogenous condition was the global crackdown on terrorist financing after the 9/11 attacks. Soon after 9/11, President Kumaratanga published an op-ed in the *Washington Post* entitled "We Know Terrorism," in which she called upon the international community to ratify the International Convention for the Suppression of Financing of Terrorism.[83] Although the treaty was drafted in 1999, only four states had ratified it by the fall of 2001. By April 2002, however, a sufficient number of countries had ratified the Terrorist Financing Convention, and it went into force. This treaty quickly led to the freezing of millions of dollars of LTTE funding, including $12 million annually from Canada.[84] When the EU designated the LTTE as a proscribed organization in 2006, this further froze their assets in 25 countries.[85] Thus the post-9/11 effect was less a function of global norms and more the result of specific efforts to choke off the LTTE's access to material resources.

Finally, the emergence of multipolarity, especially the rise of China, created a geopolitical marketplace of opportunity for a number of states to extract aid and attract material support for their domestic priorities.[86] With its geographic position on the Indian Ocean, Sri Lanka was perfectly suited to position itself within this "great game" of states vying for position in the Indian Ocean. For example, after the EU curtailed assistance to Sri Lanka because of human rights concerns, China quickly provided substantial aid, foreign direct investment, and military assistance at relatively cheap prices.[87] During Eelam IV, China is estimated to have provided $1 billion in military and financial aid annually.[88] Intelligence, military assistance and training during the campaign also came from Pakistan, Russia, Iran, Libya, Israel, Ukraine, and Indonesia.[89] India – which was motivated by geopolitical competition from China in the Indian Ocean[90] – remained politically constrained from visible support, but it facilitated substantial covert political, military, and diplomatic assistance, including naval intelligence and a $2.5 billion IMF loan.[91] Even the United States contributed with intelligence, training, communications networks setup, coastal radar systems, and possibly even interdiction of the LTTE's transnational arms supply network.[92]

Even absent a massive growth in Sri Lanka's military, the exogenous shocks and slow-moving structural changes peaking just prior to Eelam IV, converged to make conditions exceptionally ripe for a GoSL military victory. When these exogenous factors combined with the astonishing and unprecedented and growth in GoSL military size and firepower, the LTTE was no match.

A quantitative analysis of material preponderance

A series of simple quantitative tests seem to confirm the material preponderance story. Cross-national studies of counterinsurgency often treat a conflict as a single observation and code the dependent variable of "outcome" as a "win," "loss," or "draw." This approach would not allow for a quantitative analysis to evaluate the relative factors of success unless the conflict was broken into different periods like campaigns, or even conflict-years. Dividing up the conflict into observations of conflict-year and using a dependent variable of an annual loss-exchange ratio allows for a more fine-grained examination of success over time. A loss-exchange ratio (LER) modified for counterinsurgency is the number of insurgent fighters to incumbent troops killed. The LER enables users to not only capture magnitude with a continuous rather than dichotomous variable but also "focuses more closely on questions of battlefield capability per se, with less risk of misrepresenting an inferior political stake as inferior military capacity."[93]

To examine LER over time required assembly of a new data-set. LER was estimated for each year over 30 years (1980–2009) based on secondary sources including personal narratives, journalistic accounts, government statements, and older partial data-sets.[94] Data for other independent and control variables were also collected.

To examine the effect of material preponderance on LER, annual data were collected on government force size, rebel force size, size of contested geographic space of area of operations (in square kilometers), and population size within area of operations. From this, raw force levels, force to insurgent ratios, force to space ratios, and force to population ratios were assessed.[95] Scatter plots and correlation tests of LER with various measures of material preponderance (total force levels, force-to-rebel ratio, force-to-space ratio, and force-to-population ratios) all suggest a robust relationship between these two variables. Strong, positive correlations are shown in Figure 15.6.[96] Since force-to-rebel ratio proved the strongest, these variables were utilized for subsequent regression analysis.

The next step was to further probe the relationship between material preponderance while controlling for some other fine-grained explanatory variables and accounting for random error. An additional proxy of material preponderance is the state's gross domestic product (GDP) per capita, which the extant literature suggests captures something about state capacity or level of development.[97] To control for other explanations of military effectiveness that could account for annual change, data was collected on the capitalization of the military as a measure of skill,[98] primary completion rates and enrollment rates (lagged by eight years) as a measure of human capital,[99] and democracy and polity scores to measure regime type.[100] Additionally, in the years 1987–1989, data on the features of the incumbent were drawn from India since it fielded the IPKF against the LTTE. Table 15.2 offers descriptive statistics of all the variables.

Following this, the data was divided into two samples for high versus low force-to-rebel ratios. T tests were performed to assess whether the difference in two samples' difference in means of the dependent variable was statistically significant. The T tests proved significant at the 0.01 level ($p = 0.0098$) and remained statistically significant even after repeatedly redefining the threshold, dividing high and low force-to-rebel ratios at different levels. This evidence suggests that a scenario with a high force-to-rebel ratio is likely to have a different loss-exchange ratio than one with a low force-to-rebel ratio. As the boxplot in Figure 15.7 shows, high force-to-rebel ratios tend to have significantly higher loss-exchange ratios, even when accounting for outliers.

Initial bivariate regressions accounting for the error term generally confirmed the relationship between these measures of material preponderance – whether measured as total government force size or a force-to-rebel ratio – and military effectiveness ($p < 0.001$).

The results of multiple linear regression analysis (see Table 15.3) using ordinary least squares (OLS) on various models confirm the relationship between

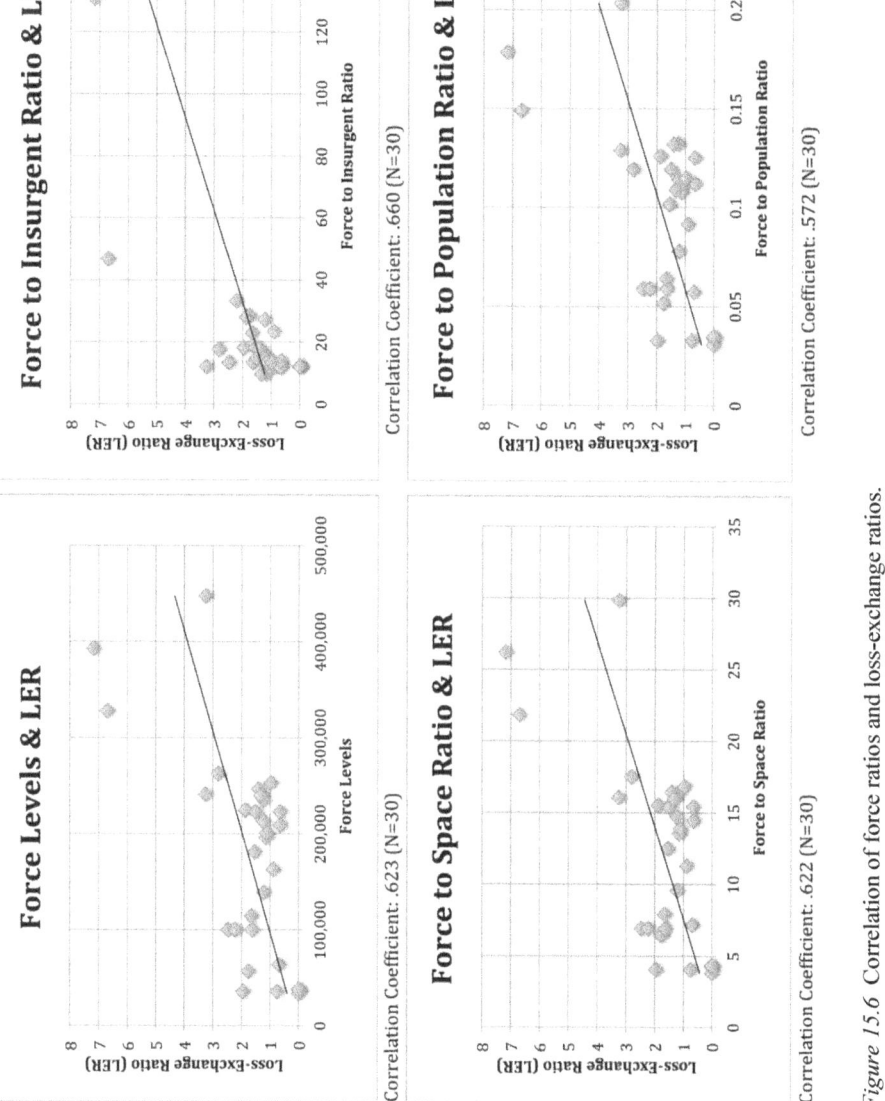

Figure 15.6 Correlation of force ratios and loss-exchange ratios.

Table 15.2 Descriptive statistics of variables

Variable	Mean	Standard deviation	Minimum	Maximum
Loss-exchange ratio	1.77	1.63	0	7.17
Govt. force to rebel ratio	25.94	32.16	9.58	149.2
Total govt. forces	177,831.50	106,804.87	33,940.00	447,600.00
Rebel size	9,803.33	6,775.59	2,000.00	25,000.00
Govt. force to space ratio	12.53	6.58	3.82	29.84
Govt. force to pop. ratio	0.1	0.04	0.03	0.2
GDP/capita	2,948.33	1,068.11	1,125.00	5,011.01
Military capitalization	3,656.82	1,861.15	777.48	8,970.78
Primary completion rate	83.73	18	42.36	107.26
Primary enrolment rate	99.65	10.51	81.19	111.36
Democracy	0.37	0.49	0	1
Polity score	5.57	0.94	5	8

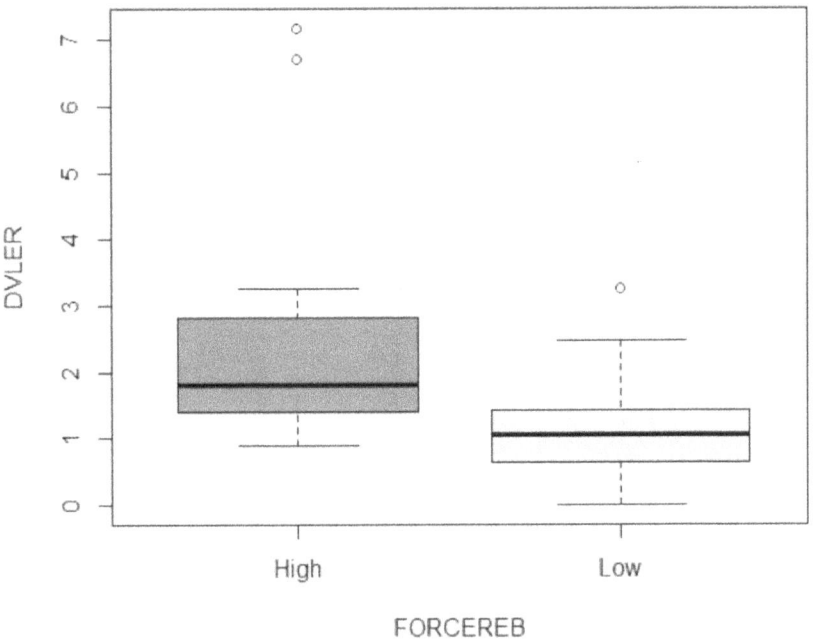

Figure 15.7 Boxplot of loss-exchange ratios (DVLER) from high force-to-rebel ratios (FORCEREB) versus low force-to-rebel ratios.

Table 15.3 Linear regression analysis

	No controls	Force + rebel + 1 control				Force + rebel + primecomp + 1 control			Force + rebel + 3 or 4 controls		
	Model 0	Model 1	Model 2	Model 3	Model 4	Model 5	Model 6	Model 7	Model 8	Model 9	Model 10
Govt force size	0.00001245*** (0.000002132)	0.00001198* (0.000004639)	0.00001253*** (0.000002123)	0.00001714*** (0.000003045)	0.00001227*** (0.000002101)	0.00001234** (0.000004176)	0.00001706*** (0.000003429)	0.00001648*** (0.000003476)	0.00001654*** (0.000003708)	0.00001095* (0.000004876)	0.000006837 (0.000005684)
Rebel size	−0.0001115** (0.00003361)	−0.0001121** (0.00003462)	−0.000105* (0.00003396)	−0.00008554* (0.00003419)	−0.00009858** (0.00003434)	−0.00008071* (0.00003329)	−0.00008565* (0.00003491)	−0.00008438* (0.0003486)	−0.00008424* (0.00003567)	−0.00008116* (0.00003376)	−0.00007464* (0.00003355)
GDP/capita		0.0005431 (0.000468)				0.0008342 (0.0005126)				0.0009286 (0.0005451)	0.001285* (0.0005979)
Military capitalization			0.0001271 (0.0001133)				0.000007794 (0.0001315)		−0.000007709 (0.0001389)	0.00007624 (0.000133)	0.00004974 (0.0001322)
Primary completion				−0.03992† (0.01944)		−0.06122* (0.02295)	−0.03915 (0.02374)	−0.0348 (0.02322)	−0.03537 (0.02584)	−0.05605* (0.02495)	−0.04967^ (0.02498)
Polity score					0.3147 (0.227)			0.1095 (0.2606)	0.1137 (0.2763)		0.3872 (0.2873)
Constant	0.6453 (0.4369)	0.5755 (0.7482)	0.1023 (0.6506)	2.9* (1.173)	−1.201 (1.399)	3.03* (1.14)	2.823 (1.766)	1.968 (2.518)	2.008 (2.673)	2.291 (1.731)	−0.6855 (2.789)
N	30	30	30	30	30	30	30	30	30	30	30
Multiple R-squared	0.5652	0.5654	0.5853	0.6259	0.5951	0.6617	0.626	0.6285	0.6286	0.6663	0.6907
Adjusted R-squared	0.533	0.5153	0.5374	0.5827	0.5484	0.6076	0.5661	0.5691	0.5512	0.5968	0.6101

Notes
*** $p < 0.001$;
** $p < 0.01$;
* $p < 0.05$;
† $p < 0.10$.

material preponderance and military effectiveness, even when controlling other variables. Previous studies have found factors like capitalization of military forces to capture skill, human capital, and democracy improve military effectiveness. Controlling for these variables in various models, I find that only measures of material preponderance, such as total government force size, and rebel force size, consistently had a statistically significant effect on Sri Lanka's military effectiveness during its war with the Tamil Tigers.

Keeping the force-to-rebel ratio as two separate variables for government and rebel force levels, several multiple linear regression models were generated to control different factors and as robustness checks. In models 1–4, only one control was tested at a time; the government force size coefficient was found to be significant at the 0.05 level in all cases. Rebel force size, another measure of the material balance of power, was also found to be significant, but it was likely endogenous to the previous year's loss-exchange. In models 5–7, different combinations of control factors were tested, all including primary education completion rate. This variable was chosen because of its highest values of multiple R-squared and adjusted R-squared compared to the other models. Out of these models, government force size and rebel force size continued to be significant at the 0.01 and 0.05 level, respectively. Force size continued to retain the statistical significance even when adding additional control variables until model 10 in which rebel size remained significant and government force size did not. This is not concerning, however, in that model 10 contains more than the recommended number of independent variables for a sample size of 30 and was mainly conducted for sensitivity analysis.

A series of robustness checks was performed to confirm the relationship between material capacity and battlefield outcomes. Introducing additional control variables – such as area of operations size and population size, as well as alternative measures of human capital and regime type such as primary enrolment rate and a dichotomous coding of democracy – into various models yielded similar results. The residual plots of the single linear regressions between the dependent variable and each independent variable were analyzed to check that the residuals did not have a non-random pattern, which was confirmed. However, the effect of outlier points was noticeable, so a natural log transformation was applied on the independent variables to reduce the variation caused by those extreme values. The same ten models were run with log-transformed independent variables and produced similar results. Finally, the extreme values from the years were removed as a third check. Though there is good reason to include them, when three outlier observations from the years 2007–2009 were excluded from the analysis, for the most part government force size continued to retain statistical significance.

While confidence in any statistical analysis with 30 observations needs to be measured, the correlations, T tests, and multiple regression analysis all seem to provide quantitative support for prior qualitative assessments that incumbent material preponderance was central to Sri Lanka's military effectiveness against the LTTE. The statistical results suggest that – independent of the level of rebel size, state capacity, general military skill, human capital, or regime type – as the Sri Lankan state expanded its manpower while the insurgent force shrunk, the GoSL made gains on the battlefield it had never seen in almost three decades of conflict.

Conclusion

This article has argued that upon close examination, the GoSL's victory over the Tamil Tigers is not particularly surprising. The inherent material advantage states possess over violent non-state actors increased manifold for Sri Lanka due to a number of fortunate events and unfolding structural processes just prior to the conflict that strengthened the state and weakened the LTTE.

Findings

Force employment, insurgent attributes, and leadership are perceived as far more interesting explanations of the LTTE's defeat than the banality of material preponderance because they accord agency to key actors in the conflict. Nevertheless, these features of the conflict were not sufficient for victory and may not have even been necessary either. When considering the counterfactual, military victory would have been inconceivable without material preponderance. By contrast a GoSL endowed with material preponderance but employing more unconventional strategies and tactics, facing a more fissiparous and deinstitutionalized insurgent organization, or possessing different leadership still seems likely to have achieved military victory.

Competing theories face two weaknesses: they fail to explain variation over time and the empirical record does not conform to the mechanisms of each of these theories. This article points out gaping holes in the empirical support for competing theories and their explanatory mechanisms. Close scrutiny reveals the absence of any serious innovations in the GoSL's tactics and operations on the battlefield but rather simply a higher frequency and tempo due to scale. Much praise for the GoSL strategy comes ex-poste and when the actual strategy is laid bare, it turns out to be nothing more imaginative than brute force. The criticism the GoSL has suffered for the incredibly high levels of civilians killed in the final campaign may also be a product of operational scale and tempo. Though precise casualty figures are notoriously suspect in this conflict, Table 15.1 estimates of civilian fatalities as a fraction of total casualties reveals Eelam IV's level of indiscriminate violence did not significantly stray from previous campaigns.

Even if we assume the LTTE made all the right strategic choices, it is unlikely to have made a difference when the GoSL marshaled force ratios of one to two orders of magnitude above the LTTE. Contrary to related criticisms, insurgent organization appears to have been quite resilient even to the very end. Despite intense military pressure, the absence of fracturing or defection other than Karuna in 2004 indicates supreme resolve and cohesion.

Finally, praise for leadership neglects how events actually played out. Little in Rajapaksa's track record pre- and post- conflict suggests he exercised good judgment and made smart, hard, and objective decisions. By contrast, plenty of evidence has been unearthed in the years since the conflict to indicate he relied on a corrupt, personalist regime composed of family and sycophants prone to groupthink. One can imagine that material preponderance would have dictated the outcome of the conflict with the LTTE even in the absence of Rajapaksa.

Implications

Sri Lanka's military victory over the LTTE offers valuable lessons for practitioners and scholars. First, under certain conditions, material factors may matter more than strategy and, at extreme levels, may be able to compensate for other vulnerabilities. If the host nation possesses sufficient motivation to defeat a militant organization, material support in the form of money and materiel to support resource-intensive counterinsurgency may prove useful.

Second, the GoSL victory underscores the importance of examining the experiences of smaller, non-Western domestic counterinsurgents. In this case, the GoSL did not even try to protect the population, collect local intelligence, win hearts and minds, or reform political institutions, and still they proved successful. This may suggest that in contrast to foreign incumbents, domestic incumbents face a different set of possibilities and constraints and may evaluate success on different terms.

Third, counterinsurgency scholarship often assesses success and failure based on some eye-test of whether incumbents follow counterinsurgency "best practices," but these criteria would have missed the successes of the GoSL's Eelam IV campaign. If, however, counterinsurgency analysts start to take into account other quasi-objective quantifiable measures, such as loss-exchange ratios, they might be able gain greater purchase on estimating the degree of success or failure to better inform policymakers.

Last, the GoSL defeat of the LTTE serves as one of the few outright victories in modern counterinsurgency warfare, but the nature of this victory should underscore the sheer difficulty of counterinsurgency. The degree of material asymmetry required to achieve a brute force military victory is one that very few foreign or domestic incumbents will be able to muster. For decades, the GoSL struggled with over a 10:1 ratio of government to rebel forces. Only when it achieved force ratios of between 50:1 and 150:1 did the state achieve decisive material gains. That scale seems almost impossible to replicate for most incumbents facing strong and sizable insurgent organizations.

Acknowledgments

I am grateful to the excellent research assistance of Vinod Kannuthurai and Hamza Shad.

Notes

1 Anderson, "Death of the Tiger," 41.
2 FBI, "Taming the Tamil Tigers"; Indian Express "LTTE most Dangerous Extremist Outfit: FBI."
3 Layton, "How Sri Lanka Won the War."
4 Swamy, *The Tiger Vanquished*, 105.
5 Pape, "The Strategic Logic of Suicide Terrorism"; Bloom, "Ethnic Conflict, State Terror and Suicide Bombing"; Hoffman, "The First Non-state Use of a Chemical Weapon."
6 Hashim, *When Counterinsurgency Wins*, 34.
7 Ramachandran, "Any Country Facing Terrorism."
8 Biswas, "The Challenges of Conflict Management"; Nieto, "A War of Attrition," 573, 584; Clarke, "Conventionally Defeated but Not Eradicated," 157–188; Weaver and Chamberlain, "Sri Lanka Declares End to War."
9 Hashim, *When Counterinsurgency Wins*, 2.
10 Staniland, *Networks of Rebellion*, 155–156; Gunaratna, *Indian Intervention in Sri Lanka*, 89–134.
11 Kydd and Walter, "The Strategies of Terrorism."
12 Staniland, "Between a Rock and Hard Place."
13 Staniland, *Networks of Rebellion*, 169.
14 Mampilly, *Rebel Rulers*, 108–115.
15 Staniland, *Networks of Rebellion*, 172.
16 Staniland, *Networks of Rebellion*, 149–152, 156–158.
17 FBI, "Taming the Tamil Tigers."
18 Vaughn, *Sri Lanka*.
19 Mearsheimer, *Conventional Deterrence*.
20 Arreguín-Toft, "How the Weak Win Wars"; Pape, *Bombing to Win*.
21 Biddle, *Explaining Victory and Defeat in Modern Battle*.
22 Shashikumar, "Lessons from Sri Lanka's War"; Anderson, "Death of the Tiger."
23 Singh, "Endgame in Sri Lanka," 8; Mehta, "Sri Lanka's Ethnic Conflict," 11.
24 Byman, "'Death Solves All Problems'"; Lyall, "Does Indiscriminate Violence Incite Insurgent Attacks."
25 Karunatilake, "The Tigers Return."
26 Layton, "How Sri Lanka Won the War."
27 Chandraprema, *Gota's War*, 197–201; Gunaratna, *Indian Intervention in Sri Lanka*, 438–448.
28 Mehta, "Sri Lanka's Ethnic Conflict," 12.
29 Mehta, "Sri Lanka's Ethnic Conflict," 13.
30 Shashikumar "Winning Wars: Political Will is the Key," 18.
31 Pape, "The Strategic Logic of Suicide Terrorism"; Krause, "The Political Effectiveness of Non-State Violence."
32 Kydd and Walter, "The Strategy of Terrorism."
33 Weinstein, *Inside Rebellion*.
34 Huang, "Rebel Diplomacy in Civil War"; Mampilly, *Rebel Rulers*.
35 Arreguín-Toft, *How the Weak Win Wars*.
36 Fortna, "Do Terrorists Win?"
37 Staniland, *Networks of Rebellion*; Weinstein, *Inside Rebellion*; Johnston, "The Geography of Insurgent Organization."
38 Cunningham et al., "It Takes Two ."
39 Mehta, "Sri Lanka's Ethnic Conflict," 17; de Silva, *Defeat of the LTTE*, 203.
40 Oakford, "For Years After a Tamil Defeat."
41 Mehta, "Sri Lanka's Ethnic Conflict," 18; Hashim, *When Counterinsurgency Wins*, 35.

42 Hashim, *When Counterinsurgency Wins*, 193.
43 Harihan, "Why LTTE Failed."
44 Marks, "Sri Lanka and the Liberation Tigers of Tamil Eelam," 511.
45 There is also a recency bias in terms of data collection that tends to undercount historic incidents since there was far less media coverage.
46 Moyar, *A Question of Command*; for a critique of this perspective, see Rovner, "The Heroes of COIN."
47 Singh, "Endgame in Sri Lanka?"
48 Chandraprema, *Gota's War*, 244.
49 Mehta, "Sri Lanka's Ethnic Conflict," 15.
50 Hashim, *When Counterinsurgency Wins*, 35, 122–123, and 191.
51 Staniland, *Networks of Rebellion*, 163–168.
52 Mehta, "India's Counterinsurgency Campaign in Sri Lanka," 161; Krause, "The Political Effectiveness of Non-State Violence."
53 Smith, "The LTTE: A National Liberation and Oppression Movement," 101.
54 Smith, "Understanding Sri Lanka's Defeat of the Tamil Tigers," 44. See also Rajagopalan, *Fighting Like a Guerilla*, 99; Mehta, "Sri Lanka's Ethnic Conflict," 17.
55 Friedman, "Manpower and Counterinsurgency."
56 Beckley, "Economic Development and Military Effectiveness."
57 Sullivan and Karreth, "The Conditional Impact of Military Intervention."
58 Interview with a local journalist, Colombo, February 24, 2015.
59 Shashikumar, "Winning Wars: Political Will is the Key," 18.
60 Hashim, *When Counterinsurgency Wins*, 187.
61 Other sources are less generous but still report the army doubled to 200,000, and total military strength increased from a previous level of 179,000 to 350,000. Manpower jumped 80 percent in 2008 alone. The SIPRI data shows a lower (but still significant) rate of growth of 14 percent over this period.
62 See IISS, *The Military Balance 2005*, 246–248; Hashim, *When Counterinsurgency Wins*, 188; Moorcraft, *Total Destruction of the Tamil Tigers*, 71.
63 The GoSL estimated the LTTE had 25,000 by 2006. Mehta estimated this dropped to 15,000 by 2006 following the losses from the Karuna defection, the Tsunami, and low-level conflict. See Waduge, "Q&A with Facts about the LTTE"; Mehta, "Sri Lanka's Ethnic Conflict," 17. See also, Buerk, "Tamil Tiger 'Forced Recruitment'."
64 Mehta, "Sri Lanka's Ethnic Conflict."
65 Gunaratna, *Indian Intervention in Sri Lanka*, 258.
66 Lawson-Tancred, "Army Says It Has Hold on Tamil Tigers."
67 Singh, "Endgame in Sri Lanka."
68 Interview with Sri Lankan defense analyst, Colombo, February 25, 2015.
69 "The Battle Progress"; Jayasundera, "Government's War of 'Attrition'."
70 Hashim, *When Counterinsurgency Wins*, 42.
71 This statement is based on comparing average annual data between Eelam IV and previous campaigns Eelam I–III. Sri Lankan military annual casualty rates estimated based on data reported by Ferdinando, "A Battering for the Luckless People of B'loa." LTTE fatality rates derived from data reported in Gunaratna, *Indian Intervention in Sri Lanka*, 465; Bhatt and Mistry, *Cost of Conflict in Sri Lanka*, 18; and "Sri Lankan Army Deaths Revealed."
72 de Silva, *Defeat of the LTTE*, 102; Interview with Sri Lankan scholar, Colombo, March 3, 2015.
73 DeSilva-Ransinghe, "Maritime Counter-Terrorism and the Sri Lanka Navy."
74 Specifically, the number of naval personnel increased from 10,000 to 15,000 and the number of patrol boats from 39 to 130. For more on this, see IISS, *Military Balance*, 2009.
75 DeSilva-Ransinghe, "Maritime Counter-Terrorism and the Sri Lanka Navy," 32.
76 Marks, "Sri Lanka and the Liberation Tigers of Tamil Eelam," 516.

77 Smith, "Understanding Sri Lanka's Defeat of the Tamil Tigers."
78 Wickremesekera, *The Tamil Separatist War*, 178; Harihan, "Why LTTE Failed"; Smith, "A National Liberation and Oppression Movement," 99; Balachandran, "Karuna's Defection Reduced LTTE's Manpower."
79 Staniland, *Networks of Rebellion*, 174.
80 Harihan, "Why LTTE Failed."
81 "Indian Ocean Tsunami Anniversary"; Nizam, "Sri Lankans Paid Tribute to Over 40,000 Persons Lost."
82 Hashim, *When Counterinsurgency Wins*, 122 and 172.
83 Kumaratunga, "We Know Terrorism."
84 Smith, "Understanding Sri Lanka's Defeat of the Tamil Tigers," 42.
85 de Silva, *Sri Lanka and the Defeat of the LTTE*, 181.
86 National Intelligence Council, "Mapping the Global Future."
87 For additional information on the strategic implications of the Indian Ocean, see Kaplan, *Monsoon: the Indian Ocean*, 195–196, 207–208.
88 Smith, "Understanding Sri Lanka's Defeat of the Tamil Tigers," 43.
89 Kaplan, *Monsoon: the Indian Ocean*, 207; Mehta, "Sri Lanka's Ethnic Conflict," 14; SIPRI Arms Transfer Database (www.sipri.org/databases/armstransfers).
90 Destradi, "India and Sri Lanka's Civil War," 595–616.
91 Mehta, "Sri Lanka's Ethnic Conflict," 20–21; Author Interview with Ashok Mehta, New Delhi, India, March 10, 2015.
92 Smith, "The LTTE: A National Liberation and Oppression Movement," 106; Layton, "How Sri Lanka Won the War."
93 Biddle, *Military Power: Explaining Victory and Defeat in Modern Battle*.
94 Full list available upon request.
95 Friedman, "Manpower and Counterinsurgency."
96 The correlations are driven in part by three observations from 2007–2009 which register extreme values. There is good reason not to exclude them as outliers as they are not random but all observations from a period of deliberate material buildup. Nevertheless, as a robustness check, excluding these from the analysis still generates moderate positive correlation coefficients greater than 0.3 for all four measures of force.
97 Fearon and Laitin, "Ethnicity, Insurgency, and Civil War"; Beckley, "Economic Development and Military Effectiveness."
98 Stam, *Domestic Politics and the Crucible of War*.
99 Biddle and Long, "Democracy and Military Effectiveness," 525–546.
100 Reiter and Stam, *Democracies at War*.

References

Anderson, Jon Lee. "Death of the Tiger: Sri Lanka's Brutal Victory Over Its Tamil Insurgents." *The New Yorker*, (January 17, 2011) available at www.newyorker.com/magazine/2011/01/17/death-of-the-tiger.

Arreguín-Toft, Ivan. "How the Weak Win Wars: A Theory of Asymmetric Conflict." *International Security* 26 (1) (2001), 93–128.

Balachandran, P. K. 2016. "Karuna's Defection Reduced LTTE's Manpower by Half Paving Way for Defeat." *The New Indian Express*, May 22.

"Battle Progress Map." Sri Lanka Ministry of Defence, (May 18, 2009) available at www.defence.lk/orbat/.

Beckley, Michael. "Economic Development and Military Effectiveness." *Journal of Strategic Studies* 33 (1) (2010), 43–79.

Bhatt, Semu and Devika Mistry. *Cost of Conflict in Sri Lanka*. Mumbai: Strategic Foresight Group, 2006.

Biddle, Stephen D. *Military Power: Explaining Victory and Defeat in Modern Battle.* Princeton, CA: Princeton University Press, 2004.

Biddle, S. and S. Long. "Democracy and Military Effectiveness: A Deeper Look." *Journal of Conflict Resolution*, 48(4) (2004), 525–546. https://doi.org/10.1177/0022002704266118.

Biswas, Bidisha. "The Challenges of Conflict Management: A Case Study of Sri Lanka." *Civil Wars* 8 (1) (2006), 46–65.

Bloom, Mia M. "Ethnic Conflict, State Terror and Suicide Bombing in Sri Lanka." *Civil Wars* 6, (1) (2003), 54–84.

Buerk, Roland. "Tamil Tiger 'Forced Recruitment'." *BBC News*, (July 30, 2007) available at http://news. bbc.co.uk/2/hi/south_asia/6916291.stm.

Buncombe, Andrew. 2010. "Up to 40,000 Civilians 'Died in Sri Lanka Offensive'." *Independent*, (February 12, 2010) available at www.independent.co.uk/news/world/asia/up-to-40000-civilians-died-in-sri-lanka-offensive-1897865.html.

Byman, Daniel. " 'Death Solves All Problems': The Authoritarian Model of Counterinsurgency," *Journal of Strategic Studies* 39 (1) (2016), 62–93.

Chandraprema, C. A. *Gota's War: The Crushing of the Tamil Tiger Terrorism in Sir Lanka.* Colombo: Ranjan Wijeratne Foundation, 2012.

Clarke, Ryan. "Conventionally Defeated but Not Eradicated: Asian Arms Networks and the Potential for the Return of Tamil Militancy in Sri Lanka." *Civil Wars* 13 (2) (2011), 157–188.

Cunningham, David E., Kristian Skrede Gleditsch and Idean Salehyan. "It Takes Two: A Dyadic Analysis of Civil War Duration and Outcome," *Journal of Conflict Resolution* 53 (4) (2009), 570–597.

de Silva, K. M. *Sri Lanka and the Defeat of the LTTE.* New Delhi: Penguin, 2012.

DeSilva-Ranasinghe, Sergei. "Reflections on the Tigers." *The Diplomat*, (May 21, 2010) available at http://thediplomat.com/2010/05/reflections-on-the-tigers/?allpages=yes.

DeSilva-Ranasinghe, Sergei. "Maritime Counter-terrorism and the Sri Lanka Navy." *Asia Pacific Defence Reporter* 35 (9) (2009), 32–33.

Destradi, Sandra. "India and Sri Lanka's Civil War." *Asian Survey* 52 (3) (2012), 595–616.

Eckhardt, William. "Wars and War-related Deaths, 1900–1995." In Ruth Leger Sivard (ed.) *World Military and Social Expenditures 1996*, 17–19. Washington, DC: World Priorities, 1996.

Fearon, James D. and David Laitin. "Ethnicity, Insurgency, and Civil War." *American Political Science Review* 97 (1) (2003), 75–90.

Ferdinando, Shamindra. "A Battering for the Luckless People of B'loa: War on Terror Revisited." *The Island*, (January 20, 2013) available at www.island.lk/index.php?page_cat=article-details&page=article-details&code_title=70797.

Fortna, Virginia Page. "Do Terrorists Win? Rebels' Use of Terrorism and Civil War Outcomes." *International Organization* 69 (3) (2015), 519–556.

Friedman, Jeffrey A. "Manpower and Counterinsurgency: Empirical Foundations for Theory and Doctrine." *Security Studies* 20 (4), (2011), 1–36.

Gunaratna, Rohan. *Indian Intervention in Sri Lanka: The Role of India's Intelligence Agencies.* Colombo: South Asian Network on Conflict Research, 1993.

Harff, Barbara and Ted Robert Gurr. "Toward Empirical Theory of Genocides and Politicides: Identification and Measurement of Cases since 1945." *International Studies Quarterly* 32 (3) (1988), 359–371.

Harihan, R. "Why LTTE Failed," *Frontline*, (May 9, 2009) available at www.frontline.in/static/html/fl2610/stories/20090522261001200.htm

Harihan, R. "A Tale of Two Interventions." *The Hindu*, July 28, 2012 available at www.thehindu.com/opinion/lead/a-tale-of-two-interventions/article3693348.ece.

Hashim, Ahmed S. *When Counterinsurgency Wins: Sri Lanka's Defeat of the Tamil Tigers*. Philadelphia, PA: University of Pennsylvania Press, 2013.

Hoffman, Bruce. "The First Non-state Use of a Chemical Weapon in Warfare: The Tamil Tigers' Assault on East Kiran." *Small Wars and Insurgencies* 20 (3–4) (2009), 463–477.

Huang, Reyko. "Rebel Diplomacy in Civil War." *International Security* 40 (4) (2016), 89–126.

"Indian Ocean Tsunami Anniversary: Memorial Events Held." *BBC News*, (December 26, 2014) available at www.bbc.com/news/world-asia-30602159.

International Institute for Strategic Studies (IISS) *The Military Balance 2005*. London: Routledge, 2005.

International Institute for Strategic Studies (IISS) *The Military Balance 2009*. London: Routledge, 2009.

International Institute for Strategic Studies (IISS) *The Military Balance 2013*. London: Routledge, 2013.

Jayasundera, Ranjith. "Government's War of 'Attrition'." *The Sunday Leader*, (November 2, 2008).

Jha, Prashant. "A Leader Falls: Why Sri Lanka Ousted Mahinda Rajapaksa," *Hindustan Times*, (January 10, 2015) available at www.hindustantimes.com/world/a-leader-falls-why-sri-lanka-ousted-mahinda-rajapaksa/story-qKUYmWXpOkuzC1BVQ494PN.html.

Johnston, Patrick. "The Geography of Insurgent Organization and Its Consequences for Civil Wars: Evidence from Liberia and Sierra Leone," *Security Studies* 17 (1) (2008), 107–137.

Kaplan, Robert D. *Monsoon: The Indian Ocean and the Future of American Power*. New York: Random House, 2011.

Karunatilake, Waruna. "The Tigers Return," *Outlook*, (July 31, 1996) available at www.outlookindia.com/magazine/story/the-tigers-return/201850.

Krause, Peter. "The Political Effectiveness of Non-state Violence: A Two-level Framework to Transform a Deceptive Debate." *Security Studies* 22 (2) (2013), 259–294.

Kumaratunga, Chandrika Bandaranaike. "We Know Terrorism." *Washington Post*, (November 7, 2001) A29.

Kydd, Andrew H. and Barbara F. Walter. "The Strategies of Terrorism." *International Security* 31 (1) (2006), 49–80.

Lawson-Tancred, Alastair. 2000. "Army Says It Has Hold on Tamil Tigers." *Guardian*, (July 14, 2000) available at www.theguardian.com/world/2000/jul/15/1.

Layton, Peter. "How Sri Lanka Won the War," *The Diplomat*, (April 9, 2015) available at http://thediplomat.com/2015/04/how-sri-lanka-won-the-war/

"LTTE Most Dangerous Extremist Outfit: FBI." *Indian Express*, (January 11) available at http://archive.indianexpress.com/news/ltte-most-dangerous-extremist-outfit-fbi/260483/.

Lyall, Jason. "Does Indiscriminate Violence Incite Insurgent Attacks? Evidence from Chechnya." *Journal of Conflict Resolution* 53 (June 2009), 331–362.

Mampilly, Zachariah Cherian. *Rebel Rulers: Insurgent Governance and Civilian Life during War*. Ithaca, NY: Cornell University Press, 2011.

Mapping the Global Future: Report of the National Intelligence Council's 2020 Project. Washington, DC: National Intelligence Council, 2004.

Marks, Thomas A. "Sri Lanka and the Liberation Tigers of Tamil Eelam." In Robert J. Art and Louise Richardson (eds.), *Democracy and Counterterrorism: Lessons From the Past*, 483–530. Washington, DC: USIP, 2007.

Mearsheimer, John J. *Conventional Deterrence*. Ithaca, NY: Cornell University Press, 1985.

Mehta, Ashok "India's Counterinsurgency Campaign in Sri Lanka." In Sumit Ganguly and David P. Fidler (eds.), *India and Counterinsurgency: Lessons Learned*. New York: Routledge, 2009.

Mehta, Ashok. "Sri Lanka's Ethnic Conflict: How Eelam War IV Was Won." Paper for Centre for Land Warfare Studies, (2010) available at www.srilankaguardian. org/2010/05/sri-lankas-ethnic-conflict-how-eelam.html.

Moorcraft, Paul. *Total Destruction of the Tamil Tigers: The Rare Victory of Sri Lanka's Long War*. Barnsley: Pen and Sword Books, 2012.

Moyar, Mark. *A Question of Command: Counterinsurgency from the Civil War to Iraq.* New Haven, CT: Yale University Press, 2010.

Nieto, W. Alejandro Sanchez "A War of Attrition: Sri Lanka and the Tamil Tigers," *Small Wars and Insurgencies* 19 (4) (2008), 573–587.

Nizam, A. A. M. "Sri Lankans Paid Tribute to over 40,000 Persons Lost in the 2004 Tsunami," *Asian Tribune*, (December 28, 2011) available at www.asiantribune.com/ news/2011/12/27/sri-lankans-paid-tribute-over-40000-persons-lost-2004-tsunami.

Oakford, Samuel. "Four Years after a Tamil Defeat, the Diaspora Regroups." *Inter Press Service*, (October 25, 2013) available at www.ipsnews.net/2013/10/four-years-after-a-tamil-defeat-the-diaspora-regroups/.

Pape, Robert A. *Bombing to Win: Air Power and Coercion in War*. Ithaca, NY: Cornell University Press, 1996.

Pape, Robert A. "The Strategic Logic of Suicide Terrorism." *American Political Science Review* 97 (3) (2003), 1–19.

"Paper: 20,000 Killed in Sri Lanka Conflict." 2009. *CNN*, (May 29) available at www. cnn.com/2009/WORLD/asiapcf/05/29/srilanka.death.toll/.

Rajagopalan, Rajesh. *Fighting Like a Guerrilla: The Indian Army and Counterinsurgency*. New Delhi: Routledge, 2008.

Ramachandran, Ramesh. "Any Country Facing Terrorism Should Follow Sri Lankan Model." *The Asian Age*, (September 15, 2010) available at http://jdsrilanka.blogspot. com/2010/09/any-country-facing-terrorism-should.html.

Reiter, Dan and Allan C. Stam. *Democracies at War*. Princeton, CA and Oxford: Princeton University Press, 2002.

Rovner, Joshua. "The Heroes of COIN," *Orbis* 56 (2) (2012), 215–232.

Shashikumar, V. K "Lessons from Sri Lanka's War." *Indian Defence Review* 24 (3) (2009) available at www.indiandefencereview.com/spotlights/lessons-from-the-war-in-sri-lanka/0/.

Shashikumar, V. K. "Winning Wars: Political Will is Key," *Indian Defence Review* 25, (2) (2010), 16–26.

Singh, Ajit Kumar. "Endgame in Sri Lanka." *Faultlines* 20 (2011), 131–170.

Smith, Chris. "The LTTE: A National Liberation and Oppression Movement." In Christophe Jaffrelot (ed.), *Armed Militias of South Asia: Fundamentalists, Maoists and Separatists*, 91–111. London: Hurst, 2009.

Smith, Justin O. "Maritime Interdiction in Sri Lanka's Counterinsurgency," *Small Wars and Insurgencies* 22 (3) (2011), 448–466.

Smith, Niel A. "Understanding Sri Lanka's Defeat of the Tamil Tigers," *Joint Force Quarterly* 59 (4) (2010), 40–44.

South Asia Terrorism Portal. "Fatalities in Terrorist Violence in Sri Lanka 2002–2016," Institute for Conflict Management, available at www.satp.org/satporgtp/countries/shri-lanka/database/annual_casualties.htm (accessed August 30, 2016).

"Sri Lankan Army Deaths Revealed." *BBC News* (May 22, 2009).

Stam, Allan C. *Win, Lose, or Draw: Domestic Politics and the Crucible of War*. Ann Arbor, MI: University of Michigan Press, 1996.

Staniland, Paul. "Between a Rock and a Hard Place: Insurgent Fratricide, Ethnic Defection, and the Rise of Pro-State Paramilitaries," *The Journal of Conflict Resolution*, 56 (1) (2012), 16–40.

Staniland, Paul. *Networks of Rebellion: Explaining Insurgent Cohesion and Collapse*. Ithaca, NY: Cornell University Press, 2014.

Sullivan, Patricia L., and Johannes Karreth. "The Conditional Impact of Military Intervention on Internal Armed Conflict Outcomes," *Conflict Management and Peace Science* 32 (3) (2015), 269–288.

Swamy, M. R. Narayan. *The Tiger Vanquished: The LTTE Story*. New Delhi: Sage, 2010.

"Tamil Rebel Leader Ends Isolation," *BBC*, (April 9, 2002) available at http://news.bbc.co.uk/2/hi/south_asia/1919144.stm.

"Taming the Tamil Tigers." Federal Bureau of Investigation, (January 10, 2008) available at www.fbi.gov/news/stories/2008/january/tamil_tigers011008.

"The Battle Progress." Sri Lanka Ministry of Defence, (February 16, 2009) available at https://web.archive.org/web/20090216101817/http://defence.lk/PrintPage.asp?fname=20080623_02.

Vaughn, Bruce. *Sri Lanka: Background and U.S. Relations*. Washington, DC: Congressional Research Service, 2011.

Waduge, Shenali. "Q&A with Facts about the LTTE." Sri Lanka Ministry of Defence and Urban Development. (March 14, 2012) available at https://web.archive.org/web/20120810023959/www.defence.lk/new.asp?fname=Facts_about_the_LTTE_20120221_02 (accessed September 10, 2014).

Weaver, Matthew, and Gethin Chamberlain. 2009. "Sri Lanka Declares End to War with Tamil Tigers," *Guardian*, (May 19, 2009) available at www.theguardian.com/world/2009/may/18/tamil-tigers-killed-sri-lanka.

Weinstein, Jeremy. *Inside Rebellion: The Politics of Insurgent Violence*. New York: Cambridge University Press, 2007.

Wickremesekera, Channa. *The Tamil Separatist War in Sri Lanka*. New York: Routledge, 2016.

16 Sri Lanka

State response to Liberation Tigers of Tamil Eelam

Thomas A. Marks

Sri Lanka has until recently been the site of one of the most complex conflicts in recent history, that of the Liberation Tigers of Tamil Eelam (LTTE).[1] An insurgency that privileged terrorism as a method of action yet ultimately fielded land, air, and sea regular forces, rounded out by powerful special operations and information capabilities, LTTE grew in capacity until it was capable of forcing the government to agree to a February 2002 ceasefire and the de facto existence of a Tamil state, or *Tamil Eelam*. It was this victory of sorts that produced a host of unforeseen consequences and led to the July 2006 resumption of hostilities that resulted in May 2009 total victory in the field for Colombo.

What makes the case particularly salient for examination is that it actually contains a number of distinct conflicts, generally labeled: Eelam I (1983–1987), Eelam II (1990–1995), Eelam III (1995–2002), and Eelam IV (2006–2009). These dates are negotiable given realities on the ground. The gap between Eelam I and II saw the interlude of the Indian Peace Keeping Force (IPKF), which clashed bitterly with LTTE; while the gap between Eelam III and IV saw the effective rule of the *Tamil Eelam* state in areas of the north and east. This was accompanied by an uneasy ceasefire. In fact, each of the *Eelam* conflicts involved periods of negotiation and cessation of hostilities, though all were problematic in implementation and intent (certainly upon the part of LTTE). All involved foreign participation. Further complicating the picture, the IPKF years saw Sri Lanka fully committed to suppression of JVP II, the second upsurge of the original Maoist *Janatha Vimukthi Peramuna* uprising (JVP, People's Liberation Front), which had erupted and been crushed in 1971 (i.e., JVP I). In aggregate, casualty figures are subject to considerable disagreement but cannot be less than 120,000 dead.[2]

Regardless, the end of *Eelam*, when it came, was as spectacular as all other facets of LTTE's three decades in existence. Having grown from a veritable band of angry young men to an impressive guerrilla group, then to a full-fledged army, the self-proclaimed flag-bearer of Tamil nationalism found itself caught in the same position as the South in the 1861–1865 American Civil War: outmobilized and outfought. Its sometime foreign supporters, notably neighboring India, had deserted it, and even a pronounced global shift of attitude with respect to what was acceptable in warfighting could not turn outrage to tangible pressure upon Colombo before the Tiger end came. A force that at one point was thought to field as many as 35,000 combatants, found its maneuver-space squeezed by the inexorable advance of government columns using innovative, punishing tactics. A last stand on a narrow stretch of northeastern beach saw total decimation, with considerable collateral damage among population forced to accompany LTTE as human shields.

LTTE itself admitted defeat on May 17, 2009, with its major figures overwhelmingly killed in action, to include the near-legendary leader, Velupillai Prabhakaran, who had emerged in the late 1970s as the group's head and ruthlessly held the position throughout the conflict. Ironically, the book has not yet been closed, as a shift in the political winds caused any number of governments, led by European states and the United States, to turn on their former Sri Lankan partner and join cause-oriented groups in seeking sanction through international humanitarian and human rights law for what they charged was callous (and illegal) indifference to civilian casualties in the final period of struggle. An outraged Sri Lanka became estranged from those democratic nations with which it had the most in common and reoriented its foreign policy to new regional forces, notably China. Even the January 2015 upset win of an opposition coalition headed by a former ruling party intimate, Maithripala Sirisena, did not result in a shift in the direction desired by those who sought, in a sense, to mandate that war be something other than what it is, barbarous and cruel.

Initial reaction to "terrorist" challenge

It cannot be said that Sri Lanka's conflict ended in a fashion much different from historical instances of major combat. What is more unique is the sheer savagery of the war that developed over the three decades mentioned above. Significant was the complexity of the threat faced. Liberation Tigers of Tamil Eelam (LTTE) was labeled "terrorist" by any number of governments. In reality, it was an insurgency in intent and methodology. It had, however, gone from using terror as a tool for mass mobilization to using it as the premier element in its approach to gaining *Tamil Eelam.*

The problem for security forces is that, early on, armed challenges to the government's writ appear more or less the same. A systemic response centered about use of force, to the near exclusion of other facets, may be inappropriate in counterterrorism, complicating the effort, but in counterinsurgency it can often be disastrous. Most commonly, abuse of the populace creates a new dynamic which allows an operationally astute insurgent challenger for state power to mobilize additional support. This is precisely what occurred in Sri Lanka.

A less likely setting for conflict would be hard to imagine, as the West Virginia-sized island was (and remains) a tropical paradise in its physical aspects. In terms of human terrain, though, British colonialism (1815–1948) had left unresolved issues as to the meaning of independence and societal composition.[3] The Buddhist, Sinhala-speaking majority – 10,979,561 of 14,846,750 as per the 1981 census, or 73.95 percent – dominated the British-inspired parliamentary democracy, yet the

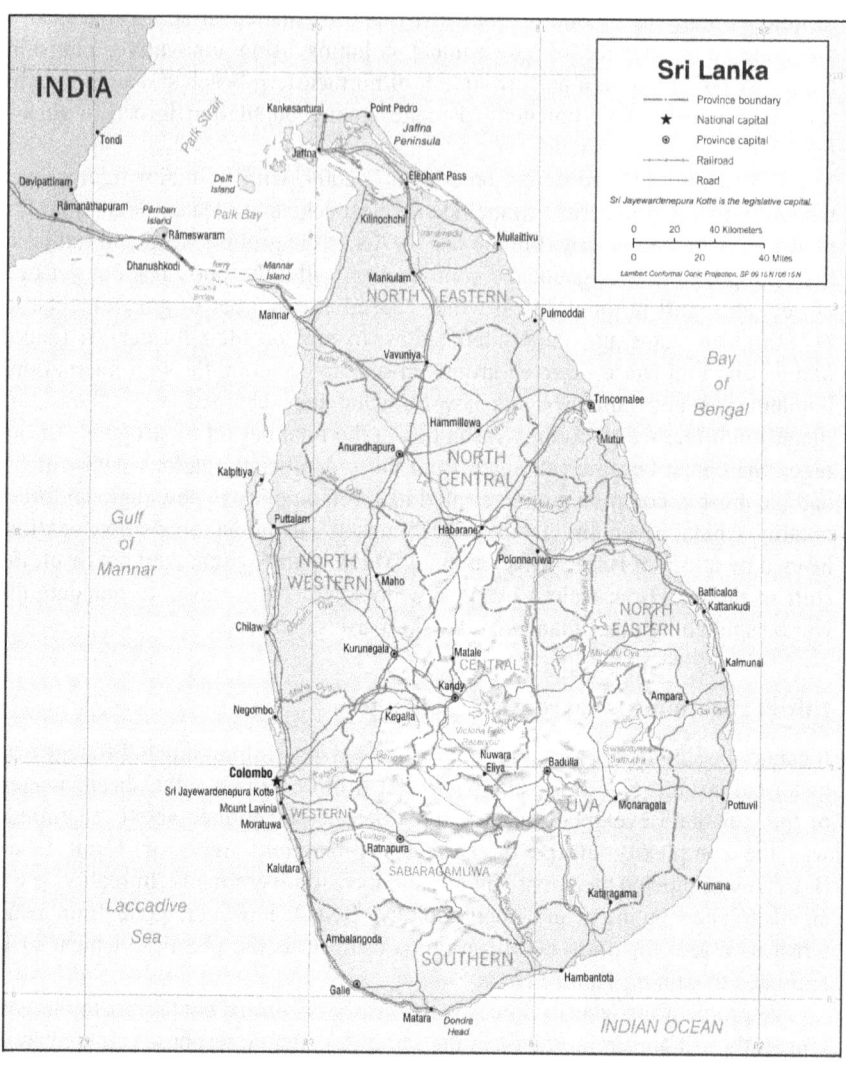

Figure 16.1 Sri Lanka political map.

principal minority group, overwhelmingly Hindu (with Christian pockets) and Tamil-speaking, 12.71 percent (1,886,872; hereafter referred to simply as "Tamils"), had maneuvered within the British imperial structure to achieve a position of relative advantage in commerce and the professions. This was much resented by the majority, which increasingly resorted to inequitable measures to improve its standing (e.g., making Sinhala the language of the civil service).[4]

Two other Tamil-speaking populations inhabited the island, the 1,046,926 "Moors" (7.05 percent, also known simply as "Muslims") and the 818,656

Indian Tamils (5.51 percent), who were the remainder of a larger migrant population recruited overwhelmingly in Tamil Nadu by the British to work on, initially, the coffee, later tea (and rubber), plantations.[5] These, though, had their own parochial issues and did not generally participate in the increasingly raw political battle between the two largest communities, a battle which saw a Tamil protest movement demanding justice – ironically, part of a larger, island-wide social movement demanding the same end – move increasingly from street action to proto-insurgency.

In the decades after achieving independence from Britain in 1948, Sri Lanka was a state remarkably unprepared to deal with even substantial overt protest action much less subversion and its challenges, whether terrorism or guerrilla action. Following the country's annexation by Britain in the three Kandyan Wars (1803–1805, 1815, and 1817–1818),[6] its martial heritage had effectively ended. In 1971, when JVP I occurred,[7] the principal armed capacity of the state was to be found in just 10,605 policemen, armed at best with the venerable .303 Lee Enfield rifle and scattered in small stations amidst a population of 12.5 million. The military was in a similar state: small (the army numbered but 6,578 in five battalions) and indifferently equipped. These forces grew but little in the decades that followed, even as the population reached first 15 million and later roughly 18 million.

Political efforts to raise the position of the majority that was Sinhala-speaking Buddhists increasingly clashed with the efforts of the Tamil minority to retain their position. Particularly resented by the Tamils were government efforts, carried out with international assistance, to open up unused lands in the north and east, through irrigation and resettlement, in areas claimed as traditional homelands by Tamils (this despite the reality that no less than one-third of all Tamils lived amidst the majority).[8] Small groups of radical Tamil youth formed, both at home and abroad, strongly influenced by Marxism. Their solution to their "oppression" was to call for "liberation," that is, the formation of a separate socialist or Marxist Tamil state, *Tamil Eelam*. Numbers were in the aggregate perhaps 200. Later, as primary group dynamics gave way to secondary group considerations, leaders sought to ideologically indoctrinate youthful manpower. This proved problematic as both socio-economic-political grievances and the desire for revenge (in response to instances of state violence) lent themselves more readily to an embrace of communalism as opposed to confusing Marxist-Leninist ideological assessment.[9]

Neither, at this point in events, were the Tamil people, whatever their plight, much interested in giving their support to would-be revolutionaries. Whatever its flaws, Sri Lanka remained a functioning democracy. Absent a mass base, the insurgents could do little more than to plan future terrorist actions. Police and intelligence documents speak of small, isolated groups of a half-dozen or so would-be liberationists meeting in forest gatherings in Sri Lanka to plot their moves. Actions that occurred, bombings and small-scale attacks upon government supporters and police positions, were irritating (albeit horrific) but dismissed as the logical consequence of radicalism. There was method to the upstart schemes, though. By 1975, contacts had been made with the Palestine Liberation Organisation [*sic*] through PLO representatives in London. Shortly thereafter, Tamils began to train in the Middle East. At home, LTTE initiated its armed struggle with a reported April 7, 1978 ambush, when four members of a police party were killed and their weapons captured. This was followed by hit-and-run attacks which led to the banning of the "Liberation Tigers" on May 19, 1978 by Parliament.

Though the police bore the brunt of LTTE activities, the army was also committed early on. This was carried out through the normal, legal procedures of representative government. The burden for implementation of very correct dictates and prohibitions, modeled in the event after those of the former British colonial power, fell upon a post-colonial security apparatus which proved inadequate to the task. By July 11, 1979, the government claimed 14 policemen had been killed by LTTE. Hence a state of emergency was declared on that date in Jaffna and at the two airports located in the Colombo vicinity. It was soon extended to the entire country and remained in force for 28 years (renewed at monthly intervals).[10] A week later (July 20, 1979), Parliament passed a "Prevention of Terrorism Act," which, though modeled after British legislation, contained a number of controversial provisions, such as the authority to detain for 18 months (in 6, renewable 3-month increments) without trial those suspected of activities connected with "terrorism." In the same context (i.e., terrorism), murder, kidnapping, and abduction were made punishable by life imprisonment. Members of the security forces acting within the scope of the Act were granted blanket immunity.[11]

Nevertheless, the situation continued to deteriorate. In Jaffna, Charles Anton, LTTE "military wing" commander, had been killed in a firefight with Sri Lankan military personnel on July 15, 1983. In retaliation, on July 23, an ambush executed by an LTTE element left 13 soldiers dead. Their subsequent funeral in Colombo ignited widespread rioting and looting directed against Tamils, at least part of which was planned and led by elements of the political spectrum. At least 400 persons were killed and 100,000 left homeless; another 200,000–250,000 fled to India. Police stood by, and in many cases members of the security forces participated in the violence. The episode proved a critical turning point in the conflict.[12]

This communal spasm served to traumatize the Tamil community and provided LTTE with an influx of new manpower. Thus the ascendancy of radical leadership in the struggle for *Tamil Eelam* was complete.[13] Though there may at one point have been more than three dozen different groups active, they were dominated by just five: LTTE; People's Liberation Organisation of Thamileelam (PLOT, also frequently rendered as PLOTE in the Western press, a variance caused by use of *Tamil Eelam* rather than *Thamileelam* as adopted by the group itself for its formal communications); Tamil Eelam Liberation Organisation (TELO); Eelam People's Revolutionary Liberation Front (EPRLF); and Eelam Revolutionary Organisation (EROS). LTTE emerged as the dominant force by ruthless application of terror against its rivals.[14] For funding, criminality (to include apparent involvement in the drug trade) was quickly surpassed by donations (both actual and forced), increasingly from Tamil Nadu (both private and public sources) but ultimately in the main from the Tamil diaspora.[15] Materially, groups also looked to India.

Clearly a product of Sri Lankan internal contradictions, the groups nevertheless existed within the larger strategic realities of the Cold War. Sri Lanka was, at least under the United National Party (UNP) administration so central to events described here, a Western-oriented democracy with a market economy. In contrast, neighboring India was closely linked to the Soviet Union, was a democracy with a socialist economic approach, and had a geo-strategic view that called for domination of its smaller South Asian neighbors. Apparently to gain information on developments concerning the Sri Lankan port of Trincomalee, which New Delhi feared was coveted by the West as a base,[16] Indian Prime Minister Indira Gandhi agreed, in 1982, to a plan by the Research and Analysis Wing (RAW – India's equivalent of the CIA) to establish links with a number of the Tamil terrorist organizations above. India was not interested in the ideology of those who received its training. It sought to safeguard its regional position while calming aroused pro-Tamil communal passions within its own borders.[17]

Consequently, an extensive network of bases in Tamil Nadu, the 55 million Tamil-majority Indian state directly across the narrow Palk Strait from Sri Lanka, was allowed to support the clandestine counter-state that was formed within Sri Lanka itself.[18] This allowed dramatic expansion of insurgent actions, and by the end of 1984, insurgent activity had grown to the point that it threatened government control of Tamil majority areas in northern Sri Lanka. The security forces had increased in size and quality of weaponry, but a national concept of operations was lacking. The result was a steadily deteriorating situation and hundreds of dead; mostly civilians killed through terroristic acts.

The extent to which insurgent capabilities had developed was amply demonstrated in a well-coordinated and executed attack on November 20, 1984 in which a Tamil force of company size used overwhelming firepower and explosives to demolish the Chavakachcheri police station on the Jaffna peninsula (east of Jaffna city) and kill at least 27 policemen defending it. There followed continued ambushes of security forces, as well as several large massacres of Sinhalese civilians living in areas deemed by the insurgents to be "traditional Tamil homelands." Use of automatic weapons, mortars, and RPG-7 rocket launchers was reported.

Even as these developments took place, it became clear to the authorities that a drastic upgrading of security force capabilities was needed, a task which was accomplished in remarkably short order. Oxford-educated Lalith Athulathmudalai, a possible successor to President Junius R. Jayewardene, was named head of a newly created (March 1984) Ministry of National Security, as well as Deputy Defense Minister (Jayewardene himself was Defense Minister). This effectively placed control of the armed services and counterinsurgency operations under one man. Inter-service coordination improved under a Joint Operations Center (JOC; formed February 11, 1985), commanded by a recalled veteran of the 1971 JVP I conflict and a former commander in Jaffna, Cyril Ranatunga (who was promoted from brigadier to lieutenant general).[19] New manpower, formations, and equipment resulted in better discipline and force disposition. To relieve pressure on the military, a new police field unit, Special Task Force (STF), was raised under the tutelage of ex-SAS (Special Air Services) personnel employed by KMS Ltd. STF took over primary responsibility for security in the Eastern Province in late 1984, freeing the army to concentrate on areas of the Northern Province (which included Jaffna).[20] The army itself stood up new special forces and commando units.

Nevertheless, the situation continued to worsen. Terrorism not only was destructive in its own terms but prompted further communal strife. Attacks upon Muslims in April 1985, for instance, sparked Muslim-Tamil riots and significant population displacement. Bombs were discovered in the capital even as attacks hit trains, buses, and other modes of transportation. On April 29, a parcel bomb damaged several buildings in the army headquarters complex in Colombo. On May 14, though, occurred an outrage that caused others to pale. Disguised as security force personnel, LTTE combatants on May 14 used a bus to enter the confines of one of Sri Lanka's most sacred shrines, the *Sri Maha Bodhi*, a *bo* tree grown from a sapling under which Buddha attained enlightenment. Attacking the worshipers indiscriminately, LTTE ultimately murdered some 180 pilgrims. All too predictably, there followed communal riots. In the field, a quickening tempo of guerrilla attacks displayed rapidly improving insurgent numbers and capabilities.

India's covert role has already been detailed. In July–August 1985, it endeavored to be more constructive by hosting peace talks between all major *Eelam* groups, to include the non-insurgent Tamil United Liberation Front (TULF), and representatives of the Sri Lanka state in Thimpu, the capital of Bhutan. Already, at this point, it was clear that LTTE was the most intransigent of the groups, and eventually its leadership had to be coerced by New Delhi to continue the discussions. Though various principles were agreed upon, LTTE's intent was to escape what had become virtual imprisonment in Thimpu, hence to return to its chosen course of action, armed struggle. This it did.[21] As 1986 commenced, attacks, massacres, and bombings occurred in what seemed an endless succession. On May 3, 1986, an Air Lanka flight from London to Colombo, continuing to the Maldives, was delayed in its Colombo departure long enough for what was intended to be a mid-flight bomb to explode while the plane was still on the ground; 21 passengers died, 41 were injured.

At this point, despite the substantial steps that had been taken, the situation was clearly out of control. Tactically, the changes made had been reasonably effective; government response was crippled, though, by the state's inability to set forth a viable political solution within which stability operations could proceed. Focusing on "terrorism" rather than insurgency that used terror as but one weapon, Colombo ordered its military leaders to go after the armed threat groups and to stamp out the violence. There was little movement toward political accommodation that would have isolated the insurgent hardcore from the bulk of the movement, the followers.

At heart, this stemmed from an unresolved debate as to just what independent Sri Lanka was: multi-ethnic nation-state or last bastion of a religion which at one point had dominated much of South Asia, Buddhism. Tamils and other Sri Lankan minorities, whatever their precise historical role on the island, could only participate as equals in the first society, while the second, though by no means the dominant choice amongst socio-economic-political elites, many of whom were trilingual and had schooled together in elite institutions (with English the link-language throughout the island), became increasingly more salient as Tamil response to state violence took on many of the chauvinistic characteristics it purported to be struggling against. This was particularly the case as concerned LTTE, which even during its flirtation with Marxist-Leninism was dominated by the chauvinism if not racism of Prabhakaran.

The result, therefore, was that the struggle, which was framed in the language of counterinsurgency and counterterrorism, was more accurately a clash of contending nationalisms,[22] with an increasingly beleaguered element of the national elite seeking to champion the fluid boundaries which saw communities mix and intermarry. The situation was further complicated by the reality that, while language and community were at the core of each national conception, the Sinhalese essence was defined by a Buddhism that was also a central element about which resistance to colonialism had rallied. In contrast, Tamils had not only generally embraced the opportunities afforded by colonialism but were divided into the two communities discussed earlier, the indigenous Sri Lanka Tamils (further differentiated by region) and the Indian Tamils. Ideologically, whereas the Sinhalese increasingly used political Buddhism as a tool for mobilization,[23] the *Eelam* movement was informed either by the secular ideology of Marxism[24] or by the raw emotions of communalism, but a communalism that rejected the traditional structures of Tamil society (whether pertaining to caste or gender).[25]

Seeking a way forward, President Jayewardene, an experienced politician (he would lead the country from 1977 to 1989) who was in a sense a representative from an earlier era (he had been born in 1906), increasingly used his immediate family and a small circle of trusted associates in an effort to determine how to proceed and who within the military leadership could best deal with the fluid situation. A strategic plan requested in mid-1986 highlighted the imperative that, used correctly, military action served to implement a political solution through elimination of grievances, area domination, increased international support, and astute diplomacy with respect to India – as opposed to the reactive posture that dominated. The actual mechanics of implementation, particularly the tangible steps necessary to restore government writ to all areas of the country, were quite straightforward and adhered closely to what was finally done successfully in 2006–2009. In the event, political realities dictated that the military facets of response continued to dominate, and the politically necessary steps were not taken.[26]

Transformation of threat

Though it did not put together the necessary *national* campaign plan, Colombo did come up with an approach for the *military* domination of insurgent-affected areas. Pacification in the east and near-north left only Jaffna as an insurgent stronghold as 1987 began. As their position in Jaffna peninsula collapsed, the Tigers became more fanatical. They adopted the suicide tactics normally associated with violent radical Islamist movements. Individual combatants were issued with cyanide capsules to use rather than be captured. Suicide attacks began, both using individuals and vehicles. A "Black Tigers" suicide commando was formed to carry these out. There remains considerable debate as to precise inspiration for this shift, but the result was not in doubt – the character of LTTE's terror and its level of violence became much more lethal.[27] Surprisingly, it was not these tactics that rescued LTTE but democratic India.

When, in May 1987, Sri Lankan forces launched *Operation Liberation* and appeared on the verge of delivering a knockout blow,[28] New Delhi, responding to domestic pressure, entered the conflict directly as an Indian Peacekeeping Force (IPKF), thus bringing to a conclusion what is termed Eelam I. Sri Lankan forces returned to barracks, and India assumed responsibility for overseeing implementation of a to-be-agreed-upon cessation of hostilities. After an initial honeymoon period, during which all *Eelam* groups save LTTE chose to align themselves with Indian expectations, hostilities commenced between IPKF and LTTE.[29] The Indian presence, while having some tactical advantages, was strategically disastrous. It not only reinforced the nationalist aspects of the *Eelam* appeal amongst the Tamil mass base but also provoked a Sinhalese nationalist reaction in the south that absorbed virtually all of security force attention.[30]

As the Indians attempted to deal with the Tamil insurgents, the Sri Lankans were forced to move troops south to deal with Sinhalese Maoist insurgents of the JVP, a body that had led a 1971 insurgency (JVP I, as termed earlier) that was crushed at the cost of some thousands of dead and at least 16,500 youth detained. In this, its second effort to seize state power (i.e., JVP II), the JVP insurgents gained influence far beyond their numbers by exploiting nationalist passions and using terrorism to murder those who did not comply with their demands. The industrial sector, thoroughly cowed by a spate of well-selected assassinations, was functioning at a mere 20 percent capacity. Such economic paralysis, in turn, fed the JVP cause. Sri Lanka staggered.

A change in leadership, with Ranasinghe Premadasa replacing the retiring President Jayewardene, brought a government approach that again turned the tide. Crucial to this was the employment of the very techniques that had gradually come to be standard in dealing with the Tamil insurgency. Particularly salient was the command and control structure that had evolved. A prerequisite for everything was the continuing evolution of the military, especially the army. Having become a more effective, powerful organization, its 76 battalions were now deployed to areas where, among other things, it spoke the language of the inhabitants and had an excellent intelligence apparatus.

Superimposed upon the tactical organization of the army was the counterinsurgency structure itself. Administratively, Sri Lanka's nine provinces were already divided into districts, 22 in all, each headed by a Government Agent (GA) who saw to it that services and programs were carried out. To deal with the insurgency, these GAs were paired with Coordinating Officers (COs), whose responsibility it became to handle the security effort in the district. Often, to simplify the chain of command, the CO would be the commander of the battalion in the district.

Only as the conflict progressed did the army place its battalions under permanent, numbered brigades – though these remained continually changing in composition – and its brigades under divisions. In theory, there was a brigade for each of Sri Lanka's nine provinces. These were grouped under three division headquarters, only two of which were operational at this point in time, because the third was designated for the area under Indian occupation. The brigade commanders acted as Chief Coordinating Officers (CCO) for their provinces and reported to Area Commanders (the divisions). Areas 1 and 2 divided the Sinhalese heartland into southern and northern sectors, respectively; Area 3 was the Tamil-populated zone under IPKF control.

Used historically with considerably effect by any number of security forces, particularly the British, this system had the advantage of setting in place security personnel whose mission was to win back their areas. They could be assigned assets, military and civil, as circumstances dictated. COs controlled all security forces deployed in their districts; they were to work closely with the GAs to develop plans for the protection of normal civilian administrative and area development functions. For this work, they were aided by a permanent staff whose job it was to know intimately the area. In particular, intelligence assets remained assigned to the CO headquarters and guided the employment of operational personnel. They did not constantly rotate as combat units came and went.

The framework culminated in the JOC. This, though, never really hit its stride as a coordinating body. Instead, manned by senior serving officers, it usurped actual command functions to such an extent that it *became* the military. The service headquarters, in particular the army, were reduced to little more than administrative centers.

Though often lacking precise guidance from above, local military authorities nonetheless fashioned increasingly effective responses to JVP insurgency. This was possible, because the CO and operational commanders, older and wiser after their tours in the Tamil areas, proved more than capable of planning their own local campaigns. Decentralization, in a state lacking communications and oversight capabilities, led some individuals and units to dispense with the tedious business of legal process. Those suspected of subversion were simply eliminated. Under the combined authorized and unauthorized onslaught, the JVP collapsed.[31]

In ending the second Sinhalese insurgency, the security forces were able to return their attention again to the Tamil campaign when India withdrew in January–March 1990 (after almost three years and casualties of 1,155 dead and 2,984 wounded). New Delhi's involvement remains highly controversial to date, with considerable disagreement concerning the achievements of its counterinsurgency effort.[32] Ultimately, relations between India and Sri Lanka were so strained that the latter was apparently actually assisting the various Tamil insurgent groups in their resistance. At this point, the Indians knew it was time to leave. Ominously, it was a greatly strengthened LTTE that awaited Colombo in Eelam II.

LTTE power grew during a round of post-IPKF negotiations, which the Tigers used to eliminate their Tamil insurgent rivals. The talks collapsed when LTTE demanded that police stations in Eastern Province be vacated, then massacred more than 300 policemen who had been ordered to accept guarantees of safety.[33] There followed widespread terrorism and a leap from guerrilla to mobile warfare – the insurgents attacking in massed units, often of multiple battalion strength, supported by a variety of heavy weapons. Deaths numbered in the thousands, reaching a peak in July–August 1991 in a series of set-piece battles around Jaffna. The 25 days of fighting at Elephant Pass, the land bridge that connected the Jaffna peninsula with the rest of Sri Lanka, saw the first insurgent use of self-manufactured armor, which was supported by artillery and a sealing off of the battle space by extensive concrete-reinforced siege works protected by thick anti-aircraft belts. The battalion was in danger of being overrun when one of the LTTE armored bulldozers, followed by infantry, breached the perimeter, but the assault was turned back in intense hand-to-hand fighting. Relief came overland through difficult terrain after landing on the eastern coast but not before casualties on both sides totaled several thousand.[34]

Elsewhere, terror bombings and assassinations became routine. Even national leaders, such as Rajiv Gandhi of India and Sri Lanka's President Premadasa, fell to LTTE bomb attacks (on May 21, 1991 and May 1, 1993, respectively), as did numerous other important figures, such as Lalith Athulathmudalai (April 1993) and members of the JOC upper echelons.[35] Heavy fighting in Jaffna in early 1994, as the security forces attempted to tighten their grip around Jaffna City, resulted in government casualties approaching those suffered by LTTE in the Elephant Pass action. The conflict had been reduced to a tropical replay of the First World War trenches.

At the national command level, Premadasa's position as president was assumed on May 7, 1993 by Dingiri Banda Wijetunga, who had been prime minister from March 3, 1989. He would lead the country until November 12, 1994. Ironically, in Sri Lanka's mixed system, wherein the president dominates and, if his party controls Parliament, all but names the prime minister, Wijetunga had been selected due to his "old school" grace and lack of further political ambitions. Yet he was experienced and well versed in the security situation, his preparation having included in-depth discussions in mid-August 1991 with security experts who highlighted the imperative that armed action serve to facilitate a political program beyond return to the status quo. Nevertheless, he was unable to reorient the counterinsurgency approach in the brief time he was in office.[36]

Only with the election of a coalition headed by the Sri Lanka Freedom Party (SLFP) in August 1994, followed by the November presidential victory of SLFP leader Chandrika Bandaranaike Kumaratunga, was politics again introduced into the debate on state response to insurgent challenge. The SLFP sweep ended 17 years of UNP power and led to a three-month ceasefire, during which Colombo sought to frame a solution acceptable to the warring sides. This hiatus is recognized as the end of Eelam II. The effort came to an abrupt halt when LTTE again – as it had in every previous instance – unilaterally ended the talks by surprise attack on government forces. Eelam III had begun.

Significantly, the wave of assaults highlighted the military side of the conflict. LTTE techniques included the use of underwater assets to destroy navy ships, as well as the introduction, somewhat later (April 1995) of surface-to-surface missiles, which were eventually used to destroy five aircraft.[37] In the field, LTTE regularized guerrilla formations (i.e., those fighting as regular military units albeit as light infantry) proved capable of engaging with government forces on more or less even terms. What had begun as a campaign by terrorists had grown to the force-on-force warfare augmented by terrorism as a tactic and guerrilla action.

These new circumstances dictated a review of approach by Colombo. In mid-1995, therefore, the government held a series of meetings designed to settle upon a revised national strategy for ending the conflict. On the political side, as directed by the president, a plan was articulated for devolution which came close, in all but name, to abandoning the unitary state in favor of a federal system. On the military side, Chandrika, as she was normally called, relied upon an arrangement similar to that of President Jayewardene; she was the Minister of Defense, while a trusted associate (reported as her uncle), parliamentarian Anuruddha Ratwatte, was Deputy Minister and hence effectively Minister. He had reached the rank of lieutenant colonel (LTC) while a mobilized reservist and had military experience but none at higher levels of command. This was to prove a key factor, because the strategic review rapidly became a fierce battle of opposing positions.

All participants in the debate basically agreed that LTTE would have to be dealt with militarily in order for a political solution to be implemented, but there was considerable disagreement on the plan of operations. On the one side were those who favored a *military-dominated* response. Opposed were those who favored a *counterinsurgency* campaign that would use force as the shield behind which restoration of government writ would occur. The first, in a sense, called for strike operations, the second for the classic "oil spot" approach, the systematic domination of areas which were then linked in a steadily expanding flow. Essentially, it was this approach that had emerged during the Wijetunga presidency as the default security forces posture. It was not favored by Ratwatte, who sought something more decisive, in particular the liberation of Jaffna peninsula, which had been held by LTTE for a decade.[38]

Contextually, there were grounds for favoring such an approach. With the end of the Cold War, LTTE had quietly dropped all talk of Marxism, though it continued to portray itself as socialist. Its links with the diaspora had matured, but its rupture with New Delhi was complete. For its part, India, though still closely linked to (now) Russia, had seen its Soviet patron collapse and cautiously reached out to establish more normal relations with the U.S. and other supporters of Colombo. There was no objection raised when the U.S. agreed, in mid-1994, to begin a series of direct training missions conducted by special operations elements.[39] (Washington would designate LTTE as a Foreign Terrorist Organization [FTO] on October 8, 1997.[40]) These steps facilitated the already mature Sri Lankan relations that existed with Britain and India. Further military ties and assistance were developed with Pakistan and China, to a lesser extent with Israel (always a controversial proposition in Sri Lankan politics due to the views of the Sri Lankan Muslim population and the large number of expatriate workers employed in the Middle East). The upshot was that the military seemed in relatively good shape internally, with strong external linkages to provide a steady stream of assistance and support.

Operation Riviresa ("Rays of Sunlight") was consequently launched in October 1995 to retake Jaffna, a goal it accomplished by December 2. Strong leadership was able to overcome an array of personnel and operational difficulties, but the victory left the occupying forces in a perilous position, cut off by the extensive territory to the south and east that remained in LTTE hands. It had been a conventional response to an unconventional problem, executed successfully but a "bridge too far," leaving multiple brigades stranded in Jaffna, where they could be supplied only by sea or air. LTTE adroitly used a combination of main-force and guerrilla units, together with special operations, to isolate exposed government units and then overrun them. These included headquarters elements, with even brigade and division headquarters being battered. In the rear area, LTTE used a suicide truck bomb to decimate the financial heart of Colombo in February 1996, killing at least 75 and wounding more than 1,500.

A pressing need for further force development led Colombo to approach the American firm, Military Professional Resources Incorporated (MPRI), which had impressed with its apparent success in overseeing the modernization and training of the Croatians as on display in the successful Croatian summer 1995 offensive (Operation Storm) against Serb-supported forces. An assistance plan was developed with U.S. acquiescence, but this went no further than the proposal stage when a reaction set in amongst the Sri Lankan officer corps.[41]

Much worse was to come, as overextension of forces and an inability to handle the complexities of conventional operations left the Sri Lankan military badly deployed. Debacle was not long in coming, and on July 17, 1996, an estimated 3,000–4,000 LTTE combatants isolated and then overwhelmed an under-strength brigade camp at Mullaitivu in the northeast, inflicting at least 1,520 dead total on the security forces. This exceeded the 1,454 total death toll for 1994 and shattered army morale. Desertion, already a problem, rapidly escalated, even as the isolation of Jaffna tightened. The linkup effort, *Operation Jayasikurui* ("Certain Victory") kicked off in May 1997 but quickly slowed to a crawl, as time and again LTTE demonstrated the ability to use combinations of regular and irregular action to inflict defeat in detail upon government forces. There followed stalemate.

LTTE, having only to exist as a rump counter-state that mobilized its young for combat, had demonstrated the ability to construct mechanisms for human and fiscal resource generation that defied the coercive capacity of the state. Linkages extended abroad, from whence virtually all funding came (some US$20–30 million per year); and diasporic commercial linkages allowed the obtaining of necessary weapons, ammunition, and supplies. Though the security forces could hold key positions and even dominate much of the east, they simply could not advance upon the well-prepared and fortified LTTE positions in the north and north east, which, in any case, were guarded by a veritable carpet of mines.

Political disillusionment again followed and increased as LTTE continued to pull off spectacular actions: the most sacred Buddhist shrine in the country, the Temple of the Tooth in Kandy, was attacked by suicide bomb in 1998; Chandrika herself narrowly missed following Premadasa as a presidential victim, surviving a 1999 LTTE bomb attack but losing an eye; the Elephant Pass camp that had previously held out against such odds fell in 2000; and in July 2001, a sapper attack on the international airport in Colombo left 11 aircraft destroyed. Ratwatte, who in the flush of victory occasioned by the recapture of Jaffna had been made a full general by Chandrika, was no longer in his position at that point, having been replaced in 1999.[42]

It was not altogether surprising that in the December 2001 parliamentary elections, the UNP, led by Ranil Wickremasinghe, was returned to power. This left the political landscape badly fractured between the majority UNP and its leader, the prime minister, and the SLFP's Chandrika, still the powerful president in Sri Lanka's hybrid system similar to that of France. That the two figures were longtime rivals, with considerable personal animosity, did not ease the situation.

Again, it was changes in the international arena that dealt a wild-card. The increased concern of the world with terrorism, already a factor in the new millennium but central after the "9/11" terrorist attacks in the U.S., caused additional Western countries to proscribe LTTE and move against its fundraising activities on their soil. LTTE was already banned in the United States and India (where it had been proscribed in 1992) when Britain announced its listing on February 28, 2001. This was an important step, since the Tamil diaspora in the United Kingdom ranked second in size only to that of Malaysia.[43] Ottawa finally moved on April 14, 2006,[44] and the next month, the entire European Union followed with its own common position of proscription.

Amidst these events, shortly after "9/11," for reasons that remain unclear, LTTE in February 2002 suddenly offered to negotiate with the new UNP government. Colombo accepted the offer, and an uneasy truce commenced. The cessation of hostilities was a very mixed bag.[45] LTTE used the restrictions on Sri Lankan security forces to move aggressively into Tamil areas from which it had hitherto been denied and to eliminate rival Tamil politicians. Throughout Tamil-populated areas, Tamil-language psychological operations continued to denounce the state. When, in October 2003, LTTE issued a proposal, *Interim Self-Governing Authority* (ISGA), which would have pushed beyond de facto realities to make it the *de jure* power in the Northern and Eastern Provinces, there was strong reaction in the increasingly restive, Sinhalese majority southern heartland of the island.[46] Chandrika watched uneasily and then asserted her power, in early November 2003, while Wickremasinghe was in Washington meeting with U.S. President George W. Bush. Claiming that the UNP approach was threatening the "the sovereignty of the state of Sri Lanka, its territorial integrity, and the security of the nation," she ousted the three UNP cabinet ministers most closely associated with the talks, dismissed Parliament, and ordered the army into Colombo's streets.

LTTE waited, but in the April 2004 parliamentary elections that were held as a consequence of talks between the dueling Sinhalese parties, SLFP unexpectedly swept back into power at the head of a United People's Freedom Alliance (UPFA). The Tigers withdrew from negotiations but did not renew active hostilities, for they were preoccupied with what had once seemed unthinkable, a split within the movement. Long chafing under the domination of LTTE by Northern Tamils, Eastern cadre under the leadership of longtime LTTE cadre Vinayagamoorthy Muralitharan (more commonly known as Colonel Karuna Amman) had revolted in March. Though they were crushed in intense fighting followed by a wholesale vetting and purge of Eastern cadre and combatants,[47] the fracture remained permanent, with alienated Eastern Tamils, represented by Karuna's *Tamil Makkal Viduthalai Pulikal* (TMVP or Tamil Peoples Liberation Tigers), increasingly making common cause with the government. This was to prove a key development in the events that followed.

As events on the ground served to strain the cessation of hostilities, contingency entered the picture with the devastating December 26, 2004 Indian Ocean tsunami, which left more than 35,000 dead. Tamil areas were hit particularly hard. International aid poured in, but the issue of how it was to be distributed stripped the last fig leaf from the unstated understandings that had given LTTE its *Eelam*. When it demanded that aid be channeled through its own counter-state bureaucracy, with the original *Interim Self-Governing Authority* (ISGA) proposal taking on all the trappings of statehood, the strained ceasefire effectively collapsed.[48]

Progressive deterioration of the situation followed, though LTTE was careful not to move too aggressively. The cease-fire served as the ideal cover for the elimination of all the group saw as standing in its way. This included even the Sri Lankan foreign minister, Lakshman Kadirgamar, an ethnic Tamil, assassinated in August 2005; Sarath Ambepitiya, the judge who had sentenced Prabhakaran to 200 years in jail in absentia for the 1996 bombing of Colombo; and literally hundreds of Tamil politicians and activists (including the misidentified) opposed to LTTE. For LTTE remained committed to *Eelam*, whatever the verbiage connected with the peace process, and behaved as such. In his annual November 27, speech, delivered on LTTE Heroes Day, "President and Prime Minister of Eelam" (as he was billed by Tamil media) Prabhakaran warned that LTTE intended to renew hostilities if the government made no tangible moves toward "peace."[49]

In what was seen at the time as but a tactical error, yet one that ultimately proved fatal, LTTE ordered a boycott of a presidential election hastily held in November 2005 after a Supreme Court decision ruled that Chandrika's presidential term had run its course. Hard-line SLFP Prime Minister Mahinda

Rajapaksa eked out a narrow victory against the UNP's Ranil Wickremasinghe on some 73 percent turnout. Wickremasinghe would by all historical accounts have been the better option for LTTE's plans, as its continued ceasefire violations dwarfed in number those of the government and steadily steeled the new Rajapaksa administration for what was to come. A number of prominent LTTE suicide attacks, which included an effort to kill the army head, LTG Sarath Fonseka, and a successful targeting of the army number three, pushed the situation beyond redemption. Last gasp efforts by Norway, the lead facilitator of the attempted settlement, came to naught.[50] As fighting became more general, suicide attacks hit even targets in the deep south, such as Galle. By August 2006, Sri Lanka was again at war, Eelam IV.

Transformation of response

What followed, as has been noted in the introductory section of this chapter, was unlike what had gone before. Often touted as a victory for "counterinsurgency," the crushing of LTTE could be deemed such only in the sense that any civil war must by definition begin with an act of rebellion. What ended LTTE's three decades of struggle was an operational clash of arms not unlike the American Civil War in ferocity albeit quite distinct in its tactics and parochial features. Operationally, what occurred was a signal illustration of military adaptation executed in concert with national mobilization even as LTTE proved unable to do likewise. Examined more strategically and theoretically, the vanquishing of LTTE as an illicit power structure and the post-war conflict it unleashed serve to illustrate the profound changes that globalization has brought about in everything from the manner in which insurgency is waged to what is permissible in response.

Precisely what occurred operationally is easier to describe than to explain, for even at nearly a decade's remove, considerable disagreement continues as to just who initiated key aspects of the reform and was responsible for a series of astute combat decisions. The basis for renewed combat obviously lay in national mobilization, accomplished in the first instance as a consequence of a veritable declaration of Buddhist holy war by the powerful *Sangha*; in the second instance, by the Rajapaksa government, with Mahinda as president and his army veteran brother, Gotabhaya, as Defence Secretary, marshaling the financial support and determination necessary to rearm and re-equip an expanded military.[51]

In the field, General (following promotion) Fonseka insisted upon a free hand that allowed him to field the finest galaxy of combat leadership that had surfaced until that time. Innovative deployment of entire battalions as squad and even fire-team (i.e., "brick") sized units schooled in light infantry (i.e., commando) tactics and able to call in supporting fires served to dramatically multiply the defensive demands for an LTTE now reduced to struggling to defend its pseudo nation-state. Its governance, though innovative in some respects, had remained grounded in coercion, which made problematic the willingness of the populace to mobilize in defense of *Eelam*. Indeed, one of the most contradictory aspects of the entire conflict was that throughout, best evidence demonstrated, a substantial proportion of Tamils, as well as nearly the entire Indian Tamil and Tamil-speaking Muslim populations, remained within government-controlled areas.[52]

First steps to seal off the battle-space and strangle LTTE's supply lines came with a successful high seas campaign that hunted down and destroyed LTTE's ocean-going merchant navy. Simultaneously, the development of coastal high speed craft and tactics succeeded in neutralizing LTTE's hitherto formidable swarm of maritime suicide craft. The air force, though faced even in the final phase of the struggle with LTTE suicide efforts to attack Colombo, used overhead imagery and ground patrol coordination of targeting to eliminate the threat air arm.

On the ground, the progression followed was that laid out originally in the 1986 planning documents. Seizure of the Eastern Province by July 2007, with help of the defecting Eastern Tamil elements of the TMVP (perhaps a majority of the most effective combatants), allowed converging columns to draw an ever tighter noose around LTTE forces trapped in the northeast coastal area – even as the first provincial elections were held to foster legitimacy for political reincorporation.[53] Mannar District, in the west, fell by August 2008, with the forces then able to move east. Other units cleared Jaffna Peninsula and pushed south. The LTTE administrative center of Kilinochchi was abandoned and fell to the government in early January 2009. By early 2009, the remaining LTTE forces, with perhaps 30,000 civilian hostages as human shields, were trapped in the coastal area of the Nanthi Kadal Lagoon, north of Mullaitivu. There, five divisions, a force conventionally put at some 50,000, crushed them by mid-May 2009. In all aspects save the innovative tactics used by the Sri Lankan infantry,[54] the conflict had been major combat, featuring everything from heavy artillery to rocket launchers to extensive minefields and suicide attacks.[55]

This final point highlights that the end-game, coming as they did at the conclusion of three decades of ever more intense conflict that progressively brutalized all facets of Sri Lankan life,[56] most resembled the island battles of the Second World War's Pacific Theater with the very real emphasis that, like Okinawa, Sri Lanka was heavily populated. It was this reality that increasingly galvanized human rights advocacy groups, which became shriller as the climax rose to a crescendo. When Colombo refused to heed calls from advocacy groups and certain Western governments, among them the United States and Britain, to allow some form of humanitarian intervention, advocacy gave way to outright opposition and siding with the defeated insurgents. This posture has continued in many respects to the present, though the change of power in Colombo has altered to a large extent the shrill tone of state critique.

It is in this sense that the Sri Lankan case transcends the mere facts on the ground. The tangible conflict, horrific as it was, nevertheless was fought by a democracy that maintained throughout the rule of law (albeit with very sharp elbows). That major combat places the rule of law under severe strain is a reality that should be readily recognized, particularly given the trajectory of American warfighting since Sherman's "March to the Sea" during the Civil War. There appear to be no credible sources that claim Sherman gratuitously inflicted harm upon the innocent; but there are few sources that do not recognize his intense determination to embrace the very horror of war for the purpose of bringing it to a conclusion, a stance that delivered victory to democratic order, however flawed it might be. This was the position in which Sri Lanka found itself. The war simply had to end if the country was to survive.

Lessons in an era of "New War"

If we endeavor to use the near-decade since the end of the ground war to draw lessons from this most complex case of a counter-state that challenged a state-that-erred – and in doing so exposed its ample flaws – some thoughts may be offered.

On the one hand, LTTE was an insurgency which struggled to transcend its origins as a traditional rebellion in order to leverage the new possibilities inherent to a post-Cold War world. This it did, both physically and virtually, integrally linking its struggle to regional and global Tamil communities – the Tamils of southern India and the Sri Lankan Tamil global diaspora, respectively – in such manner that it could retain the strategic advantage long enough to achieve its goal of *Eelam*. In the process, it became near-legendary for its melding of commitment to destruction with its imagery of a new world emerging. With its suicide bombers and the cyanide capsules worn by its combatants, many of both being women,[57] it set the Sri Lankan state back on its heels time and again even as the dictatorial *Eelam* world it created was hailed for ostensibly giving a people dignity and freedom, driving off not only the communal Sinhalese oppressors but in the process shattering the Tamil bonds of caste and gender inequity.[58]

On the other hand, the conflict waged by the state, which indeed began as ineffective counterinsurgency and gradually grew to equally ineffective civil warfighting, put on display something much more. At each stage in the conflict, Sri Lanka struggled to comprehend just what it was involved in and came up short. Initially, it treated proto-insurgency as emerging terrorism, emphasizing kinetic response when it should have been addressing roots of conflict. Later, having mastered counterinsurgency's martial facets, it neglected the necessity of holistic response, resulting in Indian intervention. In the post-Indian context, the emergence of compound war was mistaken for conventional conflict, resulting in devastating defeats and the emergence of LTTE's temporary victory. Finally, in the renewed 2006–2009 fighting, a new civil-military team engaged in the functional equivalent of national mobilization and delivered a virtuoso display of integrating strategic, operational, and tactical levels of combat to come off the canvas and deliver a knockout punch.

LTTE's end, when it came, had all the characteristics of the Second World War's denouement in Berlin or the ashes of Japan's incinerated cities. Ecstatic at its triumph, Colombo simply could not comprehend that it had again missed the larger picture, the fundamental shift to an age of "new war" which saw an orientation and a structure of law devoted to enforcing a sanitized version of combat which was designed, both practically and ideologically, to make impossible the total war of past eras. Sri Lankan warfighting adaptation had been almost completely in the application of kinetic power, without reforms in human rights and legal components necessary to engage in combat within a global fishbowl. It was quite ignorant of (certainly unprepared for) the corresponding growth of new global norms, notably the responsibility to protect (R2P) and the right to intervene, together with the accompanying demands of what has been termed "the liberal peace." Indeed, it would be difficult to understate the mounting intensity with which both state and non-state actors sought to slow if not end Colombo's final push to LTTE's elimination, and hence the growing sense of betrayal that ultimately erupted in Colombo.[59]

In the events outlined above, a pathway led from the traditional world of war to what has been called "new war" or "postmodern war" or "post-Heroic war or even (though from a different theoretical angle) "4th Generation Warfare" (4GW).[60] Regardless of terminology, the heart of the strategic matter was that in the global arena, use of force was required to be legitimate, discriminating, and secondary to more compelling concerns (e.g., human security). If such in one sense seemed to describe the operational dictates of counterinsurgency, the fundamental distinction lay in the fact that the latter balanced kinetic and nonkinetic facets as a matter necessary to successful mobilization to the extent necessary for victory, while the former saw the use of kinetics as itself a symptom of a larger failure.[61] To use force to resolve the issue at hand – in this case a drive for separatism – was to forfeit legitimacy. To add to this the bloodshed and destruction inherent to total war was to cross into criminality, which is precisely what very vocal and active voices asserted in demanding legal measures be taken against the victors following the May 2009 elimination of *Eelam*.[62]

Indeed, if any one characteristic may be assessed as a linchpin for postmodern war, it is in the supremacy of framing and narrative over the normal imperatives of war. Colombo's frame of victory found itself all but overwhelmed by a shrill counter-frame of repression, and Colombo's narrative trumpeting a triumph over terrorism all but swamped by a rival narrative of communal repression and barbarism. Warfare as traditionally waged found itself struggling to deal with *lawfare*.[63] Given the astonishing level of brutality and suffering that Sri Lanka had endured for three decades, its wounded pushback was quite comprehensible. Matters were not eased by what can only be described as the profound hubris, if not chauvinism, of both state and non-state critics.

Again, we return to the value of the Sri Lankan case for an examination of conflict in the second decade of the twenty-first century. In a sense, a globalized world has empowered *netwar*[64] at the geostrategic-legal level of international relations to such an extent that it all but compels the waging of conflict in the intangible rather than tangible dimension. "Facts *on* the ground" count for far less than "facts *in* the mind," irrespective of whether "facts" are true or later proved to be false. Seeing is no longer believing; believing has become seeing, with disabling pressure from a networked world directed against the party judged to be "in the wrong," that is, the party judged to have forfeited legitimacy.

If one imagines the Chiapas conflict, which inspired the emergence of the *netwar* concept, ending not in retreat by the Mexican state but in elimination of the Zapatista challenge according to the normal procedures witnessed throughout Mexican national history (and perhaps of most states), one would be at a position not unlike that occupied post-war by Sri Lanka. Its end-state of an indivisible *Lanka*, the land of the Buddha, has been secured through achieving the objective of threat destruction, but its ways (as executed by means that included not only material but psychological national mobilization) have been found wanting. Communal chauvinism, went the critique, provided the fuel that allowed an overhauled war machine to "win," and democracy itself, together with justice, was collateral damage.[65] In such context, the reality of a counter-state that had done as much as any in the post-Second World War era to earn the label "evil" was rendered irrelevant.

This, too, may be seen as emblematic of the new age of war. Ultimately, the conflict had become one of dueling narratives as to the fundamental merits or demerits of Sri Lanka's democratic, market polity. In such a battle, the increasingly problematic and despicable nature of LTTE's decision making and actions were pushed aside, as though the very intensity of transgression revealed a great deal as to Colombo's structural and moral inadequacy and rather less concerning LTTE perverse agency. It is in examining this process that lessons both emerge and startle.

Notes

1 This article builds upon earlier work, most recently a chapter, co-authored with T. P. S. ("Tippy") Brar, in Michelle Hughes and Michael Miklaucic, eds., *Impunity: Countering Illicit Power in War and Transition* (Washington, DC: NDU Press, 2016), 217–240.

2 This estimate, if accurate for the *Eelam* conflict as a whole, is surely off the mark when the JVP insurgencies are included. One source, in fact, has noted that various estimates put the number killed in JVP II alone at between 20,000 and 60,000, with 40,000 being the most commonly cited figure. See Tom H. J. Hill, "The Deception of Victory: The JVP in Sri Lanka and the Long-Term Dynamics of Rebel Reintegration," *International Peacekeeping* 20, no. 3 (June 2013), 357–374.

3 See Harshan Kumarasingham, *"The Jewel of the East Yet Has its Flaws" – the Deceptive Tranquility Surrounding Sri Lankan Independence*, Heidelberg Papers in South Asian and Comparative Politics/Working Paper No. 72 (Heidelberg: Department of Political Science, June 2013); available at: http://archiv.ub.uni-heidelberg.de/volltextserver/15148/1/Heidelberg%20Papers_72_Kumarasingham.pdf (accessed March 9, 2018).

4 This background is covered in depth in Thomas A. Marks, *Maoist Insurgency Since Vietnam* (London: Frank Cass, 1996), 174–252; and Tej Pratap Brar, "Sri Lanka's Civil War," paper presented at The Radcliffe Institute for Advanced Study conference, "Postcolonial Wars: Current Perspectives on the Deferred Violence of Decolonialization," October 30–31, 2008, Harvard University, Cambridge, MA.

5 In 1983, tea had regained its position as Sri Lanka's leading source of foreign exchange, slightly ahead of remittances from employment abroad. See John Richardson, *Paradise Poisoned: Learning About Conflict, Terrorism and Development From Sri Lanka's Civil Wars* (Kandy, Sri Lanka: International Centre for Ethnic Studies

[ICES], 2005), 441. On the communities mentioned, an extensive literature exists on the Indian Tamils; useful summary is Ilyas Ahmed H, "Estate Tamils of Sri Lanka – a Socio-Economic Review," *International Journal of Sociology and Anthropology* 6, no. 6 (June 2014), 184–191, which may be supplemented by the superb Valentine Daniel, *Charred Lullabies: Chapters in an Anthropology of Violence* (Princeton, NJ: Princeton University Press, 1996). For the Moors, see the excellent work by Amer Ali, "The Genesis of the Muslim Community in Ceylon (Sri Lanka): A Historical Summary," *Asian Studies* XIX (April-December 1981), 65–82; Dennis B. McGilvray and Mirak Raheem, *Muslim Perspectives on the Sri Lankan Conflict* (Washington, DC: East–West Center, 2007); and Dennis B. McGilvray, *Crucible of Conflict: Tamil and Muslim Society on the East Coast of Sri Lanka* (Durham, NC: Duke University Press, 2008). As indicated in the text, the general convention is to refer to Sri Lanka Tamils (previously termed Ceylon Tamils), our subject in this chapter, simply as "Tamils." In contrast, Indian Tamils or Estate Tamils (for they remain clustered on the plantations in the south) are invariably identified as such. Moors are now generally termed "Muslims."

6 See Geoffrey Powell, *The Kandyan Wars: The British Army in Ceylon 1803–18* (Barnsley, South Yorkshire, UK: Pen & Sword, 1973), as well as Channa Wickremesekera, *Kandy at War: Indigenous Military Resistance to European Expansion in Sri Lanka 1594–1818* (New Delhi: Manohar, 2004).

7 Good overview may be found in A. C. Alles, *Insurgency 1971*, 3rd ed. (Colombo: Mervyn Mendis, 1976).

8 Good discussion may be found in Chelvadurai Manogaran, *Ethnic Conflict and Reconciliation in Sri Lanka* (Honolulu: University of Hawaii Press, 1987). For particular discussion of the settlement issue, see Chapter 3, "Tamil Districts: Conflict Over Traditional Homelands, Colonization, and Agricultural Development," 78–114. A basic overview is Sumantra Bose, *Contested Lands: Israel-Palestine, Kashmir, Bosnia, Cyprus, and Sri Lanka* (Cambridge, MA: Harvard University Pres, 2007), 6–54.

9 It remains noteworthy that Marxist-Leninist ideology, as the prism through which societal realities (especially state violence) were interpreted by the *Eelam* leadership, is simply absent from all major treatments of the conflict. This is curious given the extent to which the various groups in their formative years, to include LTTE, embraced Marxist-Leninism for both vocabulary and analytical constructs, a conclusion I base upon fieldwork, to include interviews with group leadership figures and members (both combatants and prisoners), as well as examination of ample documentation. See e.g., the mimeographed publication by LTTE's eventual number two, Anton S. Balasingham, *On the Tamil National Question* (London: Polytechnic of the South Bank, 1978), which derives its content completely from the standard Marxist figures, especially Lenin (as did all such analyses, regardless of group); on this point, compare with the contents of Michael Löwy, "Marxists and the National Question," *New Left Review* I/96 (March–April 1976), 81–100. Just where the tension between ideologically driven leadership and grievance-produced manpower would have led for the *Eelam* movement as a whole was a hypothetical never put to the test, since LTTE, even as it established its dominance, increasingly embraced communalism. The academic result is that any treatment of the classic query – who joins, who stays, who leaves? – is a dynamic affair dependent upon point of temporal entry and consideration of the group concerned. For discussion of mobilization, a number of works may be accessed, paying due heed to the considerations just mentioned: Bryan Pfaffenberger, "Ethnic Conflict and Youth Insurgency in Sri Lanka: The Social Origins of Tamil Separatism," in Joseph V. Montville, ed., *Conflict and Peacemaking in Multiethnic Societies* (NY: Lexington, 1991), 241–257; Siri T. Hettige, "Economic Policy, Changing Opportunities for Youth, and the Ethnic Conflict in Sri Lanka," in Deborah Winslow and Michael D. Woost, eds., *Economy, Culture, and Civil War in Sri Lanka*

(Bloomington: Indiana University Press, 2004), 115–130; G. H. Peiris, "Sri Lanka: Youth Unrest and inter-group Conflict," *Faultlines* 19 (2008), 127–156; and Margaret Trawick, *Enemy Lines: Warfare, Childhood, and Play in Batticaloa* (Berkeley: University of California Press, 2007).

10 The emergency was formally lifted on August 25, 2011. See Stehphanie [*sic*] Nolan, "Sri Lanka Announces End of 28-Year State of Emergency," *The Globe and Mail*, August 25, 2011; available at: www.theglobeandmail.com/news/world/sri-lanka-announces-end-of-28-year-state-of-emergency/article595949/ (accessed March 9, 2018).

11 For full text: www.sangam.org/FACTBOOK/PTA1979.htm (accessed March 9, 2018); for further discussion, to include comparison between legislation used to address both JVP and *Eelam* upheavals, N. Manoharan, *Counterterrorism Legislation in Sri Lanka: Evaluating Efficacy* (Washington, DC: East–West Center, 2006); for insight into the struggle to cope with the rapidly evolving situation, Paul Sieghart, *Sri Lanka: A Mounting Tragedy of Errors* (London: International Commission of Jurists and Justice, 1984).

12 Though short, one of the best discussions of the 1983 riots is provided by the noted Sri Lankan scholar, Stanley J. Tambiah in his *Leveling Crowds: Ethnonationalist Conflicts and Collective Violence in South Asia* (Berkeley: University of California Press, 1996), Chapter 4: "Two Postindependence Ethnic Riots in Sri Lanka," 82–100. This is profitably read in conjunction with his earlier work, *Sri Lanka: Ethnic Fratricide and the Dismantling of Democracy* (Chicago, IL: The University of Chicago Press, 1986). My own review of the latter lauds its strengths while questioning the absence of any discussion of insurgent Marxist-Leninist ideology. This issue became moot, of course, when LTTE decimated its rivals, who were strongly committed to Marxist-Leninism even as LTTE but flirted with it; see Marks, "Book Review–Stanley J. Tambiah, *Sri Lanka: Ethnic Fratricide and the Dismantling of Democracy*" (Chicago: University of Chicago Press, 1986), *Issues and Studies* 23, no. 9 (September 1987), 135–140; for discussion of *Eelam* Marxism, see Marks, "Marxist Tamils Won't Stop at Separatism," *The Asian Wall Street Journal*, May 8, 1986, 6; reprinted as "Tamil Rebels Aim Beyond Autonomy," *The Asian Wall Street Journal Weekly*, May 26, 1986, 12; and "The Ethnic Roots of Sri Lanka's Ideological Struggle," *The Asian Wall Street Journal*, August 12, 1987, 8; abridged version under same title in *The Asian Wall Street Journal Weekly*, August 31, 1987, 12.

13 Strong treatment is Christine Sixta Rinehart, *Volatile Social Movements and the Origins of Terrorism: The Radicalization of Change* (Boulder, CO: Lexington Books, 2013), 109–137 (Chapter 4, "Ceylon Tigers: Creation and Radicalization of the Liberation Tigers of Tamil Eelam [LTTE]").

14 Still the single best treatment of LTTE prior to its destruction is M. R. Narayan Swamy, *Tigers of Lanka: From Boys to Guerrillas* (Delhi: Konark Publishers, 1994 [with two subsequent editions, 1996 and 2002]); for Swamy on the LTTE leader himself see *Inside an Elusive Mind: Prabhakaran – the First Profile of the World's Most Ruthless Guerrilla Leader* (Colombo: Vijitha Yapa Publications, 2003); for a collection of Swamy's articles on the conflict, with an excellent introduction, see *The Tiger Vanquished: LTTE's Story* (New Delhi: Sage, 2010). These may be usefully augmented by consideration of S. Murari, *The Prabhakaran Saga: The Rise and Fall of an Eelam Warrior* (New Delhi: Sage, 2012) and Raj Mehta, *Lost Victory: The Rise and Fall of LTTE Supremo, V. Prabhakaran* (New Delhi: Pentagon Press, 2009). Also excellent is the more recent Samanth Subramanian, *This Divided Island: Life, Death, and the Sri Lankan War* (NY: St. Martin's Press, 2014).

15 There is yet no single work that can be pointed to as a source for information on funding of the *Eelam* groups, later LTTE alone. Although written a considerable time after the events under discussion, reference may be made to the strong treatment by Anthony (Tony) Davis, "Tamil Tiger International," *Jane's Intelligence Review* (October 1996), 469–473. Useful for its in-depth consideration of the key support

provided by the Tamil population of Canada is Paul Kaihla, "Banker, Tiger, Soldier, Spy," *Maclean's* (August 5, 1996), 28–32.

16 Good discussion of the geostrategic position of Trincomalee and Sri Lanka itself may be found in John Clancy, *"The Most Dangerous Year of the War": Japan's Attack on the Indian Ocean, 1942* (Philadelphia, PA: Casemate, 2015).

17 I discuss this Indian effort in a number of contemporaneous works, to include: "India is the Key to Peace in Sri Lanka," *The Asian Wall Street Journal*, September 19–20, 1986, 8; reprinted under same title in *The Island* [Colombo], October 5, 1986, 8; abridged version under same title in *The Wall Street Journal*, September 22, 1986, 25; and "Peace in Sri Lanka," *Daily News* [Colombo], 3 Parts, July 6–8, 1987: I. "India Acts in its Own Interests," July 6, 6; II. "Bengali Solution: India Trained Personnel for Invasion of Sri Lanka," July 7, 8; III. "India's Political Solution Narrow and Impossible," July 8, 6; published under the same titles in *The Sri Lanka News*, July 15, 1987 (cont.), 6–7; in *The Island* as "India's Covert Involvements," June 28, 1987, 8 and 10. For this body of work, access to numerous prisoners and captured documentation was supplemented by fieldwork in Tamil Nadu, where members of all groups were quite forthcoming concerning the assistance they were receiving from New Delhi and Tamil Nadu state (which was running its own foreign policy of sorts). Rumored but known only later was that RAW's station chief in Madras, K. V. Unnikrishnan, had been compromised by the CIA and for two years reported upon Indian support to LTTE before being arrested. For details see Sandeep Unnithan, "Madras Café Brings Back Uncomfortable Memories of the CIA's Honey Trap," *India Today*, August 29, 2013; available at: http://indiatoday.intoday.in/story/madras-cafe-madras-honey-trap-john-abraham-cia-ltte-raw/1/304302.html (accessed March 9, 2018). The same matter is touched upon briefly in the memoirs of the former RAW Counter-Terrorism Division head, B. Raman, *The Kaoboys of R&AW: Down Memory Lane* (New Delhi: Lancer, 2007), 207–213.

18 For background to the role of Tamil Nadu, consult G. Palanithurai and K. Mohanasundaram, *Dynamics of Tamil Nadu Politics in Sri Lankan Ethnicity* (New Delhi: Northern Book Centre, 1993).

19 Ranatunga's memoirs are quite useful in their discussion of everything from his service in 1971 to that in Jaffna, all such postings leading ultimately to his being named as General Officer Commanding JOC; see *Adventurous Journey: From Peace to War, Insurgency to Terrorism*, two editions (Colombo: Vijitha Yapa Publications, July 2009 and February 2010).

20 For details, see Tom Marks, "Sri Lanka's Special Force: Professionalism in a Dirty War," *Soldier of Fortune* 13, no. 7 (July 1988), 32–39. For negative assessment of the KMS role, as well as that of Britain itself, see the highly skewed but useful in parts work by Phil Miller, *Britain's Dirty War Against the Tamil People – 1979–2009* (Bremen: International Human Rights Association, June 2014); available at: www.tamilnet.com/img/publish/2014/07/britains_dirty_war.pdf (accessed March 9, 2018).

21 Extensive documentation, to include specifics of all proposals and statements, may be found at "Conflict Resolution: Tamil Eelam-Sri Lanka – Thimpu Talks – July/August 1985"; available at: http://tamilnation.co/conflictresolution/tamileelam/85thimpu/thimpu00.htm (accessed June 22, 2015). Further discussion is in P. Venkateshwar Rao, "Ethnic Conflict in Sri Lanka," *Asian Survey* XXVIII, no. 4 (April 1988), 419–436.

22 For discussion of the Tamil dimension, consult (among a number of options) A. Jeyaratnam Wilson, *Sri Lanka Tamil Nationalism: Its Origins and Development in the 19th and 20th Centuries* (London: C. Hurst, 2000) and Chelvadurai Manogaran and Bryan Pfaffenberger, eds., *The Sri Lankan Tamils: Ethnicity and Identity* (Boulder, CO: Westview, 1994); for the Sinhalese dimension (among a number of options), see Tessa Bartholomeusz, "First Among Equals: Buddhism and the Sri Lankan State," in Ian Harris, ed., *Buddhism and Politics in Twentieth-Century Asia* (NY: Continuum, 1999), 173–193

and Neil DeVotta, *Sinhalese Buddhist Nationalist Ideology: Implications for Politics and Conflict Resolution in Sri Lanka* (Washington, DC: East–West Center, 2007); an excellent unpublished work is Shyamika Jayasundara-Smits, *In Pursuit of Hegemony: Politics and State Building in Sri Lanka*, PhD dissertation, International Institute if Social Studies (Netherlands), 2013; available at: www.google.com/url?sa=t&rct=j&q=&esrc=s&source=web&cd=1&ved=0CB8QFjAA&url=http%3A%2F%2Frepub.eur.nl%2Fpub%2F40137%2FThesisPDC%5B1%5D.pdf&ei=y4-IVc7RPIrp-QGI0oCIBQ&usg=AFQjCNGguNzx4YTQopm9sd2rGRf-_5f0HQ&sig2=vmhUwWpZk5gNYwkbc7WQHg&bvm=bv.96339352,d.cWw (accessed March 9, 2018).

23 Noteworthy contributions to this discussion are Patrick Grant, *Buddhism and Ethnic Conflict in Sri Lanka* (Albany: State University of New York Press, 2009); Tessa J. Bartholomeusz, *In Defense of Dharma: Just-war Ideology in Buddhist Sri Lanka* (NY: RoutledgeCurzon, 2002); and Mahinda Deegalle, ed., *Buddhism, Conflict and Violence in Modern Sri Lanka* (NY: Routledge, 2006). For impact upon other communities, see Tessa J. Bartholomeusz and Chandra R. de Silva, eds., *Buddhist Fundamentalism and Minority Identities in Sri Lanka* (Albany: State University of New York Press, 1998); for negative treatment written as Eelam IV commenced, *Sri Lanka: Sinhala Nationalism and the Elusive Southern Consensus*, Asia Report No. 141 (Brussels: international Crisis Group, November 7, 2007). An equally negative approach to the subject, its title illustrating the content, is the more recent Mahinda Deegalle, "Sinhala Ethno-nationalisms and Militarization in Sri Lanka," in Vladimir Tikhonov and Torkel Brekke, eds., *Buddhism and Violence: Militarism and Buddhism in Modern Asia* (NY: Routledge, 2013), 15–36.

24 One of the most revealing illustrations of this position is Satchi Ponnambalam, *Sri Lanka: The National Question and the Tamil Liberation Struggle* (London: Zed Books, 1983).

25 Discussed at length and in some detail in Marks, "People's War in Sri Lanka: Insurgency and Counterinsurgency," *Issues and Studies* 22, no. 8 (August 1986), 63–100.

26 Additional details may be found in Marks, "Counterinsurgency and Operational Art," *Low Intensity Conflict and Law Enforcement* [incorporated into *Small Wars and Insurgencies*] 13, no. 3 (Winter 2005), 168–211.

27 Detailed discussion may be found in R. Ramasubramanian, *Suicide Terrorism in Sri Lanka* (New Delhi: Institute of Peace and Conflict Studies, August 2004). For an overview, see Mia Bloom, *Dying to Kill: The Allure of Suicide Terror* (New York: Columbia University Press, 2005), 45–75; Jeffrey William Lewis, *The Business of Martyrdom: A History of Suicide Bombing* (Annapolis, MD: Naval Institute Press, 2012), 86–112.

28 See Channa Wickremesekara, "Operation Liberation: 25 Years On," *Long Reads*, May 28, 2012; available at: http://groundviews.org/2012/05/28/operation-liberation-25-years-on/ (accessed March 9, 2018).

29 For particulars of the situation, see Marks, "Sri Lankan Minefield: Gandhi's Troops Fail to Keep the Peace," *Soldier of Fortune* 13, no. 3 (March 1988), 36–45, 74–75; as well as Marks, "Handling Snakes and Unfriendly Troops in Sri Lanka," *Honolulu Star-Bulletin*, September 22, 1987, A-17.

30 Certainly among the finest, most comprehensive treatments of all events discussed thus far (and continuing through the IPKF commitment) is Rajan Hoole, Daya Somasundaram, K. Sritharan, and Rajani Thiranagama [all of University of Jaffna], *The Broken Palmyra: The Tamil Crisis in Sri Lanka – An Insider Account* (Claremont, CA: The Sri Lanka Studies Institute, March 1990).

31 Most complete treatment is Rohan Gunaratna, *Sri Lanka: A Lost Revolution? The Inside Story of the JVP* (Kandy, Sri Lanka: Institute of Fundamental Studies, 1990) and C. A. Chandraprema, *Sri Lanka: The Years of Terror – The JVP Insurrection 1987–1989* (Colombo: Lake House Bookshop, 1991). These (among others) have been utilized for the interesting discussion in Mick Moore, "Thoroughly Modern

Revolutionaries: The JVP in Sri Lanka," *Modern Asian Studies* 27, no. 3 (1993), 593–642. I have also contributed to this literature, but the work that created particular interest within Sri Lanka was an article initially embargoed by the official censor in Colombo for its treatment of (what can best be termed) the "overlapping cleavages" that provided the basis for JVP mobilization of marginalized societal elements; see my "In Sri Lanka, Despair Explodes Into Violence," *The Asian Wall Street Journal*, August 16, 1989, 6; abridged version as "Sri Lanka's Despair Breeds Violence," *The Asian Wall Street Journal Weekly*, August 21, 1989, 15.

32 For details, see Rohan Gunaratna, *Indian Intervention in Sri Lanka: The Role of India's Intelligence Agencies*, two editions (Colombo: South Asian Network on Conflict Research, 1993 and 1994); for the Indian perspective, Shankar Bhaduri and Afsir Karim, *The Sri Lankan Crisis* (New Delhi: Lancer International, 1990); Harkirat Singh, *Intervention in Sri Lanka: The IPKF Experience Retold* (New Delhi: Manohar, 2007); and J. N. Dixit, ed., *External Affairs: Cross-Border Relations* (New Delhi: Roli, 2003), 47–96 ("Sri Lanka"); for a contextual treatment, N. Sathiya Moorthy, *India, Sri Lanka and the Ethnic War* (New Delhi: Samskriti, 2008) and S. D. Muni, *Pangs of Proximity: India and Sri Lanka's Ethnic Crisis* (Oslo and New Delhi: Prio/Sage, 1993).

33 Some sources put the figure as high as 600.

34 For battlefield photos, see Marks, "Sri Lanka: Reform, Revolution or Ruin?" *Soldier of Fortune* 21, no. 6 (June 1996), 35–39 (armor on p. 37). Forces in the camp numbered approximately 600, with the attackers normally counted in the thousands (a figure of 5,000 is often used). In the fighting, Lance Corporal Gamini Kularatne posthumously became the first recipient of Sri Lanka's highest award for gallantry, the *Parama Weera Vibhushanaya*, for his actions in assaulting and destroying the armored bulldozer that had broken through the defenses on July 14, 1991.

35 Predictably, it is the Rajiv assassination that has exercised international attention. He was, after all, apparently on the verge of again becoming prime minister in the ongoing Indian election campaign which afforded the opportunity for his targeting. For details, consult Rajeev Sharma, *Beyond the Tigers: Tracking Rajiv Gandhi's Assassination* (New Delhi: Kaveri Books, 2013); as well as to the fictional but well-grounded film, *The Terrorist* (1994), directed by Santosh Sivan (available through Amazon). The assassination (and a cameo for the Premadasa killing) serves as the backdrop for the film, *Madras Cafe* (2013), directed by Shoojit Sircar (available through Amazon), which though widely acclaimed was nonetheless banned in Tamil Nadu for (ironically) its accurate portrayal of the Madras-supported LTTE.

36 Fieldwork which included a August 15, 1991 meeting with Mr. Wijetunga, who was from the Kandy area, the heartland of Sinhalese nationalism. (He passed away at age 92 on September 21, 2008.)

37 Figures (with particulars) may be found in a later Ministry of Defence publication, *Humanitarian Operation Factual Analysis July 2006–May 2009* (Colombo: July 2011), 21, but they are not accompanied by discussion. The two aircraft lost on April 28 and 29, 1995 were Avro transports carrying soldiers on leave. Both were in the vicinity of Palali air base in Jaffna; 97 persons died. In July 1995, an FMA IA 58 Pucará providing close air support was also shot down, its pilot lost. Several years later, on September 29, 1998, Lionair Flight 602, using an Antonov An-24RV, was apparently downed by an LTTE SAM with all 55 aboard lost. Though the missile type(s) has not been stipulated, as early as mid-1987, I examined an SA-7 manual (translated into Tamil) in an LTTE safe house in Jaffna.

38 Fieldwork included a series of July 1995 meetings with Colonel Ratwatte (he had been promoted), who, like previous-President Wijetunga, was from the Kandy area.

39 As reflected in the Sri Lankan copy of the SOCPAC (Special Operations Command Pacific) assessment, dated July 20, 1994, American involvement was focused upon training and support functions; my fieldwork included meeting with the assessment authors prior to their deployment.

State response to Liberation Tigers of Tamil

40 Washington moved much more slowly to ban various LTTE fund-raising fronts, such as the Tamil Rehabilitation Organization (TRO), named a Specially Designated Global Terrorist (SDGT) under Executive Order 13224 on November 15, 2007. The Maryland-based Tamil Foundation (TF) was not banned (under the same authority) until February 11, 2009.

41 My fieldwork included a April 12, 1996 meeting with MPRI.

42 Precise reasons for his removal were unstated. Operational disaster was accompanied by his being implicated in a series of corruption scandals (the investigation of which remained ongoing at the time of his death) and accused of involvement in death squad activity (of which he was acquitted in January 2006). A detailed discussion of the corruption charges may be found at Frederica Jansz, "The Crooked General," *The Sunday Leader*, September 1, 2002; available at: www.thesundayleader.lk/archive/20020901/spotlight.htm (accessed March 9, 2018); as well as Frederica Jansz, "Anuruddha Ratwatte Corruption Case Re-Opened," *The Sunday Leader*, July 18, 2010; available at: www.thesundayleader.lk/2010/07/18/anuruddha-ratwatte-corruption-case-re-opened/ (accessed March 9, 2018). For the problem of Sri Lankan corruption in general, with the Ratwatte case discussed contextually, see "Corruption Probes, Sift Wheat From Chaff," *The Sunday Times*, January 25, 2015; available at: www.sundaytimes.lk/150125/editorial/corruption-probes-sift-wheat-from-chaff-131962.html (accessed March 9, 2018). For the murder charges, see "Ratwatte acquitted on Murder Case," BBC Sinhala, January 20, 2006; available at: www.bbc.com/sinhala/news/story/2006/01/060120_ratwatte.shtml (accessed March 9, 2018). He passed away at age 73 on November 24, 2011 and was honored with the unveiling by President Rajapaksa of a statue in Kandy (where he had also served as the chief lay custodian of the all-important Temple of the Tooth) on the date of his birthday, July 14, 2013.

43 When considering the role of the Tamil diaspora on the conflict, a distinction must be made between imperial legacy communities, such as the Tamils of Malaysia, who migrated or were recruited pursuant to the servicing of the British empire, and more recent migrants produced at least in part by the war in Sri Lanka itself. Though I have found no work that disaggregates these categories, available literature makes clear that large, active support communities for LTTE existed in Britain, the U.S., South Africa, and Canada, with the latter perhaps the leading source of funding; see Kaihla in note 15 above but also Nomi Morris, "The Canadian Connection: Sri Lanka Moves to Crush Tamil Rebels at Home and Abroad," *Maclean's* (November 27, 1995), 28–29.

44 As in the U.S. case discussed above, Canada was slower to proscribe LTTE front organizations. The important fundraising World Tamil Movement (WTM), for example, was not banned until June 2008. In all cases, such proscription was vehemently opposed. Insight into the worldview of those diaspora members who championed LTTE as the authentic representative of the Tamil people may be explored in Øivind Fuglerud, *Life on the Outside: The Tamil Diaspora and Long Distance Nationalism* (London: Pluto Press, 1999).

45 One of the best treatments of this period is G. H. Peiris, *Twilight of the Tigers: Peace Efforts and Power Struggles in Sri Lanka* (New Delhi: Oxford University Press, 2009).

46 Full text available from BBC: http://news.bbc.co.uk/2/hi/south_asia/3232913.stm (accessed March 9, 2018). For greater discussion, see Muttukrishna Sarvananthan, "Sri Lanka: Interim Self-Governing Authority – a Critical Assessment," *Economic and Political Weekly* 38, no. 48 (November 29–December 5, 2003), 5038–5040; available at: www.jstor.org/stable/4414338?seq=1#page_scan_tab_contents (accessed March 9, 2018); and Gamini Keerawella, *The LTTE Proposals for an interim Self-Governing Authority and Future of the Peace Process in Sri Lanka*, Discussion Paper No. 3 (Chiba, Japan: Institute of Developing Economies, May 2004); available at: https://econpapers.repec.org/paper/jetdpaper/dpaper3.htm (accessed March 9, 2018).

47 Details may be found in D. B. S. Jeyaraj, "Tiger vs Tiger Tenth Anniversary of Revolt Led by Eastern LTTE Leader "Col" Karuna," *Daily Mirror*, April 12, 2014; available

at: www.dailymirror.lk/45822/tiger-vs-tiger-tenth-anniversary-of-revolt-led-by-eastern-ltte-leader-col-karuna (accessed March 9, 2018), as well as the relevant section of Ajit Kumar Singh, "Endgame in Sri Lanka," *Faultlines* 20 (2011), 131–170; available at: www.satp.org/satporgtp/publication/faultlines/volume20/Article6.htm (accessed March 9, 2018).

48 For discussion, see Zachariah Mampilly, "A Marriage of Inconvenience: Tsunami Aid and the Unraveling of the LTTE and the GoSL's Complex Dependency," *Civil Wars* 11, no. 3 (September 2009), 302–320 and Alan Keenan, "Building the Conflict Back Better: The Politics of Tsunami Relief and Reconstruction in Sri Lanka," in Dennis B. McGilvray and Michele R. Gamburd, eds., *Tsunami Recovery in Sri Lanka: Ethnic and Regional Dimensions* (NY: Routledge, 2010), 17–39; for context and comparison, Malathi de Alwis and Eva-Lotta Hedman, eds., *Tsunami in a Time of War: Aid, Activism and Reconstruction in Sri Lanka and Aceh* (Colombo: International Centre for Ethnic Studies [ICES], 2009).

49 For detailed assessment of the annual speech methodology and content, see Kasun Ubayasiri, "An Illusive Leader's Annual Speech," *Tamilnation.org*, 2006; available at: http://tamilnation.co/ltte/vp/mahaveerar/06ubayasri.htm (accessed March 9, 2018).

50 Norway's role in the peace process became increasingly controversial as LTTE continued to up the ante in its provocations. Whatever may be said about Colombo's conduct, it did not begin to approach the wholesale violations of the Tigers, particularly since the profile of LTTE actions was dominated by assassinations. That Norway and other international actors could not bring themselves to vigorously counter LTTE atrocities led in the end to a loss of legitimacy for the mediators. For consideration, see Gunnar Søbø, Jonathan Goodhand, Bart Klem, Ada Elisabeth Nissen, and Hilde Selbervik, *Pawns of Peace: Evaluation of Norwegian Peace Efforts in Sri Lanka, 1997–2009* (Oslo: NORAD, September 2011); available at: www.oecd.org/countries/srilanka/49035074.pdf (accessed March 9, 2018); Kristine Höglund and Isak Svensson, "Mediating Between Tigers and Lions: Norwegian Peace Diplomacy in Sri Lanka's Civil War," Chapter 7 in Karin Aggestam and Annika Björkdahl, eds., *War and Peace in Transition: Changing Roles of External Actors* (Lund, Sweden: Nordic Academic Press, 2009), 147–169; Kristine Höglund and Isak Svensson, "Fallacies of the Peace Ownership Approach: Exploring Norwegian Mediation in Sri Lanka," Chapter 3 in Kristian Stokke and Jayadeva Uyangoda, eds., *Liberal Peace in Question: Politics of State and Market Reform in Sri Lanka* (London: Anthem Press, 2011), 63–75; and Hannes Siebert with Chanya Charles, "Sri Lanka: When Negotiations Fail – Talks for the Sake of Talks; War for the Sake of Peace," in Michael Lund and Steve McDonald, eds., *Across the Lines of Conflict: Facilitating Cooperation to Build Peace* (NY: Columbia University Press, 2015), 193–228. For context, see Maria Groeneveld-Savisaar and Siniša Vukovic', "Terror, Muscle, and Negotiation: Failure of Multiparty Mediation in Sri Lanka," Chapter 4 in I. William Zartman and Guy Olivier Faure, eds., *Engaging Extremists: Trade-Offs, Timing, and Diplomacy* (Washington, DC: United States Institute of Peace Press, 2011), 105–135.

51 Considerable insight into the working of the Rajapaksa approach can be gained through C. A. Chandraprema, *Gōta's War: The Crushing of Tamil Tiger Terrorism in Sri Lanka* (self-published, 2012; ninth printing, October 2013). For a discussion of mobilization of manpower (Sri Lanka never had to resort to a draft), see Michele Ruth Gamburd, "The Economics of Enlisting: A Village View of Armed Service," Chapter 7 in Winslow and Woost, op.cit., 151–167; for an effort to discern the integral role martial mobilization came to play economically in a developing economy, see Rajesh Venugopal, "Sri Lanka: Military Fiscalism and the Politics of Market Reform at a Time of Civil War," Chapter 3 in Aparna Sundar and Nandini Sundar, eds., *Civil Wars in South Asia: State, Sovereignty, Development* (New Delhi: Sage, 2014), 69–95.

52 Fieldwork, to include road-counts (vehicular, individuals) and examination of the relevant logs kept at major government checkpoints ringing LTTE-held areas,

consistently revealed flight away from *Tamil Eelam* and toward government-held areas, thus toward areas of relative safety. Efforts to do longitudinal studies on such IDP (internally displaced persons) populations achieved varying degrees of success. A solid effort, conducted before the time under discussion, may be found at H. L. Seneviratne and Maria Stavropoulou, "Sri Lanka's Vicious Circle of Displacement," Chapter 9 in Roberta Cohen and Francis M. Deng, eds., *The Forsaken People: Case Studies of the Internally Displaced* (Washington, DC: Brookings Institution Press, 1998), 359–398.

53 Registered as a political party affiliated with the ruling coalition, TMVP emerged dominant in the March 2008 elections for local councils and the provincial elections themselves in May. A split between Karuna and his deputy Sivanesathurai Chandrakanthan resulted in the latter becoming the first elected chief minister of Eastern Province. Karuna later became a deputy minister in the government and vice president of the ruling SLFP. For context and discussion, see Cathrine Brun and Nicholas Van Hear, "Shifting Between the Local and Transnational: Space, Power and Politics in War-torn Sri Lanka," Chapter 10 in Stig Madeson, Kenneth Bo Nielson, and Uwe Skoda, eds., *Trysts With Democracy: Political Practice in South Asia* (London: Anthem Press, 2011), 239–260.

54 For details, consult Ivan Welch, "Infantry Innovations in Insurgencies: Sri Lanka's Experience," *Infantry* (May–June 2013), 28–31; see also "General Sarath Fonseka Reveals Untold Story of Eelam War IV," *Daily FT*, March 10, 2015; available at: www.ft.lk/article/395488/General-Sarath-Fonseka-reveals-untold-story-of-Eelam-War-IV (accessed March 9, 2018). Also useful is Paul Clarke, "Sri Lanka and the Destruction of the Tamil Tigers," in Lawrence E. Cline and Paul Shemella, eds., *The Future of Counterinsurgency* (Denver, CO: Praeger, 2015).

55 For details, see especially K. M. de Silva, *Sri Lanka and the Defeat of the LTTE* (New Delhi: Penguin Books India, 2012); Paul Moorcraft, *Total Destruction: The Rare Victory of Sri Lanka's Long War* (Barnsley, South Yorkshire, UK: Pen & Sword, 2012); Ashok Mehta, *Sri Lanka's Ethnic Conflict: How Eelam War IV Was Won*, Manekshaw Paper No. 22 (New Delhi: Centre for Land Warfare Studies [CLAWS], 2010); Kumar Rupesinghe, "Sri Lanka: Tackling the LTTE," in Moeed Yusuf, ed., *Insurgency and Counterinsurgency in South Asia* (Washington, DC: United States Institute of Peace, 2014), 249–278; and Ahmed S. Hashim, *When Counterinsurgency Wins: Sri Lanka's Defeat of the Tamil Tigers* (Philadelphia: University of Pennsylvania Press, 2013). Also useful are: Neil DeVotta, "The Liberation Tigers of Tamil Eelam and the Lost Quest for Separatism in Sri Lanka," *Asian Survey* 49, no. 6 (November/December 2009), 1021–1051 and Syed Rifaat Hussain, "Liberation Tigers of Tamil Eelam (LTTE): Failed Quest for a 'Homeland'," in Klejda Mulaj, ed., *Violent Non-State Actors in World Politics* (London: Hurts, 2010), 381–412.

56 Though I do not agree with all particulars therein, an insightful treatment of this topic is S. E. Selvadurai and M. L. R. Smith, "Black Tigers, Bronze Lotus: The Evolution and Dynamics of Sri Lanka's Strategies of Dirty War," *Studies in Conflict & Terrorism* 36, no. 7 (2013), 547–572; for the concept of "dirty war" used to frame the discussion, see M. L. R. Smith and Sophie Roberts, "War in the Gray: Exploring the Concept of Dirty War," *Studies in Conflict and Terrorism* 31, no. 5 (2008), 377–398. See also Purnaka L. de Silva, "Combat Modes, Mimesis and the Cultivation of Hatred: Revenge/Counter-Revenge Killings in Sri Lanka," in Günther Schlee, ed., *Imagined Differences: Hatred and the Construction of Identity* (Münster, Germany: Lit Verlag, 2002), 215–239.

57 For discussion of this facet of the conflict, see Tamara Herath, *Women in Terrorism: Case of the LTTE* (New Delhi: Sage, 2012).

58 This point provides the inspiration for N. Malathy, *A Fleeting Moment in my Country: The Last Years of the LTTE De-Facto State* (Atlanta, GA: Clear Day Books, 2012).

59 To pursue this subject at length would require an article unto itself. Excellent considerations of the terminology and concepts may be found in works with which I do

not necessarily fully agree but are fine efforts nonetheless: Damien Kingsbury, *Sri Lanka and the Responsibility to Protect: Politics, Ethnicity and Genocide* (NY: Routledge, 2012); David Lewis, "The Failure of a Liberal Peace: Sri Lanka's Counterinsurgency in Global Perspective," *Conflict, Security and Development* 10, no. 5 (November 2010), 647–671; and David Keen, " 'The Camp' and 'the Lesser Evil': Humanitarianism in Sri Lanka," *Conflict, Security and Development* 14, no. 1 (2014), 1–31. For R2P itself, see: Gareth Evans, *The Responsibility to Protect: Ending Mass Atrocity Crimes Once and For All* (Washington, DC: Brookings Institution Press, 2008); Luke Glanville, *Sovereignty and the Responsibility to Protect: A New History* (Chicago: The University of Chicago Press, 2014); Anne Orford, *International Authority and the Responsibility to Protect* (NY: Cambridge University Press, 2011); and Philip Cunliffe, ed., *Critical Perspectives on the Responsibility to Protect: Interrogating Theory and Practice* (NY: Routledge, 2011).

60　These terms remain much debated but endeavor to assess the reality that globalization has wrought: a networked world in which modern communications and processes have empowered framing and narrative to such extent that traditional, kinetic-dominated warfare has been supplanted by mixed approaches that specifically target societal processes and beliefs in order to impact will. For further discussion, see Mats Berdal, "The 'New Wars' Thesis Revisited," Chapter 6 in Hew Strachan and Sibylle Scheipers, eds., *The Changing Character of War* (NY: Oxford University Press, 2011), 109–133; Paul Richards, "New War: An Ethnographic Approach," Chapter 1 in Paul Richards, ed., *No Peace No War: An Anthropology of Contemporary Armed Conflicts* (Athens, OH: Ohio University Press, 2005), 1–21; *Contemporary Security Policy* 26, no. 2 (February 2007), thematic issue, "Symposium: Debating Fourth Generation Warfare," 185–285; Avi Kober, "From Heroic to Post-Heroic Warfare: Israel's Way of War in Asymmetric Conflicts," *Armed Forces and Society* 41, no. 1 (2015), 96–122; Mary Kaldor, "Inconclusive Wars: Is Clausewitz Still Relevant in These Global Times," *Global Policy* 1, no. 3 (October 2010), 271–281; and Edward Luttwak, "Toward Post-Heroic Warfare," *Foreign Affairs* (May/June 1995), unpaginated download; available at: www.foreignaffairs.com/articles/chechnya/1995-05-01/toward-post-heroic-warfare (accessed March 9, 2018).

61　Thus emerged, in the face of such a daunting array of challenges, the poignant title of an excellent recent work by M. L. R. Smith and David Martin Jones, *The Political Impossibility of Modern Counterinsurgency* (NY: Columbia University Press, 2015).

62　See e.g., *War Crimes in Sri Lanka*, Asia Report No. 191 (Brussels: International Crisis Group, May 17, 2010) and Frances Harrison, *Still Counting the Dead: Survivors of Sri Lanka's Hidden War* (London: Portobello Books, 2012).

63　Lawfare can have any number of plain text meanings, but it is most often applied to the efforts of *sub-state* actors, both legal and illegal, to use the law as a weapon to impose their will upon others – hence the play on "warfare." A growing body of discussion and scholarship is available on the topic, to include a blog (jointly sponsored by the Lawfare Institute and Brookings) that adopts the more expansive definition; i.e., the use of law as a weapon of war (irrespective of user). See: www.lawfareblog.com/. For my own contribution to the discussion, which includes brief discussion of the Sri Lankan case, see "Lawfare's Role in Irregular Conflict," *inFocus* 4, no. 2 (Summer 2010), 12–14; available at: www.jewishpolicycenter.org/2010/05/31/lawfare-irregular-conflict/ (accessed March 9, 2018).

64　The reference is to networks of attachment and engagement (i.e., social networks), of which the internet may be a virtual form; but communications are not the issue, rather linkages which allow pressure to be brought to bear asymmetrically upon the seemingly stronger adversary. *Netwar* moved steadily from a position as but one LTTE campaign to become a central strategic imperative driven by a final position of military defeat. For *netwar* conceptual development, see John Arquilla and David Ronfeldt, *The Advent of Netwar* (Santa Monica, CA: RAND, 1996); utilized further in

David Ronfeldt, John Arquilla, Graham E. Fuller, and Melissa Fuller, *The Zapatista Social Netwar in Chiapas* (Santa Monica, CA: RAND, 1998), with a condensed version available in Ronfeldt and Arquilla, "Emergence and Influence of the Zapatista Social Netwar," Chapter 6 in *Networks and Netwars: The Future of Terror, Crime, and Militancy* (Santa Monica, CA: RAND, 2001), 171–199. Bringing the circle round, in the same work, Arquilla and Ronfeldt offer as Chapter 1, "The Advent of Netwar (Revisited)," 1–25.

65 This is the position of Gordon Weiss, *The Cage: The Fight for Sri Lanka and the Last Days of the Tamil Tigers* (London: The Bodley Head, 2011).

Bibliography

Ali, Amer. "The Genesis of the Muslim Community in Ceylon (Sri Lanka): A Historical Summary," *Asian Studies* XIX (April–December 1981), 65–82.

Alles, A. C. *Insurgency 1971*. 3rd ed. Colombo: Mervyn Mendis, 1976.

Arquilla, John and David Ronfeldt, *The Advent of Netwar*. Santa Monica, CA: RAND, 1996.

Balasingham, Anton S. *On the Tamil National Question*. London: Polytechnic of the South Bank, 1978.

Bartholomeusz, Tessa. "First Among Equals: Buddhism and the Sri Lankan State," in Ian Harris, (ed.). *Buddhism and Politics in Twentieth-Century Asia*. New York, NY: Continuum, 1999, 173–193.

Bartholomeusz, Tessa J. *In Defense of Dharma: Just-war Ideology in Buddhist Sri Lanka*. New York, NY: RoutledgeCurzon, 2002.

Bartholomeusz, Tessa J. and Chandra R. de Silva (eds.). *Buddhist Fundamentalism and Minority Identities in Sri Lanka*. Albany: State University of New York Press, 1998.

Berdal, Mats. "The 'New Wars' Thesis Revisited." In Hew Strachan and Sibylle Scheipers (eds.), *The Changing Character of War*. New York, NY: Oxford University Press, 2011, 109–133.

Bhaduri, Shankar and Afsir Karim. *The Sri Lankan Crisis*. New Delhi: Lancer International, 1990.

Bloom, Mia. *Dying to Kill: The Allure of Suicide Terror*. New York: Columbia University Press, 2005.

Bose, Sumantra. Contested Lands: Israel-Palestine, Kashmir, Bosnia, Cyprus, and Sri Lanka. Cambridge, MA: Harvard University Press, 2007.

Brar, Tej Pratap. "Sri Lanka's Civil War," paper presented at The Radcliffe Institute for Advanced Study conference, "Postcolonial Wars: Current Perspectives on the Deferred Violence of Decolonialization," October 30–31, 2008, Cambridge, MA: Harvard University.

Brun, Cathrine and Nicholas Van Hear. "Shifting Between the Local and Transnational: Space, Power and Politics in War-torn Sri Lanka." In Stig Madeson, Kenneth Bo Nielson, and Uwe Skoda (eds.), *Trysts With Democracy: Political Practice in South Asia*. London: Anthem Press, 2011, 239–260.

Chandraprema, C. A. *Sri Lanka: The Years of Terror – The JVP Insurrection 1987–1989*. Colombo: Lake House Bookshop, 1991.

Chandraprema, C. A. *Gōta's War: The Crushing of Tamil Tiger Terrorism in Sri Lanka*. Self-published, 2012; 9th printing, October 2013.

Clancy, John. *"The Most Dangerous Year of the War": Japan's Attack on the Indian Ocean, 1942*. Philadelphia, PA: Casemate, 2015.

Clarke, Paul. "Sri Lanka and the Destruction of the Tamil Tigers." In Lawrence E. Cline and Paul Shemella (eds.), *The Future of Counterinsurgency*. Denver, CO: Praeger, 2015.

"Conflict Resolution: Tamil Eelam-Sri Lanka – Thimpu Talks – July/August 1985"; available at http://tamilnation.co/conflictresolution/tamileelam/85thimpu/thimpu00.htm (accessed June 22, 2015).

"Symposium: Debating Fourth Generation Warfare," *Contemporary Security Policy* 26, (2) (February 2007), thematic issue, 185–285.

"Corruption Probes, Sift Wheat From Chaff," *The Sunday Times*, (January 25, 2015); available at www.sundaytimes.lk/150125/editorial/corruption-probes-sift-wheat-from-chaff-131962.html (accessed March 9, 2018).

Cunliffe, Philip (ed.). *Critical Perspectives on the Responsibility to Protect: Interrogating Theory and Practice*. New York, NY: Routledge, 2011.

Daniel, E. Valentine. *Charred Lullabies: Chapters in an Anthropology of Violence*. Princeton, NJ: Princeton University Press, 1996.

Davis, Anthony (Tony). "Tamil Tiger International," *Jane's Intelligence Review* (October 1996), 469–473.

De Alwis, Malathi and Eva-Lotta Hedman (eds.). *Tsunami in a Time of War: Aid, Activism and Reconstruction in Sri Lanka and Aceh*. Colombo: International Centre for Ethnic Studies [ICES], 2009.

De Silva, K. M. *Sri Lanka and the Defeat of the LTTE*. New Delhi: Penguin Books India, 2012.

De Silva, Purnaka L. "Combat Modes, Mimesis and the Cultivation of Hatred: Revenge/Counter-Revenge Killings in Sri Lanka." In Günther Schlee (ed.), *Imagined Differences: Hatred and the Construction of Identity*. Münster, Germany: Lit Verlag, 2002, 215–239.

Deegalle, Mahinda (ed.). *Buddhism, Conflict and Violence in Modern Sri Lanka*. New York, NY: Routledge, 2006.

Deegalle, Mahinda. "Sinhala Ethno-nationalisms and Militarization in Sri Lanka." In Vladimir Tikhonov and Torkel Brekke (eds.), *Buddhism and Violence: Militarism and Buddhism in Modern Asia*. New York, NY: Routledge, 2013, 15–36.

DeVotta, Neil. *Sinhalese Buddhist Nationalist Ideology: Implications for Politics and Conflict Resolution in Sri Lanka*. Washington, DC: East–West Center, 2007.

DeVotta, Neil. "The Liberation Tigers of Tamil Eelam and the Lost Quest for Separatism in Sri Lanka," *Asian Survey* 49 (6) (November/December 2009), 1021–1051.

Dixit, J. N. (ed.). *External Affairs: Cross-Border Relations*. New Delhi: Roli, 2003.

Evans, Gareth. *The Responsibility to Protect: Ending Mass Atrocity Crimes Once and For All*. Washington, DC: Brookings Institution Press, 2008.

Fuglerud, Øivind. *Life on the Outside: The Tamil Diaspora and Long Distance Nationalism*. London: Pluto Press, 1999.

Gamburd, Michele Ruth. "The Economics of Enlisting: A Village View of Armed Service." In Deborah Winslow and Michael D. Woost (eds.), *Economy, Culture, and Civil War in Sri Lanka*. Bloomington: Indiana University Press, 2004, 151–167.

"General Sarath Fonseka Reveals Untold Story of Eelam War IV," *Daily FT*, (March 10, 2015) available at www.ft.lk/article/395488/General-Sarath-Fonseka-reveals-untold-story-of-Eelam-War-IV (accessed March 9, 2018).

Glanville, Luke. *Sovereignty and the Responsibility to Protect: A New History*. Chicago, IL: The University of Chicago Press, 2014.

Grant, Patrick. *Buddhism and Ethnic Conflict in Sri Lanka*. Albany: State University of New York Press, 2009.

Groeneveld-Savisaar, Maria and Siniša Vukovic´, "Terror, Muscle, and Negotiation: Failure of Multiparty Mediation in Sri Lanka." In I. William Zartman and Guy Olivier Faure (eds.), *Engaging Extremists: Trade-Offs, Timing, and Diplomacy*. Washington, DC: United States Institute of Peace Press, 2011, 105–135.

Gunaratna, Rohan. *Sri Lanka: A Lost Revolution? The Inside Story of the JVP*. Kandy, Sri Lanka: Institute of Fundamental Studies, 1990.

Gunaratna, Rohan. *Indian Intervention in Sri Lanka: The Role of India's Intelligence Agencies*, two editions. Colombo: South Asian Network on Conflict Research, 1993 and 1994.

Harrison, Frances. *Still Counting the Dead: Survivors of Sri Lanka's Hidden War*. London: Portobello Books, 2012.

Hashim, Ahmed S. *When Counterinsurgency Wins: Sri Lanka's Defeat of the Tamil Tigers*. Philadelphia: University of Pennsylvania Press, 2013.

Herath, Tamara. *Women in Terrorism: Cas. of the LTTE*. New Delhi: Sage, 2012.

Hettige, Siri T. "Economic Policy, Changing Opportunities for Youth, and the Ethnic Conflict in Sri Lanka." In Deborah Winslow and Michael D. Woost (eds.), *Economy, Culture, and Civil War in Sri Lanka*. Bloomington: Indiana University Press, 2004, 115–130.

Hill, Tom H. J. "The Deception of Victory: The JVP in Sri Lanka and the Long-Term Dynamics of Rebel Reintegration," *International Peacekeeping* 20 (3) (June 2013), 357–374.

Höglund, Kristine and Isak Svensson. "Mediating Between Tigers and Lions: Norwegian Peace Diplomacy in Sri Lanka's Civil War." In Karin Aggestam and Annika Björkdahl (eds.), *War and Peace in Transition: Changing Roles of External Actors*. Lund, Sweden: Nordic Academic Press, 2009, 147–169.

Höglund, Kristine and Isak Svensson. "Fallacies of the Peace Ownership Approach: Exploring Norwegian Mediation in Sri Lanka." In Kristian Stokke and Jayadeva Uyangoda (eds.), *Liberal Peace in Question: Politics of State and Market reform in Sri Lanka*. London: Anthem Press, 2011, 63–75.

Hoole, Rajan, Daya Somasundaram, K. Sritharan, and Rajani Thiranagama. *The Broken Palmyra: The Tamil Crisis in Sri Lanka – An Insider Account*. Claremont, CA: The Sri Lanka Studies Institute, March 1990.

Hussain, Syed Rifaat. "Liberation Tigers of Tamil Eelam (LTTE): Failed Quest for a 'Homeland'." In Klejda Mulaj (ed.), *Violent Non-State Actors in World Politics*. London: Hurts, 2010, 381–412.

Ilyas, Ahmed H. "Estate Tamils of Sri Lanka – a Socio-Economic Review," *International Journal of Sociology and Anthropology* 6 (6) (June 2014), 184–191.

Jansz, Frederica. "The Crooked General," *The Sunday Leader*, (September 1, 2002); available at www.thesundayleader.lk/archive/20020901/spotlight.htm (accessed March 9, 2018).

Jansz, Frederica. "Anuruddha Ratwatte Corruption Case Re-Opened," *The Sunday Leader*, (July 18, 2010); available at www.thesundayleader.lk/2010/07/18/anuruddha-ratwatte-corruption-case-re-opened/ (accessed March 9, 2018).

Jayasundara-Smits, Shyamika. *In Pursuit of Hegemony: Politics and State Building in Sri Lanka*, PhD dissertation, International Institute if Social Studies (Netherlands), (2013); available at www.google.com/url?sa=t&rct=j&q=&esrc=s&source=web&cd=1&ved=0 CB8QFjAA&url=http%3A%2F%2Frepub.eur.nl%2Fpub%2F40137%2FThesisPDC% 5B1%5D.pdf&ei=y4-IVc7RPIrp-QGI0oCIBQ&usg=AFQjCNGguNzx4YTQopm9sd2r GRf-_5f0HQ&sig2=vmhUwWpZk5gNYwkbc7WQHg&bvm=bv.96339352,d.cWw (accessed March 9, 2018).

Jeyaraj, D. B. S. "Tiger vs Tiger Tenth Anniversary of Revolt Led by Eastern LTTE Leader 'Col' Karuna," *Daily Mirror*, (April 12, 2014); available at www.dailymirror.lk/45822/tiger-vs-tiger-tenth-anniversary-of-revolt-led-by-eastern-ltte-leader-col-karuna (accessed March 9, 2018).

Kaihla, Paul. "Banker, Tiger, Soldier, Spy," *Maclean's* (August 5, 1996), 28–32.

Kaldor, Mary. "Inconclusive Wars: Is Clausewitz Still Relevant in These Global Times," *Global Policy* 1, (3) (October 2010), 271–281.

Keen, David. " 'The Camp' and 'the Lesser Evil': Humanitarianism in Sri Lanka," *Conflict, Security and Development* 14 (1) (2014), 1–31.

Keenan, Alan. "Building the Conflict Back Better: The Politics of Tsunami Relief and Reconstruction in Sri Lanka." In Dennis B. McGilvray and Michele R. Gamburd (eds.), *Tsunami Recovery in Sri Lanka: Ethnic and Regional Dimensions*. New York, NY: Routledge, 2010, 17–39.

Keerawella, Gamini. *The LTTE Proposals for an interim Self-Governing Authority and Future of the Peace Process in Sri Lanka*, Discussion Paper No. 3 (Chiba, Japan: Institute of Developing Economies, May 2004); available at https://econpapers.repec.org/paper/jetdpaper/dpaper3.htm (accessed March 9, 2018).

Kingsbury, Damien. *Sri Lanka and the Responsibility to Protect: Politics, Ethnicity and Genocide*. New York, NY: Routledge, 2012.

Kober, Avi. "From Heroic to Post-Heroic Warfare: Israel's Way of War in Asymmetric Conflicts," *Armed Forces and Society* 41 (1) (2015), 96–122.

Kumarasingham, Harshan. *"The Jewel of the East Yet Has its Flaws" – the Deceptive Tranquility Surrounding Sri Lankan Independence*, Heidelberg Papers in South Asian and Comparative Politics/Working Paper No. 72. Heidelberg: Department of Political Science, (June 2013); available at http://archiv.ub.uni-heidelberg.de/volltextserver/15148/1/Heidelberg%20Papers_72_Kumarasingham.pdf (accessed March 9, 2018).

Lawfare, blog; available at www.lawfareblog.com/ (accessed March 9, 2018).

Lewis, David. "The Failure of a Liberal Peace: Sri Lanka's Counter-insurgency in Global Perspective," *Conflict, Security and Development* 10 (5) (November 2010), 647–671.

Lewis, Jeffrey William. *The Business of Martyrdom: A History of Suicide Bombing*. Annapolis, MD: Naval Institute Press, 2012.

Liberation Tigers of Tamil Eelam (LTTE). *Interim Self-Governing Authority* (ISGA), BBC available at http://news.bbc.co.uk/2/hi/south_asia/3232913.stm (accessed March 9, 2018).

Löwy, Michael. "Marxists and the National Question," *New Left Review* I/96 (March–April 1976), 81–100.

Luttwak, Edward. "Toward Post-Heroic Warfare," *Foreign Affairs* (May/June 1995), unpaginated download available at www.foreignaffairs.com/articles/chechnya/1995-05-01/toward-post-heroic-warfare (accessed March 9, 2018).

Madras Cafe (2013) directed by Shoojit Sircar (available through Amazon).

Malathy, N. *A Fleeting Moment in my Country: The Last Years of the LTTE De-Facto State*. Atlanta, GA: Clear Day Books, 2012.

Mampilly, Zachariah. "A Marriage of Inconvenience: Tsunami Aid and the Unraveling of the LTTE and the GoSL's Complex Dependency," *Civil Wars* 11 (3) (September 2009), 302–320.

Manogaran, Chelvadurai. *Ethnic Conflict and Reconciliation in Sri Lanka*. Honolulu: University of Hawai'i Press, 1987.

Manogaran Chelvadurai and Bryan Pfaffenberger (eds.) *The Sri Lankan Tamils: Ethnicity and Identity*. Boulder, CO: Westview, 1994.

Manoharan, N. *Counterterrorism Legislation in Sri Lanka: Evaluating Efficacy.* Washington, DC: East–West Center, 2006.

Marks, Thomas A. "Marxist Tamils Won't Stop at Separatism," *The Asian Wall Street Journal,* May 8, 1986, 6; reprinted as "Tamil Rebels Aim Beyond Autonomy," *The Asian Wall Street Journal Weekly,* May 26, 1986: 12.

Marks, Thomas A. "People's War in Sri Lanka: Insurgency and Counterinsurgency," *Issues and Studies* 22 (8) (August 1986), 63–100.

Marks, Thomas A. "India is the Key to Peace in Sri Lanka," *The Asian Wall Street Journal,* (September 19–20, 1986) 8; reprinted under same title in *The Island* [Colombo], (October 5, 1986) 8; abridged version under same title in *The Wall Street Journal,* September 22, 1986: 25.

Marks, Thomas A. "The Ethnic Roots of Sri Lanka's Ideological Struggle," *The Asian Wall Street Journal,* August 12, 1987, 8; abridged version under same title in *The Asian Wall Street Journal Weekly,* (August 31, 1987) 12.

Marks, Thomas A. "Handling Snakes and Unfriendly Troops in Sri Lanka," *Honolulu Star-Bulletin,* (September 22, 1987) A-17.

Marks, Thomas A. "Book Review – Stanley J. Tambiah, *Sri Lanka: Ethnic Fratricide and the Dismantling of Democracy.*" (Chicago: University of Chicago Press, 1986), *Issues and Studies* 23 (9) (September 1987), 135–140.

Marks, Thomas A. "Peace in Sri Lanka," *Daily News* [Colombo], 3 Parts, July 6–7–8, 1987: I. "India Acts in its Own Interests," July 6, 6; II. "Bengali Solution: India Trained Personnel for Invasion of Sri Lanka," July 7, 8; III. "India's Political Solution Narrow and Impossible," July 8, 6; published under the same titles in *The Sri Lanka News,* July 15, 1987 (cont.), 6–7; in *The Island* as "India's Covert Involvements," June 28, 1987, 8 and 10.

Marks, Thomas A. "Sri Lankan Minefield: Gandhi's Troops Fail to Keep the Peace," *Soldier of Fortune* 13 (3) (March 1988), 36–45, 74–75.

Marks, Thomas A. (Tom). "Sri Lanka's Special Force: Professionalism in a Dirty War," *Soldier of Fortune* 13, (7) (July 1988), 32–39.

Marks, Thomas A. "In Sri Lanka, Despair Explodes Into Violence," *The Asian Wall Street Journal,* August 16, 1989, 6; abridged version as "Sri Lanka's Despair Breeds Violence," *The Asian Wall Street Journal Weekly,* August 21, 1989: 15.

Marks, Thomas A. *Maoist Insurgency Since Vietnam.* London: Frank Cass, 1996.

Marks, Thomas A. "Sri Lanka: Reform, Revolution or Ruin?," *Soldier of Fortune* 21 (6) (June 1996), 35–39.

Marks, Thomas A. "Counterinsurgency and Operational Art," *Low Intensity Conflict and Law Enforcement* [incorporated into *Small Wars and Insurgencies*] 13 (3) (Winter 2005), 168–211.

Marks, Thomas A. "Lawfare's Role in Irregular Conflict," *inFocus* 4 (2) (Summer 2010), 12–14; available at www.jewishpolicycenter.org/2010/05/31/lawfare-irregular-conflict/ (accessed March 9, 2018).

Marks, Thomas A. and T. P. S. ("Tippy") Brar, in Michelle Hughes and Michael Miklaucic (eds.), *Impunity: Countering Illicit Power in War and Transition.* Washington, DC: NDU Press, 2016, 40.

McGilvray, Dennis B. and Mirak Raheem. *Muslim Perspectives on the Sri Lankan Conflict.* Washington, DC: East–West Center, 2007.

McGilvray, Dennis B. *Crucible of Conflict: Tamil and Muslim Society on the East Coast of Sri Lanka.* Durham, NC: Duke University Press, 2008.

Mehta, Ashok. *Sri Lanka's Ethnic Conflict: How Eelam War IV Was Won,* Manekshaw Paper No. 22. New Delhi: Centre for Land Warfare Studies [CLAWS], 2010.

Mehta, Raj. *Lost Victory: The Rise and Fall of LTTE Supremo, V. Prabhakaran*. New Delhi: Pentagon Press, 2009.

Miller, Phil. *Britain's Dirty War Against the Tamil People – 1979–2009*. Bremen: International Human Rights Association, (June 2014); available at www.tamilnet.com/img/publish/2014/07/britains_dirty_war.pdf (accessed March 9, 2018).

Ministry of Defence. *Humanitarian Operation Factual Analysis July 2006–May 2009*. Colombo: July 2011.

Moorcraft, Paul. *Total Destruction: The Rare Victory of Sri Lanka's Long War*. Barnsley, South Yorkshire, UK: Pen & Sword, 2012.

Moore, Mick. "Thoroughly Modern Revolutionaries: The JVP in Sri Lanka," *Modern Asian Studies* 27 (3) (1993), 593–642.

Moorthy, N. Sathiya. *India, Sri Lanka and the Ethnic War*. New Delhi: Samskriti, 2008.

Morris, Nomi. "The Canadian Connection: Sri Lanka Moves to Crush Tamil Rebels at Home and Abroad," *Maclean's* (November 27, 1995), 28–29.

Muni, S. D. *Pangs of Proximity: India and Sri Lanka's Ethnic Crisis*. Oslo and New Delhi: Prio/Sage, 1993.

Murari, S. *The Prabhakaran Saga: The Rise and Fall of an Eelam Warrior*. New Delhi: Sage, 2012.

Nolan, Stehphanie [*sic*]. "Sri Lanka Announces End of 28-Year State of Emergency," *The Globe and Mail*, (August 25, 2011) available at www.theglobeandmail.com/news/world/sri-lanka-announces-end-of-28-year-state-of-emergency/article595949/ (accessed March 9, 2018).

Orford, Anne. *International Authority and the Responsibility to Protect*. New York, NY: Cambridge University Press, 2011.

Palanithurai G. and K. Mohanasundaram. *Dynamics of Tamil Nadu Politics in Sri Lankan Ethnicity*. New Delhi: Northern Book Centre, 1993.

Peiris, G. H. "Sri Lanka: Youth Unrest and inter-group Conflict," *Faultlines* 19 (2008), 127–156.

Peiris, G. H. *Twilight of the Tigers: Peace Efforts and Power Struggles in Sri Lanka*. New Delhi: Oxford University Press, 2009.

Pfaffenberger, Bryan. "Ethnic Conflict and Youth Insurgency in Sri Lanka: The Social Origins of Tamil Separatism." In Joseph V. Montville (ed.) *Conflict and Peacemaking in Multiethnic Societies*. New York, NY: Lexington, 1991, 241–257.

Ponnambalam, Satchi. *Sri Lanka: The National Question and the Tamil Liberation Struggle*. London: Zed Books, 1983.

Powell, Geoffrey. *The Kandyan Wars: The British Army in Ceylon 1803–18*. Barnsley, South Yorkshire, UK: Pen & Sword, 1973.

Raman, B. *The Kaoboys of R&AW: Down Memory Lane*. New Delhi: Lancer, 2007.

Ramasubramanian, R. *Suicide Terrorism in Sri Lanka*. New Delhi: Institute of Peace and Conflict Studies, August 2004.

Ranatunga, Cyril. *Adventurous Journey: From Peace to War, Insurgency to Terrorism*, two editions. Colombo: Vijitha Yapa Publications, July 2009 and February 2010.

Rao, P. Venkateshwar. "Ethnic Conflict in Sri Lanka," *Asian Survey* XXVIII, no. 4 (April 1988), 419–436.

"Ratwatte acquitted on Murder Case," BBC Sinhala, (January 20, 2006) available at www.bbc.com/sinhala/news/story/2006/01/060120_ratwatte.shtml (accessed March 9, 2018).

Richards, Paul. "New War: An Ethnographic Approach." In Paul Richards (ed.), *No Peace No War: An Anthropology of Contemporary Armed Conflicts*. Athens, OH: Ohio University Press, 2005, 1–21.

Richardson, John. *Paradise Poisoned: Learning About Conflict, Terrorism and Development From Sri Lanka's Civil Wars.* Kandy, Sri Lanka: International Centre for Ethnic Studies [ICES], 2005.

Rinehart, Christine Sixta. *Volatile Social Movements and the Origins of Terrorism: The Radicalization of Change.* Boulder, CO: Lexington Books, 2013.

Ronfeldt, David, John Arquilla, Graham E. Fuller, and Melissa Fuller, *The Zapatista Social Netwar in Chiapas.* Santa Monica, CA: RAND, 1998; condensed version available in Ronfeldt and Arquilla, "Emergence and Influence of the Zapatista Social Netwar." In *Networks and Netwars: The Future of Terror, Crime, and Militancy.* Santa Monica, CA: RAND, 2001, 171–199.

Rupesinghe, Kumar. "Sri Lanka: Tackling the LTTE." In Moeed Yusuf (ed.), *Insurgency and Counterinsurgency in South Asia.* Washington, DC: United States Institute of Peace, 2014, 249–278.

Sarvananthan, Muttukrishna. "Sri Lanka: Interim Self-Governing Authority – a Critical Assessment," *Economic and Political Weekly* 38 (48) (November 29–December 5, 2003), 5038–5040; available at www.jstor.org/stable/4414338?seq=1#page_scan_tab_contents (accessed March 9, 2018).

Selvadurai, S. E. and M. L. R. Smith. "Black Tigers, Bronze Lotus: The Evolution and Dynamics of Sri Lanka's Strategies of Dirty War," *Studies in Conflict and Terrorism* 36 (7) (2013), 547–572.

Seneviratne, H. L. and Maria Stavropoulou. "Sri Lanka's Vicious Circle of Displacement." In Roberta Cohen and Francis M. Deng (eds.), *The Forsaken People: Case Studies of the Internally Displaced.* Washington, DC: Brookings Institution Press, 1998, 359–398.

Sharma, Rajeev. *Beyond the Tigers: Tracking Rajiv Gandhi's Assassination.* New Delhi: Kaveri Books, 2013.

Siebert, Hannes with Chanya Charles. "Sri Lanka: When Negotiations Fail –Talks for the Sake of Talks; War for the Sake of Peace." In Michael Lund and Steve McDonald (eds.), *Across the Lines of Conflict: Facilitating Cooperation to Build Peace.* New York, NY: Columbia University Press, 2015, 193–228.

Sieghart, Paul. *Sri Lanka: A Mounting Tragedy of Errors.* London: International Commission of Jurists and Justice, 1984.

Singh, Harkirat. *Intervention in Sri Lanka: The IPKF Experience Retold.* New Delhi: Manohar, 2007.

Singh, Kumar. "Endgame in Sri Lanka," *Faultlines* 20 (2011), 131–170; available at www.satp.org/satporgtp/publication/faultlines/volume20/Article6.htm (accessed March 9, 2018).

Smith, M. L. R. and David Martin Jones. *The Political Impossibility of Modern Counterinsurgency.* New York, NY: Columbia University Press, 2015.

Smith, M. L. R. and Sophie Roberts. "War in the Gray: Exploring the Concept of Dirty War," *Studies in Conflict and Terrorism* 31 (5) (2008), 377–398.

Søbø, Gunnar, Jonathan Goodhand, Bart Klem, Ada Elisabeth Nissen and Hilde Selbervik. *Pawns of Peace: Evaluation of Norwegian Peace Efforts in Sri Lanka, 1997–2009,* (Oslo: NORAD, September 2011); available at www.oecd.org/countries/srilanka/49035074.pdf (accessed March 9, 2018).

Sri Lanka: Sinhala Nationalism and the Elusive Southern Consensus, Asia Report No. 141. Brussels: International Crisis Group, November 7, 2007.

Subramanian, Samanth. *This Divided Island: Life, Death, and the Sri Lankan War.* New York, NY: St. Martin's Press, 2014.

Swamy, M. R. Narayan. *Tigers of Lanka: From Boys to Guerrillas*. Delhi: Konark Publishers, 1994 [with two subsequent editions, 1996 and 2002].

Swamy, M. R. Narayan. *Inside an Elusive Mind: Prabhakaran – the First Profile of the World's Most Ruthless Guerrilla Leader*. Colombo: Vijitha Yapa Publications, 2003.

Swamy, M. R. Narayan. *The Tiger Vanquished: LTTE's Story*. New Delhi: Sage, 2010.

Tambiah, Stanley J. *Sri Lanka: Ethnic Fratricide and the Dismantling of Democracy*. Chicago, IL: The University of Chicago Press, 1986.

Tambiah, Stanley J. *Leveling Crowds: Ethnonationalist Conflicts and Collective Violence in South Asia*. Berkeley: University of California Press, 1996.

The Terrorist (1994), directed by Santosh Sivan (available through Amazon).

Trawick, Margaret. *Enemy Lines: Warfare, Childhood, and Play in Batticaloa*. Berkeley: University of California Press, 2007.

Ubayasiri, Kasun. "An Illusive Leader's Annual Speech," *Tamilnation.org*, (2006); available at http://tamilnation.co/ltte/vp/mahaveerar/06ubayasri.htm (accessed March 9, 2018).

Unnithan, Sandeep. "Madras Café Brings Back Uncomfortable Memories of the CIA's Honey Trap," *India Today*, (August 29, 2013); available at http://indiatoday.intoday.in/story/madras-cafe-madras-honey-trap-john-abraham-cia-ltte-raw/1/304302.html (accessed March 9, 2018).

Venugopal, Rajesh. "Sri Lanka: Military Fiscalism and the Politics of Market Reform at a Time of Civil War." In Aparna Sundar and Nandini Sundar (eds.), *Civil Wars in South Asia: State, Sovereignty, Development*. New Delhi: Sage, 2014, 69–95.

War Crimes in Sri Lanka, Asia Report No. 191. Brussels: International Crisis Group, May 17, 2010.

Weiss, Gordon. *The Cage: The Fight for Sri Lanka and the Last Days of the Tamil Tigers*. London: The Bodley Head, 2011.

Welch, Ivan. "Infantry Innovations in Insurgencies: Sri Lanka's Experience," *Infantry* (May–June 2013), 28–31.

Wickremesekera, Channa. *Kandy at War: Indigenous Military Resistance to European Expansion in Sri Lanka 1594–1818*. New Delhi: Manohar, 2004.

Wickremesekara, Channa. "Operation Liberation: 25 Years On," *Long Reads*, (May 28, 2012); available at http://groundviews.org/2012/05/28/operation-liberation-25-years-on/ (accessed March 9, 2018).

Wilson, A. Jeyaratnam. *Sri Lanka Tamil Nationalism: Its Origins and Development in the 19th and 20th Centuries*. London: C. Hurst, 2000.

Conclusion

Framing effective responses and future pathways

Shanthie Mariet D'Souza

South and South East Asia remain sites of various forms of extremism and violence by nonstate actors. Commitment of inadequate resources, inappropriate force mobilization, and lack of unified effort and political will to address social and religious faultlines are among the factors that have contributed to the longevity of armed groups, a number of which have changed form substantially as the decades have rolled on. The capacity of the latter to achieve success – either through the collapse of state structure or outright seizure of power – has remained the exception, yet this had not impeded the disruption which remains a feature of life in many countries. Threat group survival has been achieved through a variety of mechanisms, ranging from splintering to temporarily lying low, from exploiting peace deals to merging with other ideologically diverse groups. The case studies of Maoists in Nepal, Taliban in Afghanistan and Pakistan, and the Naxalite movement in India are illustrative of such developments.

Once thought unlikely, these trends are actually accelerating. Even as the war in Afghanistan has dragged on – the American president had actually announced a supposed intensification to match a Taliban seizure of the strategic initiative already well underway – Southeast Asia emerged as a new front in the war on terrorism, joining the Middle East and Africa (arguably now the major theater of operations). In all theatres, episodes of seizure of an entire major city, as has recently happened in the Philippines, now tend to overshadow direct U.S. and Western commitment to combat, as in the case of Afghanistan (where it is often forgotten that U.S. allies remain engaged). Behind the headlines, as highlighted by the contributions to this volume, much more is going on that requires analysis and response. For it has profound implications, not only for immediate security of the individual states themselves and the global community but also for theory. The continued miscue of lumping together of phenomena such as terrorism and insurgency, much less radicalization and mobilization, has profound implications for the viability of response.

Ironically, in many ways, the past has become the future. It is now well understood how seizure of Afghanistan by what seemed at the time a medieval relic of religious expression could leap from local to global challenge, becoming the so-called "glocal." The phenomenon became quite clear as al-Qaeda inroads were made in South Asia, and the Afghanistan–Pakistan region was turned into a hub of insurgent and terrorist activity, built upon a foundation of violent extremism and tied into international criminality. Groups and forces that at one point seemed to demonstrate little if any ambition for global expansion, much less targeting the innocent, suddenly became the "new war" that inspired both the need for global counter and more robust, expansive body of analysis.

Assessment, perhaps not surprisingly, remains focused upon the South and Southeast Asian cases as tactical manifestations of conflict, rather than as generators of knowledge of threat dimensions. Yet the plans of violent radical challenges for integration of the local with the global demand greater inspection, most crucially so that political leaders and practitioners will understand the nature of war and the requirements for response. It seems but a moment ago that the lessons of "a decade of war" were being culled, but now we begin to approach the two-decade mark. The intervening flurry of activity has created an illusion of analytical progress, though it cannot be judged that the understanding of insurgency and terrorism has advanced in any compelling manner. On the one hand, as mentioned in the Introduction, the obsession with "counting" has infected even terminology; on the other hand, even the most lauded recent work largely restates findings of the past (though it often seems quite unaware of doing so). Finally, if the American "new strategy" in Afghanistan is an indicator, we continue to live in a policy world where "kill them all" remains the privileged approach.

This volume in addition to addressing some of the existing research gaps, underscores the need for irregular war scholars and practitioners to systematically examine the changing character of irregular movements to help devise an effective response. South and Southeast Asia provide a laboratory of sorts in addressing violent extremist challenges. These necessarily lodge, in the first instance, in existing states, even as they operate regionally, and even globally. Counter must necessarily be concerned with technique, with tactics, but it is the larger issues of strategic and operational moment that are central to policy relevance. This is particularly demonstrated in the cases examined in this volume. Too often response is drawn from the past while the threat demands challenges that build for the future.

Emerging challenges

If we look at the discussion that has filled the pages of this volume, we can identify a number of emerging trends with regard to terrorism, insurgency, and violent extremism in South and Southeast Asia.

The continuing threat of the Islamic State (IS)

Rise and spread of IS has added to the complexity of the security situation in South and Southeast Asia. Despite numerous challenges, IS has been able to gain a foothold and change the dynamics of militancy in the regions, where a number of violent radical Islamist groups have been active for decades. Although the group failed to get the widespread support it expected and suffered several setbacks, the idea of setting up regional Caliphates has appeal even as the core has been all but lost in Iraq and Syria. The weakness of local IS factions and the sustained state pressure increases the possibility of collusion and convergence between IS and its rival groups. As seen in many parts of the world, the overlap and teamwork among various violent Islamist groups enhances their effectiveness and operational capabilities to carry out deadlier attacks with relative ease. The daring seizure of parts of the Filipino city of Marawi, the country's largest Muslim-majority city, by the Maute Group, a self-styled Jihadist group, is a case in point. Despite the liberation of the city, which took five long months and enormous military effort, IS seems to have expanded its presence in the region.

IS has made a concerted effort to promote its brand in South East Asia by building on the legacy of Jemaah Islamiyah (JI). In Indonesia, the IS affiliates include Jamaah Ansharut Daulah (JAD), a network of several small radical factions that coalesced and pledged allegiance to Abu Bakr al Baghdadi. IS also constitutes the biggest threat for Malaysia, where its propaganda has inspired and promoted radicalization, violent extremism, and religious-motivated terrorism. In South Asia, IS has been able to establish control in several provincial districts in eastern Afghanistan and has taken root in Pakistan. There is a high probability that IS in Asia will try to adapt to local conditions and exploit the existing ideological and sectarian tensions and rivalries between different state actors.

Re-emerging threats

Violent armed groups in South and South East Asia have remained in business, either through opportunistic alliances with IS or on their own, taking advantage of the gaps in state responses. As pointed out earlier, the legacy of JI continues particularly in the world's largest Muslim state, even as militants in the Philippines, thought to have become of small relevance after a peace deal with the government, are back to their warring days. Though the death of 1,100 cadres during the Marawi operations provides the government some breather in formulating a systematic, holistic response, the threat is severe. In Burma/Myanmar, the Arakan Rohingya Solidarity Army (ARSA) is small, but state genocidal response has pushed it forward as possibly the only active expression of Rohingya self-defense. This, it needs to be remembered, is precisely how the marginal Tamil revolutionary groups achieved traction, when unchecked state violence thrust forward dozens of self-defense groups, of which LTTE was then but one. Thailand has for now avoided falling into the same trap in its counterinsurgency in the southern border provinces (SBP), but it was inept, brutal response in 2004 which gave the insurgency its jump-state from marginal to central threat. Hence, the danger of further error and attendant consequences is ever present. In Pakistan, smaller militant groups, such as Ansar ul-Sharia Pakistan (ASP), Jamaat-ul Ahrar (JuA), and Jandullah, have also left strong marks in recent years. In India's Kashmir, local cadres of Pakistan-based groups continue to pose grave security challenges. Unlike earlier insurgencies, the present challenge is driven not by separatism or ideology but by violent radical Islamism. It bears noting that with or without alignment with the global jihadists such as al-Qaeda and IS, none of the conflict theatres in Asia are demonstrating signs of normalization.

Changing character of insurgency

The nature of insurgency and its character have also been changing. The most apt example of this is perhaps Afghan Taliban, which has evolved as a counter-state controlling territory and population. Since 2011, the leadership has professionalized guerrilla tactics, often offering training courses in Pakistan and now Iran. Taliban has become increasingly adept at ambushes and most importantly at laying mass-produced improvised explosive devices (IEDs). The strategy changed in 2014 when it decentralized and developed the ability to deploy integrated, self-sustaining units at the regional level, fighting competently and effectively. Such fighting ability, with shifts between large-scale attacks and hybrid war, as well as the development of "expeditionary" logistics, has continued in spite of the fact that the group has been heavily burdened by infighting and factionalism. Most recently, it has even sought to win rather than compel allegiance of target populations.

Similarly, in Pakistan, Tehrik-e-Taliban (TTP) has tapped deep-seated contradictions with roots in economic, social, and political deprivation. Ready access to weapons and money has combined with the legitimacy drawn from tribal connections and professed Islamic piety to enable the group to challenge the state and impose locally its ideas of Islamist governance as a step toward eventually effecting change in the political and social fabric of Pakistan. While military operations have weakened the organization, affecting its ability to carry out operations deep inside the country, a more serious challenge comes from those extremist groups embedded within the country's heartland, the Punjab province. South Punjab is the organizational home of several of the country's leading jihadi organizations and its most powerful Islamist movements, all of which aim to replace the current order with an idealized state governed by *Sharia*. The province's mosques and madrassas, among Pakistan's largest and most influential, frequently serve up recruits for violent extremist groups, as do ostensibly non-political organizations, such as Punjab-based Tablighi Jamaat, a proselytizing movement preaching moral rearmament.

External funding, sanctuaries, and transnational linkages

In a world where religion continues to play a critical role in fueling dissent, the armed organizations have found ready access to the funds needed to sustain their movements. Although opium cultivation still contributes significantly to Taliban coffers, external funding remains a significant force multiplier. Such funding to Taliban grew steadily over the years until 2014, enabling the group to deploy as many as 40,000 mobile fighters inside Afghanistan, with substantial numbers resting or in reserve. Even a ragtag insurgency such as ARSA has grown with the support it has generated in Pakistan and Saudi Arabia, in addition to its safe haven in Bangladesh. Its own ostensible transition to moderation notwithstanding, Saudi Arabia continues to be the source of foundational fundamentalism that frequently appears later in the form of violent radical Islamist recruits moving through Saudi-funded mosques and madrassas after absorbing the intolerant Salafism on display in the likes of the Maldives and South East Asia.

Vulnerability of multi-ethnic societies and governance deficits

In addition to fledgling state institutions and governance deficits, complex and combustible tribal-clan dynamics have contributed to the emergence of the extremists. This has been apparent in the cases examined. The crisis in the southern Philippines is but the most salient recent expression putting on display the growing organizational reach of the extremists as well as the depth of democratic deficit and inadequate (especially corrupt) governance. It is the latter that provides a conducive atmosphere for radicalization and extremist mobilization. Indonesia's fractured nationalism and contested identity have set the stage for two distinct periods of violent extremism and state response in Indonesia: the rise of JI and other terrorist organizations (1999–2009), and the ascendency of decentralized terrorism inspired by the IS (2016 to the present).

Effective responses

What then should constitute an appropriate strategy for the state as it counters such armed challenges, and what are the basic concepts which need to be understood for constructing this strategy of response?

Over the last two decades, the nature of nonstate actor violence, the drivers of radicalism, and counterterrorism policy have evolved in subtle but important ways. As Clausewitz put it directly, it is imperative for political leaders and practitioners to understand the kind of war on which they are embarking. The Nepal case, mentioned several times earlier, illustrates well the collapse in "New War" of the division between peace and war, thus the continued reliance upon warfighting as waged mainly by the military.

The Nepal case, amply indicates the danger. Even as the insurgency recognized that its violent action had reached a culmination point of sorts, leading to a shift to non-violent lines of effort, the state remained focused upon kinetic action. This could have some impact but could not be strategically decisive in any manner, leaving the strategic initiative with the threat challenge. The objective of that challenge remained precisely what it had been all along, the overturning of the old-order in favor of new structural formulations.

Even where the state did adapt, as in the Sri Lankan case, the failure to recognize the essence of the matters at hand left it invariably jousting at windmills. Ultimately crushing the foe nevertheless resulted, as recently as July 2018, in

358 Shanthie Mariet D'Souza

Colombo's warriors being called out as war criminals in non-public but publicized nonetheless reports by the United Nations. The continued claim that only Sri Lanka had decisively defeated "terrorism" showcased the confusion of both terminology and analytical approaches, not to mention legal realities.

To set forth the point yet once again, using the most powerful challenge:

> If insurgency is an armed political campaign – mass mobilization of a counter-state to challenge the state for political power – then it is intuitively obvious that counterinsurgency is intended to prevent this. As a strategic category, the goal of counterinsurgency is always legitimacy.[1]

This is basic. It is thus ironic, nearly 17 years on since the declaration of "war on terrorism" by the Bush administration, the understanding of insurgency and terrorism continues to be so confused.

Certainly the above should not be construed as arguing that force has little role to play in counter. The point, rather, is that the type of force and the manner in which it is used is fundamental to resolution. This must be driven by correct assessment of challenge. So often presented as a choice between regular and irregular counter, with the former deemed "incorrect," reality dictates a matching up of response in such manner as to neutralize whatever armed capacity and capability is fielded to enable threat movement toward realization of its objective. Operating in regular formations when guerrilla warfare is called for is self-evidently incorrect, but so is clinging to guerrilla warfare when the enemy has moved to military power.

As discussed throughout this volume, it is more than obvious, all doctrinal and theoretical posturing to the contrary, that states, faced with armed challenge, default to responding with the same. One need only look to a state that was once a beacon of democratic hope, the United States, to observe the astonishing descent into the depths of militarization set in motion by the imperative of securing the homeland. This is doubly troubling as studies analyzing how terror groups end have established that although terror groups seldom achieve their objectives, rarely do military operations in and of themselves cause the decline and fall of the threat. Nonetheless, at the other extreme are prescriptions that nations must follow peace negotiations and/or politico-development strategies to address the root causes of armed challenge. This is the flip-side of blind reliance on a force-centric approach. Peace negotiations and development projects are in and of themselves unproductive strategies for dealing with violent extremism in whatever tangible manifestation. Especially with regard to Jihadist movements the likes of IS, such approaches are not viable.

If the Sri Lankan case, as well as others ranging from the Thai South to Afghanistan to Nepal, illustrates the role of attempts at negotiation can play in irregular warfare, Burma displays the impossibility of such when one side has objectives quite different from conflict resolution. Certainly there is always a need for dialogue with rebels, but the motives of the sides involved are crucial. LTTE and the Nepali Maoists, for instance, from first to last, were double-dealing. The need for peace negotiations to end insurgencies can never be over-emphasized, although the reasons why rebels opt for the negotiated path will have an important bearing on reality. The Nepal chapter details the Maoist decision to use peace as part of a strategy to continue the armed effort. It was fierce blowback at strategic betrayal which led to movement splintering and years that have seen hundreds of IEDs and casualties (dramatically underplayed by all concerned lest progress be revealed not quite as portrayed). In India, dialogue with the similarly oriented Naxalites (i.e., Maoists) in 2004 failed, because the extremists stressed their right to bear arms. Similarly, dialogue with the Naga insurgents led to the birth of numerous factions in dispute over desired goals as much as end-state. In South Waziristan, the June 2004 peace agreement with insurgents allowed the insurgents to keep their arms and committed the contending sides to a posture of non-aggression but was quickly revealed as window-dressing. The military negotiated from a position of relative weakness, and this could lead to only one conclusion: breakdown of agreements and strengthening of the insurgents.

Though a solution to the quagmire in Afghanistan seems unimaginable without a peace deal of some sort with Taliban, the heart of the matter will be dictated by who owns the peace process, the terms of a deal, and whether a fractured insurgency can fall in line to accept the Afghan constitution. Always, as in Sri Lanka, the issue remains as to what game is being played. The readiness of states to grasp the weakest of reeds in hope that indeed men of good will have decided to opt for "peace," has been shown time and again to be hope as method. In February 2018, President Ashraf Ghani yet again made a serious peace offer to Taliban. The framework offered follows many years of little domestic and international consensus on a peace agenda, and is further distorted by long-standing ambiguity and inconsistency related to the conditions and requirements for ending the war.

It is thus not helpful to conceptualize distinct security, development, and political strategies when the imperative is linkages between then. Each of these pillars constitutes an approach, a line of effort, that is part of the broad and comprehensive strategy required to defeat extremism of any form. Security will always be the initial instinctive response of the authorities to outbreaks of non-state violence. There are understandable reasons as to why this is the case, not least because security is likely to be the most urgent demand of the targeted populace. Less visible but no less important, though, are activities such as gathering intelligence, interdicting attacks, pacifying areas, and specific targeting of key violent militants. All of this collectively constitutes some of the most obvious lines of activity necessary for counter. And to be expected is the reality that even the most carefully honed security tactics can end up inviting tit-for-tat violence, contributing to the militarization of unresolved political and socio-economic grievances.

The Afghan military serves to illustrate the bottom line. Despite being mentored by the forces that have FM 3–24 *Counterinsurgency* as their ostensible guide, the Afghan state has throughout the conflict focused on conventional warfare. While it does succeed, to an extent, in protecting key assets such as cities and strategic locations, or in clearing specific areas of insurgent presence by deploying massive firepower, it is still not able to effectively engage the insurgents or to protect the populace in rural areas from insurgent control and intimidation. Likewise, the Afghan police forces largely limit their role to protecting the cities and the roads, playing a modest combat role only. In South East Asia, the Philippine military proved adept at transitioning from small unit-driven counterinsurgency, as mentored by the Americans in what was on its face a more correct approach – albeit one hollowed out by political lethargy and malpractice – to the demands of urban warfare. Yet the obvious question is how such a battle could achieve both tactical and strategic surprise. At the other extreme, no adaptation took place at all in response to pinprick ARSA assault, because the objective was never leveraging legitimacy to achieve a more inclusive state, rather genocide. This could hardly surprise given the Burmese approach since the Second World War against the ethnic separatist movements that ring the central Burman core. As events played themselves out, Rangoon produced a textbook case, as had Colombo before it, of state repression producing the very threat it imagined itself facing.

Beyond conflicting objectives and schemes, efforts at negotiation highlight the degree to which all irregular conflicts are battles of framing and narrative. This is even more so in the present globalized context, where the conflation of time and space has demonstrated the ability to alter reality as perceived by all concerned. Hence, strategic communication assumes critical importance in any strategy of counter. This is widely known but is operationalized only at the tactical level. Marawi provides a telling illustration. Despite their eventual defeat in October 2017, the IS-affiliates in the Philippines pulled off a major propaganda victory by holding off a large military, which was backed by a plethora foreign powers, ranging from the U.S. and Australia to China and Russia. It thus successfully framed tactical defeat as strategic victory. Time and again, the Nepali Maoists did the same. A group with minimal strength, such as ARSA of Burma/ Myanmar uses its website and social media channels to likewise project strength by claim responsibility for attacks, many of dubious provenance. Radicalized individuals and groups in the Maldives use the same diagnostic, prognostic, and motivational frames associated with both the Islamic State and Al Qaeda.[2] Social media is gaining a large number of followers within Malaysia, and this is inspiring home-grown terrorists to commit acts of terrorism within the country.

Banning websites and restricting access to online material remains one of the most common official strategies, although its usefulness is contested, together with the potential for abuse (illustrated by more than a few of the countries discussed in this volume). For the danger, as noted, is less the instrumental use of framing and narrative than the creation of alternative realities. This has both strategic and operational significance, with both the ability to recruit and to sustain the struggle being powerfully impacted.

Beyond such debates, the chapters on India highlight attempts to operate a multi-track response. In nearly all of its conflict theatres, New Delhi has grasped the need for an economic track. Its failure at political engagement through dialogue with the extremists stands out, but this is hardly unique to India. Opening and sustaining a political track with insurgents is difficult, as many governments around the world have experienced. Nevertheless, absent complete victory by one side or the other, engagement is normally found to be important to end a conflict. The fly in the ointment, as with so many things, is will, or, more precisely, lack thereof. The case of the Philippines is instructive. The failure of the government to implement fully the peace deal arrived at after protracted negotiations with the MILF is one of the ultimate reasons for the emergence of the political space filled by IS that in the recent past eventuated in the Marawi attack. Even with the siege ended, there seems no convincing evidence that the government has given serious thought to the proper blend of the kinetic and non-kinetic facets necessary for its approach to move forward.

In Indonesia's case, as IS continues to exploit legacy grievances against the state, it is critical that the national and local governments employ the full spectrum of hard and soft power approaches while adhering to internationally accepted human rights and the rule of law. By pursuing a strategy that includes military action as necessary, more responsive and inclusive governance, and regional and international cooperation, not only Jakarta but the other states in the region will be able to dent the influence the extremists might be seeking.

In this regard, Singapore has showcased eliciting cooperation from the community. For over a decade and half, it has implemented a model of community participation that dealt initially with the threat posed by Jemaah Islamiyah, more recently, that of IS. Countering ideological support to terrorism has been particularly important in seeking to prevent a future generation of terrorists from taking root. Government endorsement of community-based initiatives has co-existed alongside the conventional counter-terrorism measures. The role played by the Religious Rehabilitation Group (RRG), comprising a group of local Muslim clerics, as well as the social services extended by the Aftercare Group (ACG), both have been important.

If Singaporean model is the ultimate of sorts in local response to threat, at the other end is the increasing need for transnational cooperation and response as the challenge has become regional and beyond. Evolving realities have pushed governments to seek external assistance, and regional and multilateral cooperation have proved to be powerful reinforcements for national and bilateral efforts. South East Asia has been a notable example, as illustrated by the Indonesian case, even as events in the Philippines and Burma/Myanmar have illustrated that much remains to be done. This is certainly illustrated by the negative example of South Asia, where *lack* of cooperation remains a reason for the ever-present challenge of terrorism. Afghanistan and Pakistan have accused each other of providing safe haven to Taliban and TTP, respectively, while the Indo-Pakistani frozen conflict continues to spawn all manner of armed challenge.

Possibility of victory?

Nevertheless, some countries in the region have been able to claim irregular warfare successes, although these claims have been questioned in their particulars.

The concluding section of the volume examines the Sri Lankan case with just this in mind.

Sri Lanka's conflict ended in a fashion not altogether different from historical instances of major combat. The sheer savagery of the war was not unique though certainly a product of its specific time and place, with the violence exacerbated by the complexity of the threat faced. Although the Liberation Tigers of Tamil Eelam (LTTE) was labeled "terrorist" by any number of governments, it was an insurgency in intent and methodology. By defining it as terrorism, a standard government mis-analysis, Colombo overlooked the basic necessities of an effort that should have gone after the roots of the problem, with the security forces used only to enable the effort. This would have obviated the need for any discussion of whether force was used legitimately, discriminately, and with concerns for human security. As events unfolded at the end of three decades of bloodletting, "Colombo's frame of victory," notes Marks, found itself all but "overwhelmed by a shrill counter-frame of repression, and Colombo's narrative trumpeting a triumph over terrorism all but swamped by a rival narrative of communal repression and barbarism."

This was unfortunate, for Sri Lanka's military victory over LTTE, indeed, offers valuable lessons for practitioners and scholars. It underscores the importance of examining the experiences of smaller, non-Western domestic counterinsurgents, even as it underscores the danger of ignoring particularity and simply focusing on success, because so much was unique in the contest that it cannot simply be copied. The degree of material asymmetry ultimately achieved allowed victory in an actual civil war in a fashion impossible for all but a few countries. Colombo's triumph came at the conclusion of the three decades just mentioned, a period made necessary only by profound strategic miscue which was itself a factor in an equally profound reform and mobilization that resulted in operational elimination of the threat.

Perhaps unintentionally, the case does serve to highlight the ongoing and often unhelpful debate on counterinsurgency, or "COIN," that continues in the West. For Sri Lanka's struggle, labeled then and now a battle with terrorism, in reality was a counterinsurgency that needed to address terrorism as but one method used by LTTE. The reality is that most of the groups discussed in this volume also are insurgencies, albeit practitioners of often horrendous acts of terrorism. It is particularly unfortunate, therefore, that the existing literature almost universally sidesteps this fact and more often simply conflates the terms insurgency and terrorism, which is decidedly unhelpful in constructing counter.

If this is one challenge, analysts point to others, not least that the existing COIN literature can in fact lead to unhelpful thinking. Though much critique simply misses the point, conflating local counterinsurgency with expeditionary COIN, there are some useful findings. Not least is that COIN doctrine of any stripe highlights how the countering of insurgency involves breaking the bonds between insurgents and the local populace, and how this is achieved by protecting the populace to facilitate the provision of governance to the region. Doctrine, then, must guide the security forces to deploying in a proportionate and purposeful manner to facilitate popular incorporation and empowerment. It is crucial is to avoid the militarization that produces "operations overly process-focused and divorced from the wider political currents that inevitably define the strategic possibilities of any given war."[3]

Also reductionist is to focus tactically upon the process of winning hearts and minds to the point that the need for structural reform is neglected. If the former is ineffective, the latter is as time-consuming as it is challenging. Regardless, long-term transformation is impossible without an operational approach which, as Hazelton explains, focuses upon "buying-out the leaders and warlords responsible for the violence, while deploying enough coercive force to break the insurgency's will and capability to fight on."[4] In sum, the strategist must avoid a partisan approach to the tools being used. Each activity must be considered on its merits *and* in relation to other activities. Comprehending all of the relevant factors – how they interact, the trade-offs between them, and their accumulated impact – is a prerequisite for sound strategic thinking in responding to a complex insurgency.[5]

People's war remains an important analytical framework, if properly understood for what it is: the building of a new world to challenge an existing world, with all violent and non-violent ways but means to the end.[6] Far from being an anachronism, much less a kit-bag of techniques, people's war raises what has always been present in military history, asymmetric challenge from below, and fuses it symbiotically with what has likewise always been present politically, rebellion and the effort to seize power. This is on full display in the Nepal case study, where a strategic approach comprised of violent and non-violent components astutely emphasized the particular lines of effort appropriate to the moment in order to continue the revolutionary struggle even as international and national contexts shifted, the very essence of the so-called "new war."

Notes

1 Marks, "Counterinsurgency in the Age of Globalism," 22–29.
2 Della Porta and Diani, *Social Movements: An Introduction.*
3 Smith and Jones, *The Political Impossibility of Modern Counterinsurgency,* xvi.
4 Hazelton, "The 'Hearts and Minds' Fallacy," 80–113.
5 Freedman, *Strategy: A History,* 242–245.
6 Marks and Rich, "Back to the Future," 409–425.

Bibliography

Della Porta, Donatella and Diani, Mario. *Social Movements: An Introduction.* Malden, MA: Blackwell Publishing, 2006.

Freedman, Lawrence. *Strategy: A History.* New York, NY: Oxford University Press, 2013.

Hazelton, Jacqueline L. "'The Hearts and Minds' Fallacy: Violence, Coercion, and Success in Counterinsurgency Warfare," *International Security* 42 (1) (Summer 2017), 80–113.

Marks, Thomas A. "Counterinsurgency in the Age of Globalism," *The Journal of Conflict Studies* 27 (1) (Summer 2007), 22–29; available at: https://journals.lib.unb.ca/index.php/JCS/article/view/5936 (accessed March 4, 2016).

Marks, Thomas A. and Paul B. Rich, "Back to the Future – People's War in the 21st Century," *Small Wars and Insurgencies* 28 (3) (June 2017), 409–425.

Smith, M. L. R. and David Martin Jones. *The Political Impossibility of Modern Counterinsurgency Strategic Problems, Puzzles, and Paradoxes.* New York, NY: Columbia University Press, 2015.

Index

Page numbers in **bold** denote tables, those in *italics* denote figures.